# FOUNDATIONS OF THE THEORY OF ELASTICITY, PLASTICITY, AND VISCOELASTICITY

# FOUNDATIONS OF THE THEORY OF ELASTICITY, PLASTICITY, AND VISCOELASTICITY

**E.I. Starovoitov and F.B. Nagiyev**

Apple Academic Press

TORONTO NEW JERSEY

Apple Academic Press Inc.
3333 Mistwell Crescent
Oakville, ON L6L 0A2
Canada

Apple Academic Press Inc.
1613 Beaver Dam Road, Suite # 104
Point Pleasant, NJ 08742
USA

***Exclusive worldwide distribution by CRC Press, a Taylor & Francis Group***

International Standard Book Number: 978-1-926895-11-6 (Hardback)

Printed in the United States of America on acid-free paper

**Library of Congress Control Number: 2012935655**

**Library and Archives Canada Cataloguing in Publication**

Foundations of the theory of elasticity, plasticity, and viscoelasticity/edited by E.I. Starovoitov and F.B. Nagiyev.

Includes bibliographical references and index.
ISBN 978-1-926895-11-6
1. Deformations (Mechanics)–Textbooks. 2. Elasticity—Textbooks. 3. Plasticity–Textbooks.
4. Viscoelasticity—Textbooks. I. Starovoitov, E. I. II. Nagiyev, Faiq

QA931.F68 2012 531'.38 C2011-908702-2

# Contents

# List of Abbreviations

| | |
|---|---|
| A | Avogadro number |
| D-16T | Duralumin |
| $\delta(x)$ | Dirac's delta function |
| FT | Fourier's transform |
| G | Instantaneous elastic shear modulus |
| K | Instantaneous modulus |
| K | Kinetic energy |
| L | Vector-linear operator |
| $\lambda$ and $\mu$ | Lame's constants |
| Pa | Pascals |
| PTFE | Polytetrafluoroethylene |
| P | Pressure |
| Q(r) | Transverse force |
| SNSC | Silicon nitride structural ceramics |
| SSS | Stress–strain state |
| T | Temperature |
| $T_s$ | Cauchy stress tensor |
| W | Potential energy of deformation |
| V | Poisson's ratio |
| Y | Young's modulus E |
| Zircaloy-2 | Zirconium alloy |

# Preface

The purpose of the book was to reflect modern approaches to the statement and solution of the problems in the mechanics of deformed solids, which involves the theories of elasticity, plasticity, and viscoelasticity. The present monograph is intended for scientists, postgraduates, and high-grade students of engineering spheres. The authors believes that the subject like this should be presented with sufficient mathematical strictness and proofs, where required. The knowledge of corresponding sections of a standard course of mathematical analysis, linear algebra, and differential equations is a must to cope successfully with this material. Additional data on tensor analysis and generalized functions are given in the appendix.

The first six chapters are devoted to the foundations of the theory of elasticity. The theory of stress–strain state (SSS), physical relations and problem statements, variational principles, contact and 2D problems, and the theory of plates are considered. The reasoning in the tensor form is made in rectangular Cartesian coordinates. The theoretical part is accompanied by the examples of solving typical problems.

The last six chapters may be useful for postgraduates and scientists engaged in nonlinear mechanics of deformed inhomogeneous bodies. The foundations of the modern theory of plasticity (general, small elastoplastic deformations and the theory of flow), linear and nonlinear viscoelasticity are set forth. The statements of dynamic problems of the theory of elasticity (linear oscillations, waves) are given.

Beginning from the sixth chapter, we included a section on corresponding research of three-layered circular plates of various materials as a supplement to the considered statements of the problem and methods of their solving. Analytical solutions and numerical results for elastic, elastoplastic, linear viscoelastic, and viscoelastoplastic plates are given. Thermoviscoelastoplastic characteristics of certain materials needed for numerical account are presented in the 11th chapter.

A list of literature [1–81] reflects only the sources used to prepare the given manuscript.

The authors express deep gratitude to Professor Tarlakovski D.V. who has written the seventh chapter of this book. Any remarks or case comments from the readers are welcome.

**— E.I. Starovoitov**

# Introduction

The theories of elasticity, plasticity, and linear viscoelasticity are the parts of the mechanics of deformed solids. This is a domain of sciences studying stress–train state (SSS) of solid bodies. To these belongs also strength of materials and structural mechanics.

The problems solved by named sciences are similar and establish the relationship between the external loading and the deformable body behavior.

The difference between these parts of mechanics consists in the objects considered, the assumptions accepted and the methods of solving the problems posed. The course of strength of materials treats on the main bars; the theory of elasticity studies the bars, plates, shells, and arrays. Structural mechanics studies the systems consisting of rods (truss), beams (frames), plates, and shells. The theories of elasticity, plasticity, and viscoelasticity employ various physical laws establishing the relation between stresses and strains, although no deformation hypotheses are involved. Therefore, more intricate problems in contrast to strength of materials are to be solved, for which aim, more complex mathematical methods should be used.

The theory of elasticity (the theory of plasticity as well) is subdivided often into the mathematical and applied theories of elasticity.

The mathematical theory of elasticity does not employ any deformation hypotheses, and equations are solved by either exact, or approximated procedures, which allow to infinitely increase the degree of approximation to the exact solution. So, we may consider the results obtained by solving problems of the mathematical theory of elasticity, as a standard for verifying accuracy of various approximated theories and methods for similar problems.

The applied theory of elasticity differs from the mathematical one in that along with Hooke's law, several additional hypotheses of deformation character (hypothesis of plane sections for rods, normals to plates and shells, etc.) are used for solution of the problems. When solving the problems of the applied theory of elasticity, approximated solutions may be used along with the exact ones for corresponding equations. There is no clear border between the applied theories of elasticity closely related with the practical needs, and strength of materials.

The first mathematician engaged in studying resistance of solids to failure was Galilei.1 Although he considered solid bodies to be inelastic and did not apply the law relating shear and forces, his works have shown the paths to be followed by others.

The discovery of Hook's2 law in 1660 and Navier's3 general equations in 1821 are two milestones in the evolution of the theory started by Galilei. Hook's law has given necessary experimental substantiation of the theory. The general equations have nullified the problems concerning small deformations of elastic bodies in mathematical calculations.

In the period between the discovery of Hook's law and appearance of general differential equations of the theory of elasticity, the interest of researchers was directed to the problems of oscillations in beams and plates, along with stability of columns. We should relate to these, first of all, J. Bernoulli's fundamental works devoted to the elastic curve shape, and Euler's4 investigations in the field of stability of elastic systems. Following Euler's theory, Lagrange5 has applied it for the definition of most reliable forms of columns.

The mathematical theory of elasticity as a science has evolved in the first half of the nineteenth century basically thanks to the works of French engineers and scientists. For the first time, the equilibrium equations and oscillations of elastic solid bodies in the assumption of the discrete molecular structure were obtained by Navier. Using the variational calculus, along with the differential equations he deduced boundary conditions to be met on a body surface.

Lame6 and Clapeyron7 have extended Navier's theory with reference to civil engineering. They have written a special memoir on the internal equilibrium of solids, solved the problem for stresses and strains of a thick-walled pipe at axially symmetrical loading (Lame problem).

By the fall of 1822, Cauchy8 has discovered the majority of basic elements of the pure theory of elasticity. He has introduced the concept of stress and strain in a given point. He has shown that the concepts can be defined by six corresponding components. Proceeding from the hypothesis on a continuous and homogeneous structure of a solid body, Cauchy has received the equations for motion (or equilibrium). He was the first to introduce two elastic constants into the equations of the theory of elasticity in contrast to Navier's ones that contained only one. The equations relating small deformations and displacements bear his name.

Significant contribution into the development of the theory of elasticity was made by Saint-Venant.9 He suggested a new approach for solving several applied problems (Saint-Venant's semi-inverse method). The important problems for bending and torsion of non-round in section beams were solved with the help of this method. The investigations of oscillations, impact, and the theory of plasticity he also studied.

In the second half of the nineteenth century, Kirchhoff10 formulated the basic equations of the theory of thin rods thus initiating the development of calculation methods of elastic springs. He has also developed a consecutive theory of thin plates. The first attempts in this direction were made by Lagrange and Sophie Germain11 in 1814, but they failed to formulate correctly the boundary conditions of the problem.

At the end of nineteenth century Aron and Love12 gave the first variants of the equations of the modern theory of shells by applying the hypotheses of undeformability of a normal straight element. Boussinesq13 has studied the distribution of stresses in the elastic body under the action of a concentrated force. This has allowed Hertz14 to state the problem on interaction of two elastic bodies in contact.

The works of Russian scientists have played an important role in the development of the theory of elasticity. Fundamental results in the development of the principle of probable displacements, the theory of impact and integration of the equations of dynamics belong to Ostrogradski.15 A General of artillery, A.V. Gadolin,16 has investigated stresses in multilayered cylinders, having formulated thus the bases for designing barrels of artillery guns. D.I Zhuravski[17] has set forth the modern theory of beam bending. He has widely applied the methods of strength of materials in designing numerous railroad bridges. The plane problem solution of the theory of elasticity was essentially advanced by Kolosov's[18] and Muskhelishvili[19]'s works who were the first to apply the method based on the functions of a complex variable. I.G. Bubnov[20] has solved a number of problems on bending of plates.

Fundamental investigations in the theory of plates and shells, oscillations of rods with account of shear strains were fulfilled by S.P. Timoshenko.[21] A number of problems were solved by the energetic method suggested by him.

V.G. Galerkin[22] is the author of a series of investigations in the theory of bending of thin plates, thick plates, and the theory of shells. To deduce equations of the theory of shells he has for the first time applied the equations of three-dimensional (3D) theory of elasticity. P.F. Papkovich[23] was the first to suggest a solution the theory of elasticity in displacements in the form of harmonic functions and investigated the general theorems of stability of elastic systems, solved a great number of problems on bending of plates under various boundary conditions.

A considerable contribution to the development of a general theory of shells was made by V. Vlasov,[24] V. Novozhilov,[25] and Yu. Rabotnov.[26] V.Z. Vlasov has studied general equations of the theory of shells, has developed its engineering variant and the theory of semi-flexible shells, and suggested a new theory of bending and torsion of open-profile thin-walled rods. He is considered to be the founder of a new scientific discipline—structural mechanics of shells.

The theory of plasticity as an independent part of mechanics has evolved more than a 100 years ago. The first Saint-Venant's works related to 1868–1871 have formulated the bases of the theory of plastic yield. They present the description of metal processing under pressures (metal working). Having been limited in the amount of experimental data, Saint-Venant has managed (to some extent by analogy with hydrodynamics of a viscous liquid) to formulate a number of laws whose importance is preserved up till now.

At the beginning of twentieth century, T. Karman[27] (1909), M.T. Hyber (1904), von Mises[28] (1913), A.L. Nadai (1921), H. Hencky (1924), and some other scientists have put forward new concepts and theories that have not given a decision to the problem but expanded the circle of ideas. Along with the appearance of numerous variants of the theory, the experimental investigations were intensively carried out in the 1920s and 1930s. This was an important stage in the development of the theory of plasticity. However, by the end of 1930s and beginning of 1940s there was a situation when some experiments confirmed one theory denying all others and, vice versa, other experiments contradicted the first ones.

A.A. Ilyushin[29] (1943–1945) has clarified the situation by indicating the necessity of distinction between the character of deformation processes (simple and complex deformations). The analysis of experimental data has proved mutual concordance if they were obtained in conditions of simple deformation. These investigations have brought him to the development of the theory of small elastoplastic deformations.

At the beginning of 1950s various plasticity theories for a random complex loading were suggested. These approaches have sprung in the form of three theories: the modern theory of flow, the theory of sliding, and the general theory of elastoplastic deformations.

The construction of a general mathematical deformation theory of plasticity is based on the postulate of isotropy formulated by Ilyushin. Further development of the theory of flow of elastoplastic bodies is based on Drucker's postulate of hardening on non-negativity of the work of external forces in a closed-cycle plastic loading.

These directions were broadened significantly by theoretical and experimental studies of both home and foreign scientists.

## NOTES

1. Galilei Galileo (1564–1642), Italian scientist, one of the founders of modern mechanics, was the first to study durability of beams.
2. Hooke Robert (1635–1703), English naturalist, many-sided scientist, and experimenter.
3. Navier Anri (1785–1836), French mathematician and mechanical engineer, founder of the theory of elasticity, introduced the concept of stress.
4. Euler Leonard (1707–1783), mathematician, mechanical engineer; Swiss by birth, worked in Russia between 1727 and 1741, then in Berlin.
5. Lagrange Josef Louie (1736–1813), French mathematician and mechanical engineer; basic works in calculus of variations, analytical mechanics.
6. Lamé Gabriel (1795–1870), French mathematician and mechanical engineer, worked in Russia between 1820 and 1832 in mathematical physics, theory of elasticity.
7. Clapeyron Benua Pol Emil (1799–1864), French physicist and engineer, worked in Russia in 1820–1830.
8. Couchy Augustine Louie (1789–1857), French mathematician, one of the founders of the mathematical theory of elasticity.
9. Barre de Saint-Venant (1797–1886), French mechanical engineer, one of the founders of the theory of plasticity.
10. Kirchhoff Gustavus Robert (1824–1887), German physicist; worked in mechanics, mathematical physics, electricity, spectral analysis.
11. Germain Sophie (1776–1831), investigated equilibrium of plates, was awarded a premium of French Institute in 1815, her major work was published in 1821. The first variant of the equation of balance of plates was corrected by Lagrange in 1811.
12. Love Augustus Eduard Hweitt at the beginning of twentieth century has published a unique book [41] considered to be encyclopedia of the mathematical theory of elasticity.
13. Boussinesq Josef Valanten (1862–1929), French mechanic; works in hydrodynamics, optics, thermodynamics, theories of elasticity.
14. Hertz Henrich Rudolf (1857–1894), German physicist; worked in electromagnetism, theories of light, he gave a statement of the class of contact problems.
15. Ostrogradski Mikhail Vasilevich (1801–1861/62), Russian mathematician, mechanical engineer; has carried out important researches in integral calculus.
16. Gadolin Aksel Vilhelmovich (1828–1892), Russian scientist, has developed the theory of joining barrels of artillery guns; worked in physics, processing of metals.
17. Zhuravski Dmitri Ivanovich (1821–1891), Russian scientist and engineer, founder of school in structural mechanics and bridge building.

18. Kolosov Guriy Vasilievich (1867–1936), Russian mechanical engineer; has works in mathematics, theories of machines and mechanisms.
19. Muskhelishvili, Nikolaj Ivanovich (1891–1976), Russian mechanical engineer; has works in mathematics and theories of elasticity.
20. Bubnov Ivan Grigorievich (1872–1919), Russian engineer, founder of structural mechanics of ships, has built two submarines.
21. Timoshenko Stepan Prokofievich (1878–1972), Russian scientist, since 1922 lived in the USA; worked in stability of elastic systems, variational methods, vibrations.
22. Galerkin Boris Grigorevich (1871–1945), Russian engineer and scientist, one of the founders of the theory of bending of plates, contributed to introduction of mathematical methods into engineering research.
23. Papkovich Petr Fedorovich (1887–1946), Russian ship-builder; main works in structural mechanics of ships, theory of elasticity.
24. Vlasov Vasiliy Zakharovich (1906–1958), Russian scientist in the field of mechanics, worked in resistance of materials, structural mechanics, and theory of elasticity.
25. Novozhilov Valentin Valentinovich (1910–1987), Russian mechanical engineer; worked in the theory of elasticity, plasticity, calculation of shells, and ship designs.
26. Rabotnov Yuri Nikolaevich (1914–1985), Russian mechanical engineer; worked in the theory of shells, plasticity, creep, and failure of materials.
27. Karman Theodor von (1881–1963), German scientist; worked in aircraft construction and rocket production, aero-, hydro-, and thermodynamics, structural mechanics.
28. Mises Richard (1883–1953), German mathematician and mechanical engineer, since 1939 lived in the USA, one of the founders of the theory of plasticity; worked also in aeromechanics.
29. Ilyushin Aleksey Antonovich (1911–1998), Russian mechanical engineer, one of the founders of the theory of plasticity, viscoelasticity, and so forth.

# Chapter 1

## The Theory of Stress–Strain State

### THE CONCEPT OF ELASTIC CONTINUUM

It is acknowledged that the deformed solids change their dimensions and shape (*become deformed*) under the influence of external forces (*loads*). Inside them the *internal forces* may be generated, the magnitude and distribution of which depend on the load and geometry of the bodies.

The mechanical behavior of a solid varied and complex enough. Its general description is based on the theory of *continuum*. It is well known that there is no any continuum in fact, but this model can be accepted to understand the mechanical behavior of matter in macrovolumes. If the discrete structure of matter is ignored, it is assumed that the volume occupied by a body is continuously filled by a matter.

An infinitesimal volume of matter may be considered as a "particle" of continuum. An arbitrary divisibility of matter, as well as invisibility of separate particles constitutes one of the basic notions of the continuum mechanics.

The basis for adoption of the continuum model is the experience enabling experimental check of the theory under consideration. The material particles change their position in space at loading under the influence of external forces, that is, the continuum moves. It is also assumed that the movement of continuum is continuous. This means that all values defining deformation are the continuous functions of coordinates.

The description of deformation of solid bodies and fluids is a purely geometrical problem of mechanics and does not depend on the material behavior. This is a problem of *continuum kinematics*, so the nature of forces causing the deformation is of no importance. Similarly, the forces occurring in a deformed body do not depend on the material properties.

Let us consider an arbitrary solid body with applied to it end reactions. The influence of surrounding bodies is replaced by the forces known as *external*. The external loads may be divided into 3D (mass), *surface* and *concentrated* ones. The latter may be treated as a limiting case of the surface load application on a small portion of the body surface.

The elasticity theory is a part of the continuum mechanics. It deals with determination of deformations and internal forces in elastic solids under specified loads. The following fundamental hypotheses and assumptions related to the material properties, stresses, and deformation nature are accepted in this case.

1. Homogeneity and continuity hypothesis. This hypothesis makes it possible to study mechanical properties of bodies on the samples of relatively small size and to use differential calculation methods for the deformation analysis.

2. Assumption on small deformations. The deformations in the points of a body are considered to be small when they do not exert any essential effect on mutual location of the loads applied on the body. This allows considering the body geometry as invariable in equilibrium equations.
3. Cauchy–Euler stress principle. Interactions between forces take place in each imaginary cross-section inside a body similarly to the stresses distributed over the surface. So, by using the sectioning method we may accept the surface forces acting in a section instead of the effect of one part of a body on the other.
4. Axiom of hardening (frosting): A material body is considered to be absolutely solid in any fixed moment $t$, so the laws of theoretical mechanics are valid for this body, including Newton's laws.[1] In this case the axiomatic concepts of *force* and *mass* are used.

Note, please, that according to the continuity hypothesis motion of a matter is studied at so-called *macroscopic level*, that is, the elementary structure of matter is not taken into account. This is justified by the fact that in numerous practical problems this is the general state of a body that is of interest but not the behavior of a separate molecule (atom).

The validity of the assumptions introduced can be proved only empirically. Additional axioms and definitions will be introduced where necessary and their numbering will continue the one started here.

Let us suppose that an imaginary section divides a body into two areas: $V_1$ and $V_2$ (Fig. 1.1a). A surface element of section $\Delta S$ with point $A$ as a center is characterized by a unit vector of normal $\mathbf{v}$ directed toward $V_1$. The effect produced on $V_2$ part of the body by $V_1$ part in point $A$ may be presented by a vector force $\Delta\mathrm{P}$ and moment $\Delta\mathrm{M}$. In a limiting case, at shrinkage of the surface element to a point when $\Delta S \to 0$ (at fixed direction $\mathbf{v}$), the following physically substantiated assumptions may be accepted:

$$\lim_{\Delta S\to 0}\frac{\Delta P}{\Delta S}=\frac{dP}{dS}=\sigma_v,\ \lim_{\Delta S\to 0}\frac{\Delta M}{\Delta S}=0\,. \tag{1.1}$$

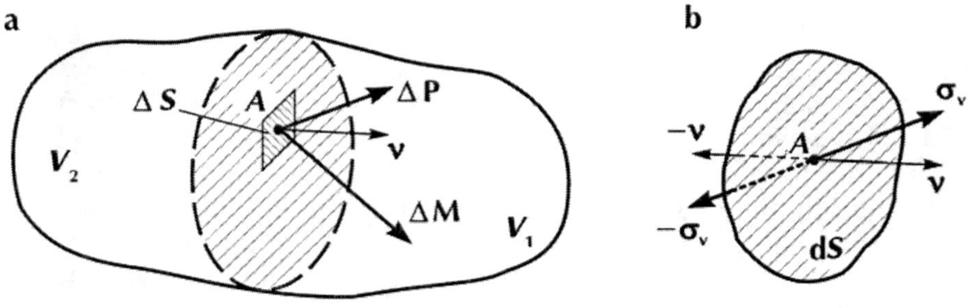


**Figure 1.1.**

Vector $\sigma_v$ found by relation (1.1) is known as a *stress vector* in point $A$. It operates on the surface element at normal direction $\mathbf{v}$ and varies with changing direction of this normal for $\Delta S$. For the stress vectors acting in the same point but directed oppositely from the element section, the following equality is valid:

$$\sigma_{-\nu} = -\sigma_{\nu} \cdot \qquad (1.2)$$

As it is seen from Fig. 1.1b it helps to describe the effect of the body part $V_2$ on $V_1$ and vice versa.

Relation (1.2) may be interpreted as a direct expression of the *third Newtonian law* (the principle of equality of action and reaction). Although, it can be also derived directly from the theorem of momentum and Cauchy stress principle. A set of stress vectors $\boldsymbol{\sigma}_{\nu}(A)$ for all directions $\boldsymbol{\nu}$ determines the stress condition in point $A$.

In a general case, stress vector $\boldsymbol{\sigma}_{\nu}(A)$ is not directed along **ν**-normal. Its projection in an arbitrary direction (defined by a unit vector) is known as a *stress vector component* in this direction. If to expand $\boldsymbol{\sigma}_{\nu}$ with respect to the normal and tangential to the area element *dS,* we obtain a so-called *normal stress* $\sigma_n$ and *tangential stress* $\tau_n$ (Fig. 1.2).

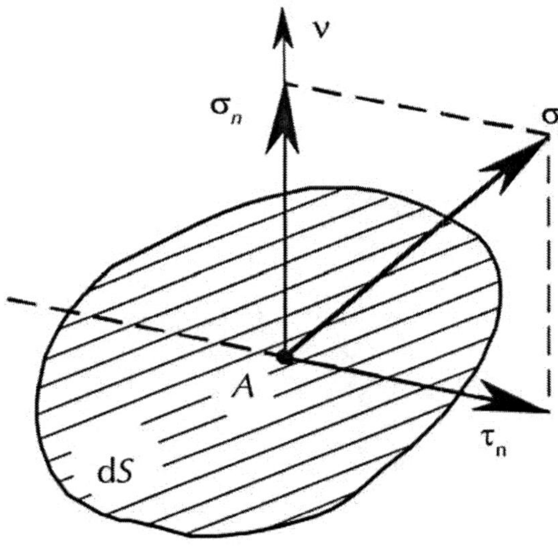


**Figure 1.2.**

Notice that the following relation will be valid for the modulus of these values:

$$\sigma_{\nu}^{2} = \sigma_n^2 + \tau_n^2$$

One should always remember that stresses $\sigma_n$ and $\tau_n$ defined in such a manner are not vector components in a common sense. Notice that tangential stresses in the plane of element *dS* are usually expanded in two coordinate directions.

## STRESS TENSOR

*Cartesian coordinates* $x_1$, $x_2$, $x_3$ will be used further are denoted henceforth by symbol $x_i$, bearing in mind that $i$ and other Latin indices take the values 1, 2, and 3. Let us designate the *base vectors* of the system of coordinates as $e_i$.

Under the effect of specified stresses in a solid body points, there occurs a *stress state*. To describe it let us cut out an imaginary elementary parallelepiped about an arbitrary point, whose edges are parallel to the coordinate axes and have short lengths $dx_i$ (Fig. 1.3a). In view of smallness of the elementary parallelepiped it may be assumed that stress vectors $\boldsymbol{\sigma}_i$ on its sides coincide with those parallel to them on the coordinate area elements drawn through the point under consideration.

We shall accept an important assumption in subsequent discussion that the *deformed and undeformed elements of the volume are identical*, which follows from the hypothesis of smallness of deformations.

Stress vectors $\sigma_i$ acting on the elementary parallelepiped faces may be expanded into *normal* $\sigma_{ii}$ (perpendicular to faces) and *tangential* $\sigma_{ij}$ $(i \neq j)$ (lying in-plane of the faces) *components* (Fig. 1.3b). The first subscript in notations indicates a normal to the area element where this stress acts. The second subscript is the axis to which the stress is parallel. The tensile normal stresses are considered positive, compressive ones are negative. Values $\sigma_{ij}$ are not the components of the vector in its common sense. They are measured in Pascals (1 Pa = 1 N/m²).

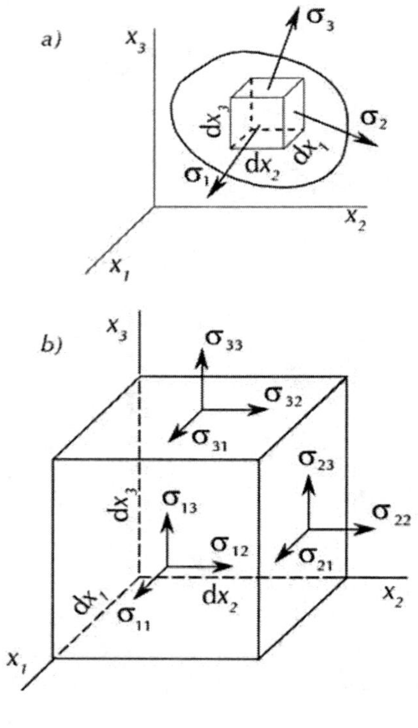


**Figure 1.3.**

The expansion of, for example, stress vector $\sigma_3$ (operating on face $x_3$ = const) along coordinate axes $x_i$ is of the form

$$\sigma_3 = \sigma_{31}e_1 + \sigma_{32}e_2 + \sigma_{33}e_3 = $$
$$= \sum_{k=1}^{3} \sigma_{3k}e_k \equiv \sigma_{3k}e_k.$$

Let us accept that every time we meet some Latin index twice in a monomial, they will be summed up from 1 to 3 over this index. The summation sign will be omitted. This index is known as a *dummy* one.

It can be substituted (like an integration variable in the integrand) by any other index. For example, $x_i y_i = x_k y_k$. If a Greek index is repeated, the summation is not made. The index found in a component only once is known as a *free* one. If a formula is written correctly, each component of the formula shall have the same free indices.

In a general case, the following equality is valid for the stress vector:

$$\sigma_i = \sigma_{ij}e_j, \tag{1.3}$$

where $e_j$—unit vector of coordinate axis $x_j$.

We are to introduce a *metric tensor* $\delta_{ij}$, which is a scalar product of the base vectors. In Cartesian coordinates $e_i \times e_j = 0$, if $i \neq j$, that is why

$$\delta_{ij} = e_i \times e_j, \; \delta_{ij} = \begin{cases} 1, & i = j, \\ 0, & i \neq j. \end{cases} \tag{1.4}$$

The matrix of this tensor is diagonal. The components of the metric tensor (1.4) are often called *Cronecker deltas.*[2]

Scalar multiplication of relations (1.3) by base vector $e_k$ *gives*:

$$\sigma_i \times e_k = \sigma_{ij}e_j \times e_k = \sigma_{ij}\delta_{kj}.$$

It follows from (1.4) that $\delta_{kj} = 1$ only at $j = k$, so we obtain:

$$\sigma_{ik} = \sigma_i \times e_k. \tag{1.5}$$

Nine components of stresses $\sigma_{ij}$ represent in aggregate a physical value known as the *Cauchy stress tensor*[3] ($T_s$). This second-order tensor (per number of indices) written in the form of a matrix looks like:

$$T_s = (\sigma_{ij}) = \begin{pmatrix} \sigma_{11} & \sigma_{12} & \sigma_{13} \\ \sigma_{21} & \sigma_{22} & \sigma_{23} \\ \sigma_{31} & \sigma_{32} & \sigma_{33} \end{pmatrix}.$$

The matrix lines contain stress tensor components acting on a common area element, the columns are those parallel to the common axis.

Leaving aside the accurate mathematical definition of the tensor, we should note that it represents an invariant object remaining unchanged at transfer from one system of coordinates to another. Only its components are changing according to a certain "tensorial" law at such transitions. Nevertheless, we shall name for briefness in our further discourse a combination of tensor components a tensor. More detailed information on the tensor concept is given in Section 14.1 or the course of continuum mechanics (Gorshkov et al., 2000, Mase, 1970, Sneddon and Berry, 1961) or the elasticity theory (Amenzade, 1976, Novackii, 1975, Williams et al., 1955).

## STRESS TENSOR PROPERTIES

Let us prove that the components of stress vector $\sigma_v$ on an arbitrary oblique area element drawn through the point considered may be expressed by the stresses on the coordinate area elements (elementary parallelepiped faces) in this point (Fig. 1.4a).

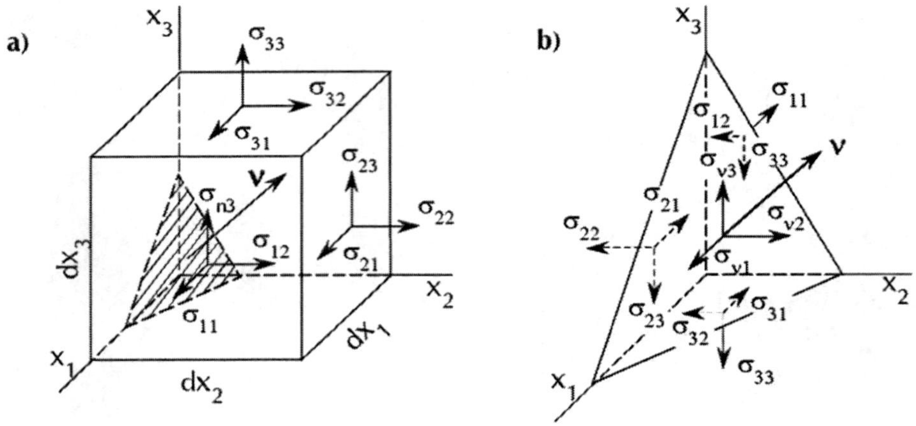


**Figure 1.4.**

We shall take now an elementary pyramid formed by the coordinate sides and the oblique area elements with a normal v (Fig. 1.4b). Here, $\sigma_{vi}$—is a projection of stress vector $\sigma_v$ on coordinate axes; $S$—area of the oblique area element; $S_i$—areas of corresponding area of coordinate faces (the subscript indicates a normal to the area element) related to $S$ via the relations:

$$S_i = Sl_i, \ (i = 1, 2, 3), \tag{1.6}$$

where $l_i = \cos(\mathrm{v}, x_i)$—are directional cosines of normal **v** to the oblique area element.

Let us show the stresses on the faces of the isolated pyramid acting on them (refer to Fig. 1.4b). We may write the *equations* for the *equilibrium of forces*, being the products of stresses on the area of corresponding faces:

$$\sigma_{\nu 1}S - \sigma_{11}S_1 - \sigma_{21}S_2 - \sigma_{31}S_3 = 0\,,$$

$$\sigma_{\nu 2}S - \sigma_{12}S_1 - \sigma_{22}S_2 - \sigma_{32}S_3 = 0\,,$$

$$\sigma_{\nu 3}S - \sigma_{13}S_1 - \sigma_{23}S_2 - \sigma_{33}S_3 = 0\,.$$

This system of equations in the tensor notations takes the form:

$$\sigma_{\nu i}S - \sigma_{ji}S_j = 0\,. \tag{1.7}$$

Notice that the summation is made over the repeated Latin index $j$. In our case, it is:

$$\sigma_{ji}S_j = \sigma_{1i}S_1 + \sigma_{2i}S_2 + \sigma_{3i}S_3\,.$$

Let us substitute relations (1.6) into equations (1.7). Canceling then by $S$, we obtain the formulas expressing the stress vector components on an arbitrary oblique area element through the coordinate stresses,

$$\sigma_{\nu i} = \sigma_{ji}l_j = 0\,. \tag{1.8}$$

Thus, it is possible to describe fully the stress state in a point using the stress tensor components on the coordinate area elements.

Stress vector $\sigma_\nu$ on an arbitrary oblique area element drawn through this point may be presented by expanding in the base vectors

$$\sigma_\nu = \sigma_{\nu i}e_i \equiv \sigma_{\nu 1}e_1 + \sigma_{\nu 2}e_2 + \sigma_{\nu 3}e_3\,. \tag{1.9}$$

Using (1.9), the modulus of this vector may be expressed by a sum of squares of its projections.

## EQUILIBRIUM EQUATIONS OF ELASTIC BODIES

If a deformed body is found in equilibrium, then any of its part arbitrary isolated from the body shall be also in equilibrium. Let us presume that the body is loaded by specified superficial forces $R_\nu$ and homogeneous bulk forces $\rho F$ ($\rho$—material density):

$$\rho F = \rho F_i e_i \equiv \rho F_1 e_1 + \rho F_2 e_2 + \rho F_3 e_3\,.$$

It is assumed that stress components $\sigma_{ij}$, as well as their first partial derivatives are continuous functions of coordinates. Deformations are considered to be small, so we may formulate the equilibrium equations for a undeformed body or its part.

For an isolated element with short edges $dx_i$ loaded similarly to Fig. 1.5, the stress components at coordinate increase by infinitesimal value $dx_i$ change, respectively, by

$d\sigma_{ij}$. For example, the changes in normal stresses $\sigma_{11}$ acting along axis $x_1$ due to only $dx_1$ increments will be as follows:

$$d\sigma_{11} = \sigma_{11}(x_1 + dx_1, x_2, x_3) - \sigma_{11}(x_1, x_2, x_3) = \frac{\partial \sigma_{11}}{\partial x_1} dx_1 \equiv \sigma_{11},_1 dx_1 .$$

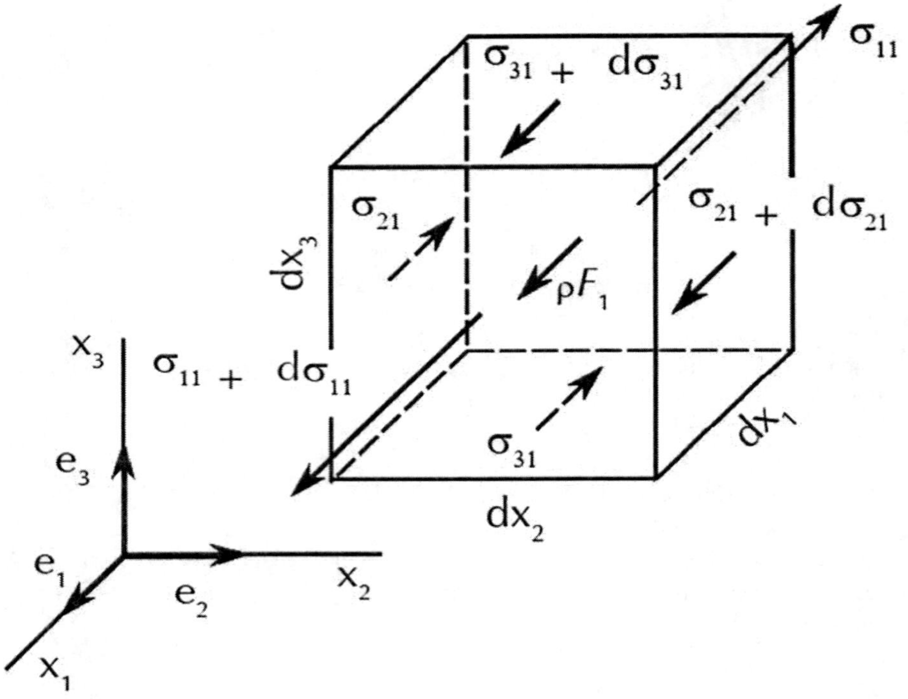


**Figure 1.5.**

Stresses $\sigma_{21}$ acting along $x_1$ axis on the other face are incrementing due to the change of coordinate $x_2$ by value $dx_2$:

$$d\sigma_{21} = \sigma_{11}(x_1, x_2 + dx_2, x_3) - \sigma_{11}(x_1, x_2, x_3) = \frac{\partial \sigma_{21}}{\partial x_2} dx_2 \equiv \sigma_{21},_2 dx_2 .$$

Stresses $\sigma_{31}$ are changing essentially in the same manner: $d\sigma_{31} = \sigma_{31},_3 dx_3$. For briefness, the partial derivation is denoted by a comma preceding the corresponding subscript:

$$\frac{\partial}{\partial x_i}(\ldots) \equiv (\ldots),_i .$$

The rest stress increments along the coordinate axes are:

$$d\sigma_{\alpha\beta} = \sigma_{\alpha\beta},_{\alpha} dx_{\alpha},$$

($\alpha$, $\beta$ = 1, 2, 3; without summation over repeated Greek index).

In provision that the sum of the moments of forces acting on the element about coordinate axes is equal to zero, we obtain the known reciprocity law for shearing stresses

$$\sigma_{ij} = \sigma_{ji} \ (i \neq j).$$

This proves that the stress tensor has only six independent components and its matrix is symmetric about the principal diagonal.[4]

Let us consider the equilibrium of forces acting on the singled-out element in direction of $x_1$ axis. For this purpose, we shall project all acting forces, including the bulk ones, onto this axis. With this aim, the stresses parallel to $x_1$ axis are multiplied by the area of the faces they are operating on, while intensity $\rho F_1$ by the volume of parallelepiped, and summed up:

$$\begin{gathered}(\sigma_{11} + \sigma_{11},_1 dx_1 - \sigma_{11})dx_2 dx_3 + (\sigma_{21} + \sigma_{21},_2 dx_2 - \sigma_{21})dx_3 dx_1 \\ +(\sigma_{31} + \sigma_{31},_3 dx_3 - \sigma_{31})dx_1 dx_2 + \rho F_1 dx_1 dx_2 dx_3 = 0 \qquad ,\end{gathered}$$

it follows that:

$$\sigma_{11},_1 + \sigma_{21},_2 + \sigma_{31},_3 + \rho F_1 = 0.$$

With the allowance for the symmetry of stress tensor ($\sigma_{ij} = \sigma_{ji}$), we obtain:

$$\sigma_{11},_1 + \sigma_{12},_2 + \sigma_{13},_3 + \rho F_1 = 0.$$

Similarly, for the rest two directions:

$$\begin{gathered}\sigma_{21},_1 + \sigma_{22},_2 + \sigma_{23},_3 + \rho F_2 = 0, \\ \sigma_{31},_1 + \sigma_{32},_2 + \sigma_{33},_3 + \rho F_3 = 0.\end{gathered}$$

In a general case, these three equations may be written in the tensorial form:

$$\sigma_{ij},_j + \rho F_i = 0. \tag{1.10}$$

They are known as *equilibrium equations of the elastic body* or *static equations*.

Notice that in the theory of generalized media, when so-called *couple stresses* arise (e.g., the theory of *Cosserat continuum*[5]), the stress tensor is no longer a symmetrical one.

## BOUNDARY EQUILIBRIUM CONDITIONS

Equilibrium equations (1.10) are valid all over the deformable body bulk, while at the boundary, that is, on the body surface, the equilibrium conditions in stresses should be met. This means that at the boundary we observe a continuous transition of the stress tensor into the surface loading it follow from (1.2):

$$R_\nu = -\sigma_{-\nu} = \sigma_\nu ,$$

where $R_\nu$ is the stress vector on the boundary given on the face with normal ν.

If we multiply this expression by $e_i$ scalar-wise and use formulas (1.8), we obtain the following *boundary conditions* for stresses:

$$R_{\nu i} = \sigma_{ij} l_j . \tag{1.11}$$

Hence, stresses on the boundary (superficial stresses) are equalized with stresses in the body bulk. In the coordinate form equalities (1.11) look like

$$R_{\nu 1} = \sigma_{11}\cos(\nu, x_1) + \sigma_{12}\cos(\nu, x_2) + \sigma_{13}\cos(\nu, x_3) ,$$

$$R_{\nu 2} = \sigma_{21}\cos(\nu, x_1) + \sigma_{22}\cos(\nu, x_2) + \sigma_{23}\cos(\nu, x_3) ,$$

$$R_{\nu 3} = \sigma_{31}\cos(\nu, x_1) + \sigma_{32}\cos(\nu, x_2) + \sigma_{33}\cos(\nu, x_3) ,$$

where the normal vector components are the directional cosines. We shall not use this subscript of the normal in our equations further.

## PRINCIPAL AXES AND PRINCIPAL VALUES OF STRESS TENSOR

Let us take a previously cut-out elementary parallelepiped with Cartesian coordinates «frozen» into it, and start rotating imaginary about the point under consideration. The values of the stress tensor components on its sides will be changing. It is acknowledged, that there is at least one position of a parallelepiped where the tangential stresses on its faces become equal to zero, and the normal stresses become limiting (Fig. 1.6). The corresponding axes of coordinates at a given point are known as the *principal axes of the stress tensor*. The stress components along these axes are expressed in terms of $\sigma_1$, $\sigma_2$, $\sigma_3$, and are called the *principal values of the stress tensor.*

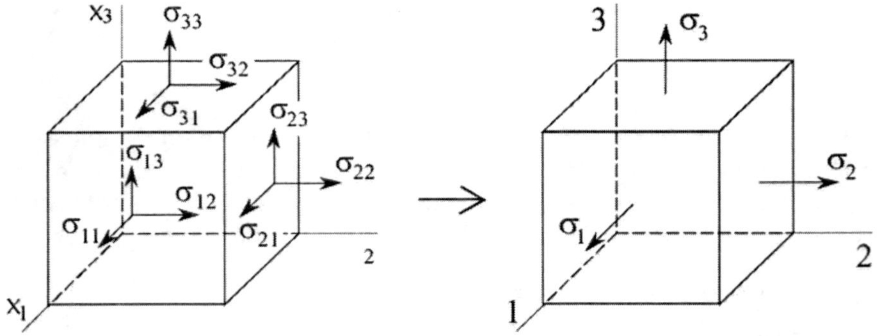


**Figure 1.6.**

The principal axes are numbered so as to meet the conditions for the principal values from the algebraic viewpoint: $\sigma_1 \geq \sigma_2 \geq \sigma_3$. The stress tensor matrix in the principal axes takes a diagonal form. The stress vector components on the oblique area element (1.8) along the principal axes (without summation over *i*) will become:

$$\sigma_{\nu i} = \sigma_i l_i . \qquad (1.12)$$

Let us consider now how to determine the values of principal stresses $\sigma_1$, $\sigma_2$, $\sigma_3$ based on known values of six coordinate components of stress tensor $\sigma_{ij}$. Remembering Fig. 1.4 and relations (1.5), let us assume that the oblique area element is the principal one. Then, the vector of combined stress $\sigma_\nu$ will be directed along normal ν to this area element since tangential stresses are absent (Fig. 1.7).

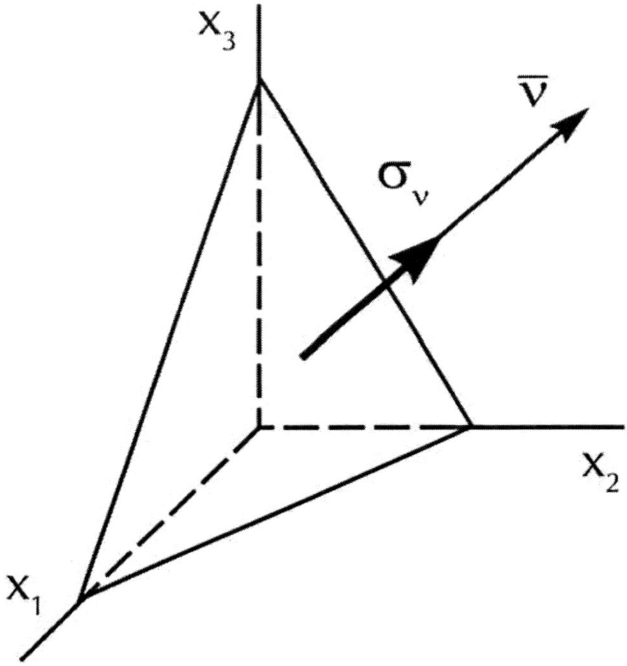


**Figure 1.7.**

We shall designate the sought value of the corresponding principal stress by σ. The projection of vector $\sigma_\nu$ on the coordinate axes will be:

$$\sigma_{\nu i} = \sigma l_i .$$

By their substitution into relations (1.8) and transposition of all items to the left side, we get a system of equations:

$$\sigma_{ij} l_j - \sigma l_i = 0 . \qquad (1.13)$$

Term $\sigma l_i$ in equations (1.13) will be of the kind

$$\sigma l_i \equiv \sigma \delta_{ij} l_j \,.$$

This manipulation ($l_i \equiv \delta_{ij} l_j$) is called «*juggling*» with indices by means of the metric tensor. We can verify this expression, for example, at $i = 1$:

$$\sigma l_1 = \sigma \delta_{1j} l_j = \sigma \delta_{11} l_1 + \sigma \delta_{12} l_2 + \sigma \delta_{13} l_3 = \sigma \delta_{11} l_1 = \sigma l_1 \,,$$

since (1.4) $\delta_{12} = \delta_{13} = 0$, $\delta_{11} = 1$.

After substitution of these relations in equations (1.13), we obtain a system of three homogeneous linear algebraic equations with respect to unknown cosines of the normal that define orientation of the principal area element in the reference coordinates:

$$(\sigma_{ij} - \sigma \delta_{ij}) l_j = 0 \quad (i, j = 1, 2, 3). \tag{1.14}$$

In the case of uniqueness of the obtained system solution, it will be equal to zero: $l_i = 0$, ($i = 1, 2, 3$). As far as the zero, the solution has no physical meaning to ground the existence of a nonzero solution it is necessary and sufficient for the system to have the determinant equal to zero:

$$|\sigma_{ij} - \sigma \delta_{ij}| \equiv \begin{vmatrix} \sigma_{11} - \sigma & \sigma_{12} & \sigma_{13} \\ \sigma_{21} & \sigma_{22} - \sigma & \sigma_{23} \\ \sigma_{31} & \sigma_{32} & \sigma_{33} - \sigma \end{vmatrix} = 0. \tag{1.15}$$

This is achieved by the appropriate selection of σ value. If condition (1.15) is met, at least one of three equations of system (1.14) will be not independent and may be linearly expressed through the other two ones. Having added the following condition to the independent equations

$$l_i l_i = 1 \,, \text{ or } l_1^2 + l_2^2 + l_3^2 = 1 \,,$$

we obtain a new system of three independent equations. It is sufficient for finding cosines of the normal to the principal area element, on which the stress equal to σ is acting.

The equation for finding *the principal stress values* is the determinant (1.15). If to expand it and to place the summands in degrees σ, we shall get a cubic equation, which is known as a *secular* one:

$$\sigma^3 - \sigma^2 J_1 + \sigma J_2 - J_3 = 0 \,.$$

Here,

$$J_1 = \sigma_{ii} \equiv \sigma_{11} + \sigma_{22} + \sigma_{33} = \sigma_1 + \sigma_2 + \sigma_3 \,,$$

$$J_2 = (\sigma_{ii}\sigma_{jj} - \sigma_{ij}\sigma_{ij}) / 2 = \sigma_{11}\sigma_{22} + \sigma_{22}\sigma_{33} + \sigma_{33}\sigma_{11} - \sigma_{12}^2 - \sigma_{23}^2 - \sigma_{31}^2 =$$

$$= \sigma_1\sigma_2 + \sigma_2\sigma_3 + \sigma_3\sigma_1$$

$$J_3 = \left\|\sigma_{ij}\right\| \equiv \begin{vmatrix} \sigma_{11} & \sigma_{12} & \sigma_{13} \\ \sigma_{21} & \sigma_{22} & \sigma_{23} \\ \sigma_{31} & \sigma_{32} & \sigma_{33} \end{vmatrix} = \sigma_1\sigma_2\sigma_3 \tag{1.16}$$

We may prove that all three roots of the secular equation are real-valued. They are the *principal values of the stress tensor* and are expressed in terms of $\sigma_1$, $\sigma_2$, $\sigma_3$. Their values are conditioned by the nature of the external load and are independent of the initial orientation of the coordinate system. Therefore, when the axes are rotated, the values of coefficients $J_1$, $J_2$, $J_3$ in the secular equation found from formulas (1.16) should remain unchangeable. In this connection, they are called the *stress tensor invariants.*

When all principal stresses are not equal to zero, the stress condition is known as a *triaxial* one. If one of them is zero, the stress state is *plane*. If two principal stresses are simultaneously equal to zero, we have a *linear* stress state.

The physical essence of the stress tensor depends on its three invariants, so far the values dependent on it (e.g., plasticity criteria) should be the functions of $J_1$, $J_2$, $J_3$. This circumstance should be taken into account in the general theories of the mechanical behavior of materials.

## MAXIMAL TANGENTIAL STRESSES

The *octahedral* area elements are known to be equally sloping to the principal axes of the stress tensor. To obtain these stress values, let us first consider an arbitrary area element in the principal axes (Fig. 1.8). Let us denote its stress vector through $\sigma_\nu$. It will be expanded into the normal ($\sigma_n$) and tangential ($\tau_n$) components:

$$\sigma_\nu^2 = \sigma_n^2 + \tau_n^2 \,. \tag{1.17}$$

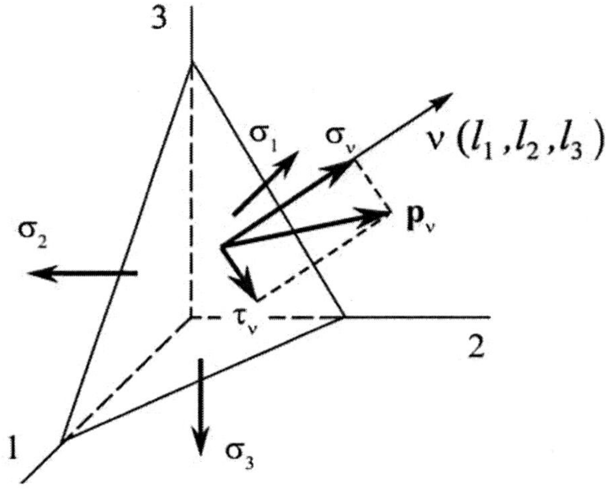


**Figure 1.8.**

The projections of vector $\boldsymbol{\sigma}_\nu$ on the principal axes according to formulas (1.12) will be:

$$\sigma_{\nu\alpha} = \sigma_\alpha l_\alpha.$$

where, $l_\alpha$ are the directional cosines of the normal to the considered arbitrarily orientated area element.

Then, the modulus of the combined stress vector on this area element may be obtained from a sum of squares of projections, and the value of the normal stress $\sigma_n$ as a sum of projections of the combined stress components on the normal:

$$\sigma_\nu^2 = \sigma_{\nu 1}^2 + \sigma_{\nu 2}^2 + \sigma_{\nu 3}^2 = \sigma_i^2 l_i^2,$$

$$\sigma_n = \sigma_{\nu i} l_i = \sigma_1 l_1^2 + \sigma_2 l_2^2 + \sigma_3 l_3^2 = \sigma_i l_i^2. \qquad (1.18)$$

Substitution of the expressions obtained for $\sigma_\nu^2$ and $\sigma_n$ in equation (1.17) and the allowance that the sum of cosine squares is $l_i l_i = 1$ gives us an *expression for tangential stresses on an arbitrary area element*:

$$\tau_n^2 = (\sigma_1 - \sigma_2)^2 l_1^2 l_2^2 + (\sigma_2 - \sigma_3)^2 l_2^2 l_3^2 + (\sigma_3 - \sigma_1)^2 l_3^2 l_1^2. \qquad (1.19)$$

Notice that $\tau_n^2$ is a positive value and tends to zero on the principal area elements. Actually, if normal ν coincides with one of the principal axes, one of the directional cosines is equal to "one," and the other two—to zero, so far, $\tau_n^2 = 0$.

Let us return to the octahedral area elements. Since they are similarly sloping to the principal axes, all their directional cosines are $l_i^2 = 1/3$. Their substitution into formulas (1.18), (1.19), gives us the *values of the normal and tangential stresses on octahedral area elements*:

$$\sigma_{\text{oct}} = \tfrac{1}{3}(\sigma_{11} + \sigma_{22} + \sigma_{33}) = \tfrac{1}{3}(\sigma_1 + \sigma_2 + \sigma_3),$$

$$\tau_{\text{oct}} = \tfrac{1}{3}\sqrt{(\sigma_1 - \sigma_2)^2 + (\sigma_2 - \sigma_3)^2 + (\sigma_3 - \sigma_1)^2} \qquad (1.20)$$

By expressing $\tau_{oct}$ through the arbitrary coordinate components of stresses we obtain:

$$\tau_{\text{oct}} = \tfrac{1}{3}\sqrt{(\sigma_{11} - \sigma_{22})^2 + (\sigma_{22} - \sigma_{33})^2 + (\sigma_{33} - \sigma_{11})^2 + 6(\sigma_{12}^2 + \sigma_{23}^2 + \sigma_{31}^2)}.$$

The octahedral stresses are invariant values and are expressed through the stress tensor invariants:

$$\sigma_{\text{oct}} = \tfrac{1}{3} J_1, \quad \tau_{\text{oct}} = \tfrac{1}{3}\sqrt{2J_1^2 - 2J_2}.$$

Namely to Nadai we owe the introduction of octahedral stresses (Nadai, 1950). They play an important role in calculations of equivalent stresses for a compound stress. They are also used to formulate the plasticity criteria.

The area elements on which the *maximal tangential stresses* arise are of particular interest. The investigation of the extremum of relation (1.19) will assist in finding the location of these area elements. Since

$$(\sigma_1 - \sigma_3) = (\sigma_1 - \sigma_2) + (\sigma_2 - \sigma_3)$$

and the square exceeds the sum of squares of its components, then

$$(\sigma_1 - \sigma_3)^2 \geq (\sigma_1 - \sigma_2)^2 + (\sigma_2 - \sigma_3)^2.$$

Hence, $\tau_n$ value reaches a maximum on the area elements where the third term of formula (1.19) is maximal. Therefore, at $l_1^{\,2} = l_3^{\,2} = 1/2$, $l_2^{\,2} = 0$, we obtain $\tau_2 = \tau_{max}$ (index 2 indicates the axis, to which the area element is parallel),

$$\tau_{max} = \tfrac{1}{2}(\sigma_1 - \sigma_3)\,. \qquad (1.21)$$

Thus, the area element with the *maximal tangential stress* is parallel to the principal axis 2 and is equally sloping to the principal area elements where the principal maximal ($\sigma_1$) and minimal ($\sigma_3$) stresses are operating (Fig. 1.9).

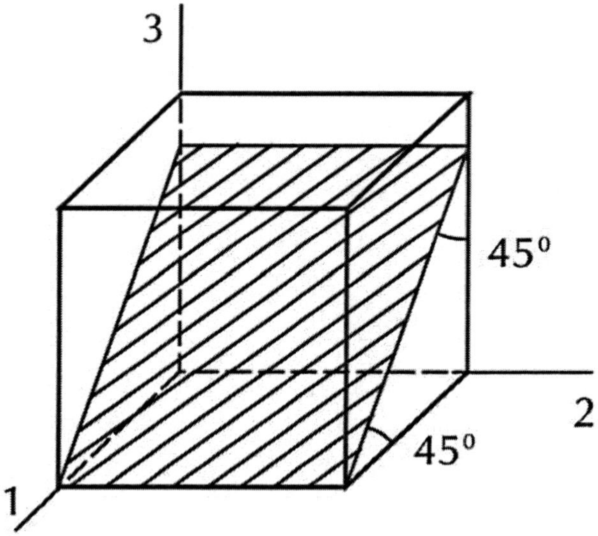


**Figure 1.9.**

If to consider two more area elements parallel to the first and third principal axes, respectively, and are similarly sloping to the rest two, we obtain two more extreme values of tangential stresses:

$$\tau_1 = \tfrac{1}{2}(\sigma_2 - \sigma_3)\,,\quad \tau_3 = \tfrac{1}{2}(\sigma_1 - \sigma_2)\,.$$

Values $\tau_1$, $\tau_2$, $\tau_3$ are known as the *principal tangential stresses*. The relation valid for them is $\tau_1 + \tau_2 + \tau_3 = 0$. The surfaces, on which the principal tangential stresses operate, are not mutually orthogonal, but create the sides of a regular dodecahedron. Its faces are not free from normal stresses.

## DEVIATOR AND SPHERICAL PART OF STRESS TENSOR

There exists its own stress tensor for each point of a solid, there is to be a field of stress tensors in a body. Let us take two features of the tensors useful in our further argumentation:

- a tensor, whose components present a sum of corresponding components of the tensor terms, that is, is a *sum of two tensors*;
- a tensor, the components of which are λ times greater than the corresponding terms of the tensor to be multiplied, are *a product of the tensor by scalar* λ.

Let us consider a stress state under which only three principal stresses σ are equal to the mean stress in a given point of the body are acting on three mutually perpendicular area elements.

$$\sigma_1 = \sigma_2 = \sigma_3 = \sigma = \sigma_i\ /3 \equiv (\sigma_1\ + \sigma_2\ + \sigma_3\ )/3\,.$$

This stress condition is described by the tensor

$$T_\sigma = \begin{pmatrix} \sigma & 0 & 0 \\ 0 & \sigma & 0 \\ 0 & 0 & \sigma \end{pmatrix}, \tag{1.22}$$

which is known as a *spherical stress tensor* (spherical part of the stress tensor).

The term «spherical tensor» is connected with a geometrical representation of the stress state in a point, and was proposed by Lame. If to draw a combined stress vector in the system of coordinates coincident with the principle axes for each area element passing through the origin of coordinates, the ends of these vectors will describe an ellipsoidal surface called the *stress ellipsoid* or *Lame's ellipsoid.*

Three semi-axes of the stress ellipsoid are equal in length to three principal stresses. Under a stress state described by the spherical tensor (1.22), all three principal stresses are equal to each other, and the stress ellipsoid transforms into a sphere.

By subtracting the spherical tensor from the stress tensor, we obtain a new tensor known as a *stress deviator*,

$$D_\sigma \equiv (S_{ij}) \begin{pmatrix} \sigma_{11} - \sigma & \sigma_{12} & \sigma_{13} \\ \sigma_{21} & \sigma_{22} - \sigma & \sigma_{23} \\ \sigma_{31} & \sigma_{32} & \sigma_{33} - \sigma \end{pmatrix}. \tag{1.23}$$

Thus, the stress tensor in each point may be presented as a sum of spherical stress tensor (σ) and stress deviator ($s_{ij}$). For their components, the following relation is met:

$$\sigma_{ij} = s_{ij} + \sigma\delta_{ij} \ (i, j = 1, 2, 3),$$

where $\delta_{ij}$ are Cronecker's deltas (1.4).

The expansion of the stress tensor into the spherical and deviatoric parts is of fundamental importance for study a behavior of the solid and plastic bodies under loading. The spherical part isolates from the stress state a uniform tension or compression that induces changes only in the volume of a given element, leaving intact its shape. The stress deviator characterizes a shear condition at which the shape of a body element is changed without changing its volume. Consequently, the stress deviator indicates a deflection (*deviation*) of the stress state under study from the uniform tension (compression) or deflection of the acquired shape of the body from its original one. As the experiments have shown, the materials react differently on the uniform compression shear stress.

By analogy with the stress tensor invariants, let us find the invariants for the tensors introduced. The first invariant of the spherical tensor coincides with the first invariant of the stress tensor:

$$J_{1s} = 3\sigma \equiv \sigma_1 + \sigma_2 + \sigma_3 .$$

The first invariant of the stress deviator is equal to zero

$$J_{1d} = s_{ii} \equiv (\sigma_{11} - \sigma) + (\sigma_{22} - \sigma) + (\sigma_{33} - \sigma) = \sigma_{11} + \sigma_{22} + \sigma_{33} - 3\sigma = 0 .$$

To determine the second invariant of the stress deviator let us use an expression for the second invariant of the stress tensor (1.16) by substituting differences $\sigma_{ii} - \sigma$ instead of stresses $\sigma_{ii}$. After simple transformations, we obtain:

$$J_{2d} = -s_{ij}s_{ij}/2 = -\left[(\sigma_{11} - \sigma_{22})^2 + (\sigma_{22} - \sigma_{33})^2 + (\sigma_{33} - \sigma_{11})^2 .\right.$$

$$\left. + 6(\sigma_{12}^2 + \sigma_{23}^2 + \sigma_{31}^2)\right]/6 = -\left[(\sigma_1 - \sigma_2)^2 + (\sigma_2 - \sigma_3)^2 + (\sigma_3 - \sigma_1)^2\right]/6$$

The third invariant of the deviator is of the form

$$J_{3d} = \sigma_{ij}\sigma_{jk}\sigma_{ki}/3 .$$

The plasticity theory often employs a notion *of the stress tensor intensity*, introduced by Ilyushin. It is formally determined through the second invariant of the deviator

$$\sigma_u^2 = -3J_{2d} = 3/2\, s_{ij}s_{ij} .$$

Consequently,

$$\sigma_u = \tfrac{\sqrt{2}}{2}\sqrt{(\sigma_{11} - \sigma_{22})^2 + (\sigma_{22} - \sigma_{33})^2 + (\sigma_{33} - \sigma_{11})^2 + 6(\sigma_{12}^2 + \sigma_{23}^2 + \sigma_{31}^2)} . \qquad (1.24)$$
$$= \tfrac{\sqrt{2}}{2}\sqrt{(\sigma_1 - \sigma_2)^2 + (\sigma_2 - \sigma_3)^2 + (\sigma_3 - \sigma_1)^2}$$

The stress tensor intensity is an invariant variable being expressed through the second invariant of the deviator. It coincides with the octahedral tangential stress (1.20) within the numerical factor. The numerical coefficient in formula (1.24) was selected so as to meet condition $\sigma_u = |\sigma_1|$ in case of a simple tension or compression ($\sigma_{11} = \sigma_1$, all other components being equal to zero).

The positive value

$$s = \sqrt{s_{ij} s_{ij}} = \frac{\sqrt{3}}{3}\sqrt{(\sigma_1 - \sigma_2)^2 + (\sigma_2 - \sigma_3)^2 + (\sigma_3 - \sigma_1)^2} \tag{1.25}$$

is known as a *modulus of stress deviator*. Let us call the stress deviator having each component divided by the deviator modulus (1.25) a *directing stress tensor*:

$$\overline{D}_\sigma \equiv (\overline{s}_{ij}) = \frac{D_\sigma}{s} = \begin{pmatrix} (\sigma_{11} - \sigma)/s & \sigma_{12}/s & \sigma_{13}/s \\ \sigma_{21}/s & (\sigma_{22} - \sigma)/s & \sigma_{23}/s \\ \sigma_{31}/s & \sigma_{32}/s & (\sigma_{33} - \sigma)/s \end{pmatrix}. \tag{1.26}$$

Since the first invariant of the deviator is equal to zero, then $\overline{s}_{ii} = \overline{s}_{11} + \overline{s}_{22} + \overline{s}_{33} = 0$. Simultaneously, $\overline{s}_{ij}\overline{s}_{ij} = 1$. Hence, the directing tensor is characterized fully by giving four numbers, since six other components have already been bound by two reduced relations. It should be noted, that the principal axes of the directing tensor coincide with the principal axes of the stress tensor and the stress tensor deviator.

With the help of the directing tensor the stress tensor components may be presented as:

$$\sigma_{ij} = \sigma\delta_{ij} + s\,\overline{s}_{ij}.$$

Two modules $\sigma$ and $s$ define scalar properties of the material. As for the vector properties of the material, they are denoted by the directing tensor ($\overline{s}_{ij}$).

The *loading* in a given point of a body is called *simple* if all components of the stress tensor change proportionally to a common variable $\lambda$, that is, $\sigma_{ij} = \lambda\sigma_{ij}^0$. In this case, the directing tensor components shall remain unchangeable, or otherwise, the loading is called *complex.*[6]

## DISPLACEMENTS AND DEFORMATIONS STRAIN TENSOR

None of materials existing in nature can be considered absolutely solid. Under the impact of external forces all bodies change their shape to a greater or lesser extent (undergo deformation). The points of a body are changing their position continuously. The solid material unbroken initially does not contain any discontinuities or cavities after deformation (no cracks are emerging).

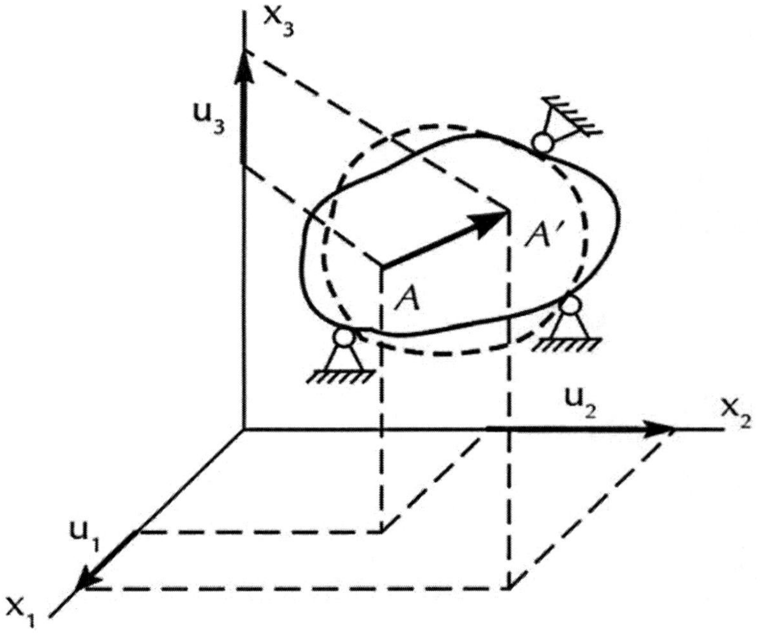


**Figure 1.10.**

Vector $u$, originating in point $A$ of an undeformed body and terminating in a corresponding point $A'$ of the deformed one is known as a *total displacement vector* of the point (Fig. 1.10).

Its projections onto the coordinate axes are:

$$u = (u_1, u_2, u_3),\ u_i \equiv u_i(x),\ x \equiv (x_1, x_2, x_3).$$

The mechanics of deformable solids examines, as a rule, invariable in kinematics respect systems, meaning that they do not allow displacements of a body in space as a rigid whole, or otherwise the constituent of this transfer shall be excluded from the displacements of all points. The displacements thus introduced are small for most of the systems considered as compared to geometrical dimensions of the body.

One of the methods of determining deformations consists in calculations of length variations of linear components and changes in the angles between two linear components proceeding from the displacements occurred. In order to relate the displacements in point $A$ with the deformation in its vicinity, let us first consider a plane problem, that is, assume that

$$u_3(x) \equiv 0,\ u_1 \equiv u_1(x_1, x_2),\ u_2 \equiv u_2(x_1, x_2).$$

Segment $AB$ with projections $dx_1$, $dx_2$ will occupy in the course of deformation position $A'B'$ (Fig. 1.11). The total displacement of point $A$ along axis $x_1$ is $u_1$. Point $B$ will move together with point $A$ to a value $u_1$ plus extra displacement $du_1$ due to the deformation of the segment along $x_1$ axis. Since the displacement increment is negligible, and $du_1 = du_1(x_1, x_2)$, so till the accuracy of the members of a higher order of smallness, we have

$$du_1 = \frac{\partial u_1}{\partial x_1}dx_1 + \frac{\partial u_1}{\partial x_2}dx_2 \equiv \frac{\partial u_1}{\partial x_j}dx_j \equiv u_{1,j}\,dx_j \ (j = 1, 2).$$

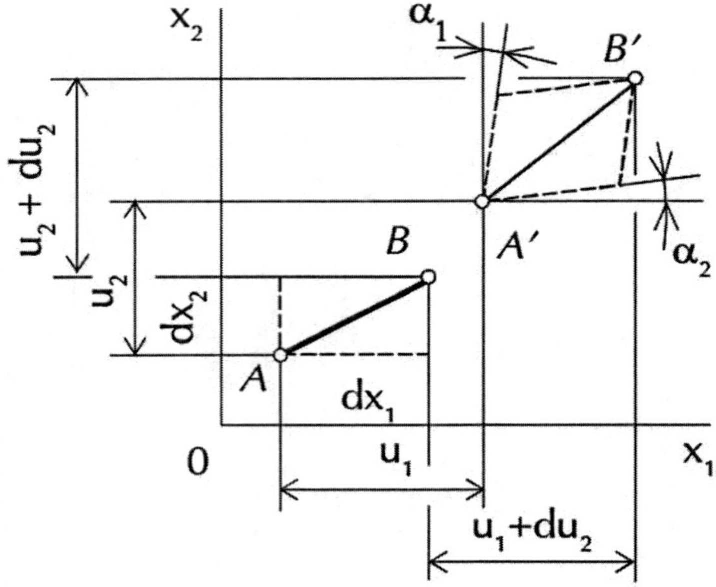


**Figure 1.11.**

Here, the first term corresponds to the elongation of component $dx_1$ along $x_1$ axis. The second term describes, in view of smallness of deformations, the displacement due to turning of the segment about axis $x_2$ by angle $\alpha_1$:

$$\alpha_1 \approx \operatorname{tg}\alpha_1 = \frac{\partial u_1}{\partial x_2} \equiv u_{1,2}.$$

When describing the displacements of point $B$ along axis $x_2$, we may obtain by analogy:

$$du_2 = \frac{\partial u_2}{\partial x_1}dx_1 + \frac{\partial u_2}{\partial x_2}dx_2 \equiv u_{2,j}\,dx_j, \alpha_2 \approx \operatorname{tg}\alpha_2 = \frac{\partial u_2}{\partial x_1} \equiv u_{2,1} \ (j = 1, 2).$$

The values describing the linear elongation along the axes are called *linear deformations*, and are expressed by:

$$\varepsilon_{11}=\frac{\partial u_1}{\partial x_1}\equiv u_{1,1},\ \varepsilon_{22}=\frac{\partial u_2}{\partial x_2}\equiv u_{2,2}.$$

To describe *shear deformations,* we use a value equal to a half of the total change of the right angle between the coordinate axes:

$$\varepsilon_{12}=\tfrac{1}{2}(\alpha_1+\alpha_2)=\tfrac{1}{2}(u_{1,2}+u_{2,1}).$$

If the deformation in the vicinity of point $A$ is cubic ($u_i(x)\neq 0$, $i = 1, 2, 3$), the displacement increment will be:

$$\mathrm{d}u_i=u_{i,j}\,\mathrm{d}x_j\ (i,j=1,2,3).$$

Besides, the following deformations are added:

$$\varepsilon_{33}=u_{3,3},\varepsilon_{32}=\tfrac{1}{2}(u_{2,3}+u_{3,2}),\varepsilon_{31}=\tfrac{1}{2}(u_{3,1}+u_{1,3}).$$

As a result, we obtain in a general case, that the deformation components are related to small displacements ($u_i^2,_j << u_{i,j}$) via the formulas called *Cauchy relations*:

$$\varepsilon_{ij}=\tfrac{1}{2}(u_{i,j}+u_{j,i})\ (i,j=1,2,3). \qquad (1.27)$$

*Note 1.* In the case of *finite deformations* (*geometrically nonlinear theory of elasticity*), when $u_i^2,_j$ and $u_{i,j}$ are comparable, the strain tensor components are related to displacements using the relations:

$$\varepsilon_{ij}=\tfrac{1}{2}(u_{i,j}+u_{j,i}+u_{k,i}u_{k,j})\ (i,j,k=1,2,3).$$

*Note 2.* In the bending problems of membranes and thin plates by transverse loading only flexures may be finite, so only $u_3^2,_j$ and $u_{3,j}$ are comparable. The equations relating deformations and displacements are of the form:

$$\varepsilon_{ij}=\tfrac{1}{2}(u_{i,j}+u_{j,i}+u_{3,i}\,u_{3,j})\ (i,j=1,2,3).$$

The strain components (1.27) create a second-order tensor, being symmetrical ($\varepsilon_{ij} = \varepsilon_{ji}$) similarly to the stress tensor. We may show that strain of an arbitrary fiber is described completely in the neighborhood of the point under consideration using this tensor.

## PRINCIPAL AXES AND PRINCIPAL VALUES OF STRAIN TENSOR

Similarly to the case considered for the stress tensor, we can study in which directions only relative deformations are present without any shear. We may prove that these deformations also correspond to the extreme values of relative deformations. These values are called the *principal relative deformations*, and they are expressed in terms of $\varepsilon_1$, $\varepsilon_2$, $\varepsilon_3$.

To determine the principal directions and principal relative deformations one may use the corresponding expressions for the stress tensor mentioned in 1.6. The following equations are valid:

$$(\varepsilon_{ij} - \varepsilon\delta_{ij})l_j = 0 \; (i, j = 1, 2, 3).$$

Resolvability conditions for these equations bring us to a *characteristic equation* for the principal values of the strain tensor:

$$\varepsilon^3 - \varepsilon^2 I_1 + \varepsilon I_2 - I_3 = 0.$$

Three invariants independent of the coordinate system orientation will be the coefficients in this equation:

$$I_1 = \varepsilon_{ii} \,{}^{\circ}\, \varepsilon_{11} + \varepsilon_{22} + \varepsilon_{33} = \varepsilon_1 + \varepsilon_2 + \varepsilon_3,$$

$$I_2 = (\varepsilon_{ii}\varepsilon_{jj} - \varepsilon_{ij}\varepsilon_{ij})/2 = \varepsilon_{11}\varepsilon_{22} + \varepsilon_{22}\varepsilon_{33} + \varepsilon_{33}\varepsilon_{11} - \varepsilon_{12}^2 - \varepsilon_{23}^2 - \varepsilon_{31}^2$$

$$= \sigma_1\sigma_2 + \sigma_2\sigma_3 + \sigma_3\sigma_1,$$

$$I_3 = \left\| \varepsilon_{ij} \right\| \equiv \begin{vmatrix} \varepsilon_{11} & \varepsilon_{12} & \varepsilon_{13} \\ \varepsilon_{21} & \varepsilon_{22} & \varepsilon_{23} \\ \varepsilon_{31} & \varepsilon_{32} & \varepsilon_{33} \end{vmatrix} = \varepsilon_1\varepsilon_2\varepsilon_3 \qquad (1.28)$$

The strain tensor matrix in the principal axes acquires a diagonal form. The first invariant $I_1$ bears a simple geometrical meaning. This is a *relative volume change*

$$\theta = u_{i,i} = \lim_{\Delta V_0 \to 0} \frac{\Delta V - \Delta V_0}{\Delta V_0} = div u.$$

The direction of the principal strain tensor axes is obtained from the relations similar to (1.12). Within the frameworks of the elasticity theory, the principal axes of the stress and strain tensors overlap for the isotropic medium.

## DEVIATOR AND SPHERICAL PART OF STRAIN TENSOR

Strain tensor $(\varepsilon_{ij})$ may be presented as two tensor components

$$\varepsilon_{ij} = э_{ij} + \varepsilon\delta_{ij} \; (i, j = 1, 2, 3). \qquad (1.29)$$

Here, $\varepsilon = \theta/3 \equiv (\varepsilon_{11} + \varepsilon_{22} + \varepsilon_{33})/3$ is a *spherical part of the strain tensor*, $\delta_{ij}$—Cronecker's deltas (1.4), $э_{ij}$—*components of the strain tensor deviator*. They are expressed in terms of deformations by the following relation:

$$D_\varepsilon \equiv (э_{ij}) \equiv \begin{pmatrix} \varepsilon_{11} - \varepsilon & \varepsilon_{12} & \varepsilon_{13} \\ \varepsilon_{21} & \varepsilon_{22} - \varepsilon & \varepsilon_{23} \\ \varepsilon_{31} & \varepsilon_{32} & \varepsilon_{33} - \varepsilon \end{pmatrix} \qquad (1.30)$$

The spherical strain tensor (1.29) describes a 3D deformation $\theta$ in some point of a body. Its first invariant coincides with the first invariant of the strain tensor (1.28):

$$I_{1s} = 3\varepsilon \equiv \varepsilon_{11} + \varepsilon_{22} + \varepsilon_{33} .$$

The strain deviator characterizes the shape change deformation. Its first invariant is equal to zero since

$$I_{1d} = э_{ii} \equiv (\varepsilon_{11} - \varepsilon) + (\varepsilon_{22} - \varepsilon) + (\varepsilon_{33} - \varepsilon) = \varepsilon_{11} + \varepsilon_{22} + \varepsilon_{33} - 3\varepsilon = 0$$

The second invariant of the strain deviator is obtained in much the same way as the stress deviator:

$$I_{2d} = -э_{ij}э_{ij}/2 = -\left[(\varepsilon_{11} - \varepsilon_{22})^2 + (\varepsilon_{22} - \varepsilon_{33})^2 + (\varepsilon_{33} - \varepsilon_{11})^2 \right.$$
$$\left. + 6(\varepsilon_{12}^2 + \varepsilon_{23}^2 + \varepsilon_{31}^2)\right]/6 = -\left[(\varepsilon_1 - \varepsilon_2)^2 + (\varepsilon_2 - \varepsilon_3)^2 + (\varepsilon_3 - \varepsilon_1)^2\right]/6 .$$

The third deviator invariant looks like

$$I_{3d} = э_{ij}э_{jk}э_{ki}/3 .$$

The second invariant of strain deviator $I_{2d}$ plays a fundamental role in the elasticity theory. The positive value

$$э = \sqrt{э_{ij}э_{ij}} = \tfrac{\sqrt{3}}{3}\sqrt{(\varepsilon_1 - \varepsilon_2)^2 + (\varepsilon_2 - \varepsilon_3)^2 + (\varepsilon_3 - \varepsilon_1)^2}$$

is called the *strain deviator modulus*.

Using expression $I_{2d}$, let us introduce a definition for the *intensity of strain tensor* $\varepsilon_u$ according to Ilyushin:

$$\varepsilon_u^2 = -\tfrac{4}{3} I_{2d} = \tfrac{2}{3} э_{ij}э_{ij}$$

or

$$\varepsilon_u = \tfrac{\sqrt{2}}{3}\sqrt{(\varepsilon_{11} - \varepsilon_{22})^2 + (\varepsilon_{22} - \varepsilon_{33})^2 + (\varepsilon_{33} - \varepsilon_{11})^2 + 6(\varepsilon_{12}^2 + \varepsilon_{23}^2 + \varepsilon_{31}^2)}$$
$$= \tfrac{\sqrt{2}}{3}\sqrt{(\varepsilon_1 - \varepsilon_2)^2 + (\varepsilon_2 - \varepsilon_3)^2 + (\varepsilon_3 - \varepsilon_1)^2} \qquad (1.31)$$

The geometrical interpretation of the deformation intensity is: $\varepsilon_u$ coincides within the accuracy of the numerical factor with octahedral shear, that is relative shear of the area element equally sloping to the principal axes of deformation.

In the case of a simple tension or compression, $\varepsilon_{11} = \varepsilon_1$, $\varepsilon_{22} = \varepsilon_{33} = -\nu\varepsilon_1$, $\varepsilon_{12} = \varepsilon_{23} = \varepsilon_{31} = 0$ ($\nu$ – Poisson's[7] ratio). Proceeding from (1.31), it follows that $\varepsilon_u = \tfrac{2}{3}(1+\nu)|\varepsilon_1|$, and for the incompressible material, ($\nu = 1/2$) – $\varepsilon_u = |\varepsilon_1|$. According to (1.31) the intensity of deformations is an invariant value.

A deviator whose components

$$\bar{э}_{ij} = э_{ij}/э$$

is known as the *directing strain tensor*, where

$$\bar{э}_{ii} = 0, \quad \bar{э}_{ij}\bar{э}_{ij} = 1/2$$

Hence, the directing tensor is characterized completely by giving four numbers, since six components have already been related via two reduced equations. By making use of the tensor, the strain tensor components may be expressed in terms of

$$\varepsilon_{ij} = \varepsilon\delta_{ij} + э\,\bar{э}_{ij}$$

Two moduli σ and $s$ determine scalar properties, and the directing tensor $(\bar{э}_{ij})$ – vector properties of the material. Deformation in a material particle of a body is known as a *simple deformation,* if all components $(\bar{э}_{ij})$ = const during the deformation, or otherwise deformation is called complex.

## COMPATIBILITY EQUATION OF DEFORMATIONS

Using three known differentiable components of the displacement field $u_i(x)$, $(x \equiv x_1, x_2, x_3)$ it is easy to determine six independent components of the strain tensor with the help of Cauchy's formulas (1.27). The inverse operation is hampered since there is not always a coincidence between the continuous field of displacements and six continuous components $\varepsilon_{ij}(x)$. If the field exists, the deformations are called *compatible*, or otherwise *incompatible*.

The combined deformations are related with each other by the equations called *deformation compatibility equations*. They may be derived in various ways. On the one hand, they may be considered as integrability conditions for six differential equations $\varepsilon_{ij} = ½(u_{i},_{j} + u_{j},_{i})$ with respect to three unknown functions $u_i(x)$. On the other hand, they express a physical fact, that the whole body before and after deformation is continuous and solid, and the material particles are bonded with each other. Saint-Venant was the first to obtain theses relations in 1860.

To derive the first group of compatibility equations establishing the links between the deformations acting in a single plane, let us write out Cauchy's relations in components:

$$\begin{aligned}&\varepsilon_{11} = u_{1},_{1}, \varepsilon_{22} = u_{2},_{2}, \varepsilon_{33} = u_{3},_{3},\\ &\varepsilon_{12} = \tfrac{1}{2}(u_{1},_{2} + u_{2},_{1}), \varepsilon_{23} = \tfrac{1}{2}(u_{2},_{3} + u_{3},_{2}), \varepsilon_{31} = \tfrac{1}{2}(u_{3},_{1} + u_{1},_{3})\cdot\end{aligned} \qquad (1.32)$$

Using (1.32) and the property of differentiation order permutations, we can write the following equation:

$$\varepsilon_{11},_{22} + \varepsilon_{22},_{11} = u_{1},_{122} + u_{2},_{211} = (u_{1},_{2} + u_{2},_{1})_{12} = 2\varepsilon_{12},_{12},$$

because $2\varepsilon_{12} = u_{1},_{2} + u_{2},_{1}$.

By analogy, we may obtain the next two similar equations. As a result,

$$\varepsilon_{\alpha\alpha},_{\beta\beta} + \varepsilon_{\beta\beta},_{\alpha\alpha} - 2\varepsilon_{\alpha\beta},_{\alpha\beta} = 0 \ (\alpha, \beta = 1, 2, 3; \alpha \neq \beta) \qquad (1.33)$$

without summation, as previously, over the repeated Greek indices.

To derive the second group from three integrity equations relating the deformations acting in different planes, let us compose the following expression:

$$(\varepsilon_{12},_3 - \varepsilon_{23},_1 + \varepsilon_{31},_2),_1 = \tfrac{1}{2}(u_1,_{23} + u_2,_{13} - u_2,_{31} - u_3,_{21} + u_3,_{12} + u_1,_{32}),_1$$
$$= u_1,_{231} = \varepsilon_{11},_{23}$$

where from,

$$(\varepsilon_{12},_3 - \varepsilon_{23},_1 + \varepsilon_{31},_2),_1 - \varepsilon_{11},_{23} = 0$$

In analogous way, with the help of a circular permutation of indices, two more equations of a similar type can be formally obtained. As a result,

$$(\varepsilon_{\alpha\beta},_\gamma - \varepsilon_{\beta\gamma},_\alpha + \varepsilon_{\gamma\alpha},_\beta),_\alpha - \varepsilon_{\alpha\alpha},_{\beta\gamma} = 0 \ (\alpha, \beta, \gamma = 1, 2, 3; \alpha \neq \beta \neq \gamma). \quad (1.34)$$

Hence, to find a corresponding field of displacements using six continuous components of the strain tensor, six compatibility differential equations (1.33), (1.34) should be satisfied in partial derivatives with respect to six components of strain tensor $\varepsilon_{ij}$. In the case of a simply-connected domain, they are necessary and sufficient, but for the multiply-connected domain, they are only necessary (Postnov, 1977).

The geometric meaning of the compatibility equations consists in the following. Imagine that a body has been divided before deformation into a multitude of material particles of a rectangular parallelepiped shape. We admit that after being subjected to arbitrary deformation $\varepsilon_{ij}$, all material particles take the shape of skew-angular parallelepipeds, which are unable any longer to make up a solid deformed body. To avoid this, the deformation components should satisfy equations (1.33), (1.34).

Along with displacements, we shall introduce other kinematics characteristics of a deformable medium. By differentiation in time, we can define velocity vector **v** and acceleration vector w:

$$v = \frac{\partial u}{\partial t} = \dot{u}, \ w = \frac{\partial v}{\partial t} = \frac{\partial^2 u}{\partial t^2} = \ddot{u} .$$

It should be underlined in conclusion, that further consideration of a probable solution of the elasticity problem in displacements involves automatic fulfillment of compatibility equations. Of the problem solution in stresses presupposes that compatibility equations are included into the number of basic equations, which are to be satisfied.

## NOTES

1. *Newton*, Isaak (1643–1727), English mathematician, mechanical engineer, and founder of classical physics.
2. *Cronecker*, Leopold (1823–1891), German mathematician.
3. *Notion "tensor"* stems from Latin "*tensio*"—tension.
4. First has been proved by Cauchy in 1822 at deriving equilibrium equations.
5. Cosserat (brothers), Eugène Maurice Pierre and Francois were the first to introduce this kind continuum and investigated it.

6. The notions of *simple* and *complex* loading were introduced by Ilyushin.
7. *Poisson*, Simeon Denis (1781–1840), French mathematician and mechanical engineer; worked in mathematical analysis, probability theory, elasticity theory.

## KEYWORDS

- **Cauchy stress tensor**
- **Cronecker deltas**
- **Juggling**
- **Macroscopic level**
- **Stress ellipsoid**
- **Stress vector**

# Chapter 2

## Physical Relations in the Elasticity Theory

The relations derived earlier turn to be insufficient for determining stresses, strains, and displacements in a deformable solid experiencing external loading. We need also physical determinants to find the stress–strain equations making allowance for the material behavior.

Since there exists a great many different materials and respectively many physical laws describing their behavior under various conditions are possible as well. However, it is practically impossible to establish a universal physical law for all materials, since general equations would become too cumbersome and inconvenient for practical use. That is why, there is a tendency to establish such relations which could describe the key types of material behavior under certain conditions. Therefore, it is worthwhile using a mathematical model to be tested experimentally for achieving good approximation to the properties of a real material. In our further examination of the elasticity theory relations we shall assume the following hypotheses to be valid in addition to the ones accepted previously for the material considered.

1. The hypothesis of a perfect elasticity material. The perfect elasticity means capability of a body to restore its initial shape and dimensions after removing the reasons that have caused its deformation (removal of external loading, temperature, electromagnetic, and radiation fields, etc.).
2. The assumption on the linear stress–strain dependence (*Hooke's law*). It is presumed that, for most materials, the displacements arising from deformation of a body are directly proportional to the loads that have caused them.

We shall presume that in its initial state a body is free from stresses, has a constant temperature and is found in a thermodynamic equilibrium with the environment.

### STRAIN ENERGY AND ELASTICITY POTENTIAL

We shall call a body whose stress in every point is a single-valued deformation function the elastic body.

The process of a stress or strain tensor variation will be called a *load path* or a *deformation path*, respectively, depending on some monotonously increasing parameter termed "time." As a matter of fact, the real time plays no part in determining the elastic body model. When using this term we speak about a sequence of events rather than about time duration.

To illustrate the stress or strain tensor we can present them as vectors, which components are equal to the ones of corresponding tensors:

$$\sigma = \sigma(\sigma_{11}, \sigma_{22}, \sigma_{33}, \sigma_{12}, \sigma_{23}, \sigma_{31}),\ \varepsilon = \varepsilon(\varepsilon_{11}, \varepsilon_{11}, \varepsilon_{11}, \varepsilon_{11}, \varepsilon_{11}, \varepsilon_{11}). \tag{2.1}$$

Then, vectors σ and ε are the images of the stress and strain tensors in six-dimensional spaces of stresses and strains, respectively. Note, that this image is not unique. A 9D space could be introduced instead of a 6D one if to neglect the symmetry of tensors $\sigma_{ij}$ and $\varepsilon_{ij}$. A. Ilyushin has applied 5D spaces for the stress and strain deviators, since only five of their components turn to be independent.

Thus, the load and deformation paths may be represented as the curves described by the ends of vectors σ and ε in corresponding spaces. The elasticity law, that is, equations (2.1), proves in fact that the closed deformation path corresponds to the closed load path and vice versa.

Let us consider now a class of elastic materials for which the work in an elementary volume exercised via the closed loop strains and stresses is equal to zero. Nonmechanical energy losses are absent in this case and thermal effects are excluded.

Let us introduce a concept of *specific potential strain energy* as a work done per unit body volume deformation:

The internal energy variation (in view of symmetry of tensors $\varepsilon_{ij}$ and $\sigma_{ij}$) may be recorded in an expanded form:

$$dU = \sigma_{11}d\varepsilon_{11} + \sigma_{22}d\varepsilon_{22} + \sigma_{33}d\varepsilon_{33} + 2(\sigma_{12}d\varepsilon_{12} + \sigma_{23}d\varepsilon_{23} + \sigma_{31}d\varepsilon_{31})\,. \quad (2.2)$$

The zero condition of the work over an arbitrary closed loop will be:

$$\int dU = \int \sigma_{ij}d\varepsilon_{ij} = 0\,.$$

To meet this condition, the integration element should present the exact differential, that is:

$$\sigma_{ij} = \frac{\partial U}{\partial \varepsilon_{ij}}\,. \quad (2.3)$$

The specific potential strain energy $U(\varepsilon_{ij})$ is a single-valued function of deformation; which is also called the *elastic potential*. On the other hand, the expression

$$dU^* = \varepsilon_{11}d\sigma_{11} + \varepsilon_{22}d\sigma_{22} + \varepsilon_{33}d\sigma_{33} + 2(\varepsilon_{12}d\sigma_{12} + \varepsilon_{23}d\sigma_{23} + \varepsilon_{31}d\sigma_{31})$$

is the exact differential as well

$$dU + dU^* = d(\sigma_{ij}\varepsilon_{ij})\,.$$

Hence,

$$\varepsilon_{ij} = \frac{\partial U^*}{\partial \sigma_{ij}}\,. \quad (2.4)$$

Value $U^*(\sigma_{ij})$ as a *complementary specific strain energy*, is an *elastic stress potential*. The following general relations are valid for the energies introduced:

$$dU(\varepsilon)=\frac{\partial U}{\partial \varepsilon_{ij}}d\varepsilon_{ij}, dU^{*}(\sigma)=\frac{\partial U^{*}}{\partial \sigma_{ij}}d\sigma_{ij}.$$

The values of $U$ and $U^*$ for a nonlinearly elastic material may be visualized by an example of a uniaxial stress state (Fig. 2.1). The figure shows that values

$$U(\varepsilon)=\int_0^{\varepsilon}\sigma d\varepsilon, U^{*}(\sigma)=\int_0^{\sigma}\varepsilon d\sigma$$

supplement each other till a rectangle of σ·ε area under and above the stress–strain curve, respectively.

The assumption on the existence of specific potential strain energy is in one-to-one correspondence with the assumption of reversibility of the examined processes of deformation, determining thereby the elastic behavior of materials.

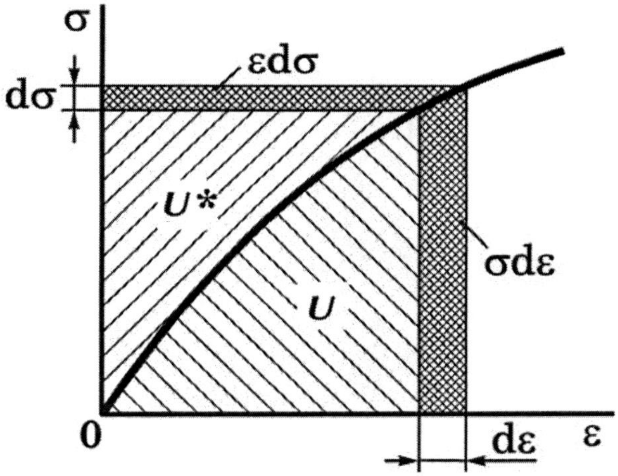


**Figure 2.1.**

This does not, however, necessarily imply a linear stress–strain relationship. Linearity is introduced only by Hooke's law.

The requirement on the single-valued solvability of equations (2.3) with regard to deformations is equivalent to the convexity condition of surfaces $U(\varepsilon_{ij})$ = const or $U^*(\sigma_{ij})$ = const in strain and stress spaces, respectively. In fact, relation (2.3) means, for example, that vector σ is normal-directed to surface $U$ = const. If this surface is strictly convex, only one point of the surface corresponds to a given direction of the normal.

The specific potential energy of deformation is a positively defined value. This feature is used, for example, to prove the uniqueness of the linear problem solution of the elasticity theory. Besides, it serves the basis for the theorems on the minimum potential energy and, respectively, that of extra energy.

## HOOKE'S LAW

Based on the experimental results with a wire, Hooke has established a linear relation between the load and total deformation. He has formulated a corresponding law stating: «Ut tensio sic vis»—«The force causes the respective deformation» and published it in 1676 in the form of an anagram "ceiiinosssttuv."[1]

The relations of the *generalized Hooke's law* for isotropic bodies are known from the course of strength of materials. In the accepted notations of the stress and strain tensor components they are:

$$\varepsilon_{11} = \frac{1}{E}\left[\sigma_{11} - \nu\left(\sigma_{22} + \sigma_{33}\right)\right],\ \varepsilon_{22} = \frac{1}{E}\left[\sigma_{22} - \nu\left(\sigma_{33} + \sigma_{11}\right)\right],$$

$$\varepsilon_{33} = \frac{1}{E}\left[\sigma_{33} - \nu\left(\sigma_{11} + \sigma_{22}\right)\right], \sigma_{12} = 2G\varepsilon_{12}, \sigma_{23} = 2G\varepsilon_{23}, \sigma_{31} = 2G\varepsilon_{31}. \tag{2.5}$$

Here, *E and G*—*Young's* and *shear moduli*[2], ν—*Poisson's ratio.* They are related by a well-known dependence $2G = E/(1+\nu)$.

When solving the problems of the elasticity theory, there arises a need for inverse relations, when stresses are expressed through strains. For this case, we have

$$\sigma_{ij} = 2\mu\varepsilon_{ij} + \lambda\theta\delta_{ij}. \tag{2.6}$$

Here,

$$\theta = \varepsilon_{11} + \varepsilon_{22} + \varepsilon_{33} = \frac{1-2\nu}{E}(\sigma_{11} + \sigma_{22} + \sigma_{33}), \tag{2.7}$$

is introduced earlier volumetric deformation; λ and μ—new constants of the material referred to as *Lame's parameters*, related with *G*, ν and *E* via:

$$\lambda = \frac{\nu E}{(1+\nu)(1-2\nu)},\ \mu = G = \frac{E}{2(1+\nu)},\ \nu = \frac{\lambda}{2(\lambda+\mu)}.$$

Consequently, only two out of five elastic constants λ, μ, *K*, ν and *E* turn to be independent. Dimensionality of values λ, μ, *K*, *E* coincides with that of stress or pressure. These parameters are positive. Poisson's ratio is a dimensionless quantity. The limitation of its possible values results from condition $K > 0$ and $\mu > 0$: $-1 \le \nu \le \frac{1}{2}$. Value $\nu = ½$ corresponds to an incompressible material. The experience suggests that for all known isotropic materials, $\nu > 0$. Numerous attempts made to prove that the lower boundary at ν equals to zero but not to a unity were in vain [53].

If it is necessary to write equations (2.6) solvable about $\varepsilon_{ij}$, we come to a reciprocal form of Hooke's law

$$\varepsilon_{ij} = \frac{1}{2G}\left[\sigma_{ij} - \frac{3\nu}{1+\nu}\sigma\delta_{ij}\right]. \tag{2.8}$$

At $\nu \to 0.5$, Lame's parameter is $\lambda \to \infty$, that is, corresponds to an incompressible material ($\theta = 0$) according to (2.7). In this case, dependencies (2.6) seem to be inadequate. It is worthwhile to write two separate relations in which volumetric deformation is more explicit. This may be achieved by, for example, using expression of Hooke's law through the deviatoric and spherical parts of stress (1.22), (1.23) and strain tensors (1.29), (1.30):

$$s_{ij} = 2G\,\text{э}_{ij}, \sigma = K\theta$$
$$K = \lambda + \frac{2}{3}\mu = \frac{E}{3(1-2\nu)} \tag{2.9}$$

where *K—volumetric deformation modulus*. For incompressible materials instead of the second of equations (2.9), the condition $\theta = 0$ is used.

It follows from Hooke's law, in line with (2.9), that the specific potential energy of deformation may be divided into two independent parts:

$$U_d = \frac{1}{2}s_{ij}\,\text{э}_{ij} = \frac{1}{4G}\left(\sigma_{ij}\sigma_{ij} - \frac{\sigma^2}{3}\right), U_v = \frac{1}{2}\sigma\theta = K\frac{\varepsilon^2}{2} \tag{2.10}$$

where $U_d$ is a *specific distortion energy*, and $U_v$ is a *specific energy of the volume change*. The specific energy of forming may be recorded also through the octahedral shear stress (1.20). In the expanded form formulas (2.10) are of the kind:

$$U_v = \frac{1-2\nu}{6E}(\sigma_{11} + \sigma_{22} + \sigma_{33})^2,$$
$$U_d = \frac{1+\nu}{6E}\left((\sigma_{11} - \sigma_{22})^2 + (\sigma_{22} - \sigma_{33})^2 + (\sigma_{33} - \sigma_{11})^2\right)$$
$$+\frac{1}{2G}\left(\sigma_{12}^2 + \sigma_{23}^2 + \sigma_{31}^2\right)$$

Both values play an important role, for example, in formulation of strength criteria and flow laws at plastic deformation. The potential energy of deformation was first introduced in 1839 by Green[3] when deducing the theory of light propagation.

## HOOKE'S LAW FOR ANISOTROPIC MATERIAL

Relations (2.6) may be generalized for an arbitrary anisotropic material by assuming a linear relation between the stress–strain tensor components in the form:

$$\sigma_{ij} = E_{ijkl}\varepsilon_{kl} \tag{2.11}$$

Equalities (2.11) may be considered as a linear expansion term in a series of general nonlinear relation $\sigma_{ij} = \sigma_{ij}(\varepsilon_{ij})$ without a constant summand due to the condition $\sigma_{ij} = 0$ at $\varepsilon_{ij} = 0$. Equality (2.11) contains in fact nine equations, each having nine summands.

Components $E_{ijkl}$ for a homogeneous material are independent of the coordinate values. However, the stresses and strains depend on orientation of the coordinates, so other elastic constants $E_{ijkl}$ should obey this relation as well. They form a matrix consisting of 81 elastic constants, being transformed as components of the fourth-rank tensor, and called the *elasticity modulus tensor*. The elastic constants are independent of the coordinate orientation only for isotropic materials (see 2.2).

In view of symmetry of tensors $\sigma_{ij}$ and $\varepsilon_{kl}$ the elasticity modulus tensor is also symmetric about indices *i* and *j*, *k* and *l*: $E_{ijkl} = E_{jikl} = E_{ijlk} = E_{jilk}$, and only 36 components remain independent. The matrix of its components is:

$$\begin{bmatrix} E_{1111} & E_{1122} & E_{1133} & E_{1112} & E_{1113} & E_{1123} \\ E_{2211} & E_{2222} & E_{2233} & E_{2212} & E_{2213} & E_{2223} \\ E_{3311} & E_{3322} & E_{3333} & E_{3312} & E_{3313} & E_{3323} \\ E_{1211} & E_{1222} & E_{1233} & E_{1212} & E_{1213} & E_{1223} \\ E_{1311} & E_{1322} & E_{1333} & E_{1312} & E_{1313} & E_{1323} \\ E_{2311} & E_{2322} & E_{2333} & E_{2312} & E_{2313} & E_{2323} \end{bmatrix}.$$

Further reduction of the number of independent components will be obtained from thermodynamic considerations if to assume the existence of specific potential strain energy. Let us substitute in turn relations (2.11) and (2.3) into the dependence for the differential of the specific potential energy (2.2). Then,

$$dU = \sigma_{ij} d\varepsilon_{ij} = E_{ijkl}\varepsilon_{kl} d\varepsilon_{ij},$$

$$dU = \sigma_{ij} d\varepsilon_{ij} = \frac{\partial U}{\partial \varepsilon_{ij}} d\varepsilon_{ij}.$$

This implies that

$$\frac{\partial U}{\partial \varepsilon_{ij}} = E_{ijkl}\varepsilon_{kl}.$$

After a repeated differentiation and taking into account possible changes in the order of differentiation, we obtain:

$$\frac{\partial}{\partial \varepsilon_{kl}}\left(\frac{\partial U}{\partial \varepsilon_{ij}}\right) = E_{ijkl}, \quad \frac{\partial}{\partial \varepsilon_{ij}}\left(\frac{\partial U}{\partial \varepsilon_{kl}}\right) = E_{klij}.$$

Consequently,

$$\frac{\partial^2 U}{\partial \varepsilon_{ij} \partial \varepsilon_{kl}} = E_{ijkl} = E_{klij}.$$

Hence, the number of independent constants for the anisotropic body becomes equal to 21. If an elastic body has the only plane of symmetry of elastic properties, the number of constants reduces to 13. For a body having three mutually orthogonal planes of symmetry (orthotropic body) the number of constants reduces to nine. The number of independent constants for an isotropic body equals to two according to above stated.

We may invert the generalized Hooke's law by expressing the deformation through the stresses. Then,

$$\varepsilon_{ij} = D_{ijkl}\sigma_{kl}. \tag{2.12}$$

The fourth-rank tensor $D_{ijkl}$ (the *elastic compliance tensor*) possesses the same properties of symmetry like the tensor of elasticity modulus $E_{ijkl}$.

It should be noted, that the number of elasticity constants involved into Hooke's law reduces only if the planes of symmetry are assumed to be the coordinate ones. In other coordinates the equations will still contain 21 constants expressed through nine independent constants.

## CLAPEYRON FORMULA

To calculate the specific potential strain energy, we shall proceed from equation (2.2). Let us consider an arbitrary element of an elastic body volume with stress–strain state (SSS) set by $\sigma_{ij}$ and $\varepsilon_{ij}$ values. We shall introduce a local system of principal axes for stresses and strains, and provide increments for the principal stresses identical to their finite values (so-called *proportional loading*). Then, the work done per unit volume will be expressed as:

$$W = \tfrac{1}{2}(\sigma_1\varepsilon_1 + \sigma_2\varepsilon_2 + \sigma_3\varepsilon_3).$$

It is independent of the load path and is equal to the specific potential strain energy $U$ according to the law of conservation of energy. Correspondingly, we have in the arbitrary coordinates:

$$U = \tfrac{1}{2}\sigma_{ij}\varepsilon_{ij}. \tag{2.13}$$

This expression is known as *Clapeyron's formula.* It may be expanded into:

$$U = \tfrac{1}{2}(\sigma_{11}\varepsilon_{11} + \sigma_{22}\varepsilon_{22} + \sigma_{33}\varepsilon_{33}) + \sigma_{12}\varepsilon_{12} + \sigma_{23}\varepsilon_{23} + \sigma_{31}\varepsilon_{31}.$$

When subject to the generalized Hooke's law in the form of (2.11) or (2.12), Clapeyron's formula (2.13) becomes:

$$U = \tfrac{1}{2}E_{ijkl}\varepsilon_{ij}\varepsilon_{kl},$$

or

$$U = \tfrac{1}{2}D_{ijkl}\sigma_{ij}\sigma_{kl}.$$

This implies that

$$\frac{\partial U}{\partial \sigma_{ij}} = \frac{1}{2}\frac{\partial}{\partial \sigma_{ij}}(D_{mnkl}\sigma_{mn}\sigma_{kl}) = \tfrac{1}{2}D_{mnkl}(\delta_{im}\delta_{jn}\sigma_{kl} + \delta_{ik}\delta_{jl}\sigma_{mn})$$

$$= \tfrac{1}{2}(D_{ijkl}\sigma_{kl} + D_{mnij}\sigma_{mn}) = D_{ijkl}\sigma_{kl},$$

or

$$\frac{\partial U}{\partial \sigma_{ij}} = \varepsilon_{ij}. \tag{2.14}$$

The relation obtained should not be confused with a one very close to it (2.4), which in contrast contains $U^*$ instead of $U$ but acknowledges validity of Hooke's law and, consequently, that of the linearly elastic behavior of the material. At the same time, relation (2.3) is deduced from the common laws of thermodynamics.

Formulas (2.3) and (2.14) present particular statements of *Castigliano's theorem*[4] we are to consider afterwards.

## TEMPERATURE EFFECTS

Let us suppose that the body under consideration is found in an inhomogeneous and nonstationary temperature field $T(x_i, t)$. The temperature shall be counted from a certain initial value $T_0$. According to *Neumann's hypothesis*[5], the total linear deformation $\varepsilon_{ii}$ consists of the strains caused by the force loading $\varepsilon'_{ii}$ and thermal expansion $\varepsilon''_{ii} = \alpha T$:

$$\varepsilon_{ii} = \varepsilon'_{ii} + \varepsilon''_{ii}.$$

Temperature increments for shear strains are equal to zero. So, *Hooke's law with regard to temperature* takes the following form:

$$\varepsilon_{ij} = \frac{1}{2G}\left[\sigma_{ij} - \frac{3\nu}{1+\nu}\sigma\delta_{ij}\right] + \alpha T\delta_{ij}. \tag{2.15}$$

Value $\alpha$ is known as a *coefficient of linear thermal expansion* of the material. By expressing the stress through deformations (2.15), we obtain an equation:

$$\sigma_{ij} = 2G\left[\varepsilon_{ij} + \frac{\nu\theta}{1\text{-}2\nu}\delta_{ij} - \frac{1+\nu}{1-2\nu}\alpha T\delta_{ij}\right].$$

The relation between the deviators of thermoelasticity remains similar to a perfect elasticity (2.9); only the relation between the spherical parts of the stress and strain tensors takes a different form:

$$s_{ij} = 2G\text{э}_{ij}, \quad \sigma = K(\theta - 3\alpha T)$$

The deformation components from (2.15) are related to displacements by Cauchy equations (1.25).

In a general case, to solve complex problems of thermoelasticity, we are to solve beforehand the problem on heat distribution. Temperature $T(x_i, t)$ in each point of the body shall meet a *thermal conductivity equation*

$$\frac{\partial T}{\partial t} = B\Delta T + \frac{Q}{c\gamma}, \tag{2.16}$$

where $B$—thermal diffusivity of the material; Δ—Laplacian operator[6]

$$\Delta = \frac{\partial^2}{\partial x_1^2} + \frac{\partial^2}{\partial x_2^2} + \frac{\partial^2}{\partial x_3^2} = (...),_{ii}$$

$Q(x_i, t)$ is the function of heat quantity generated by a thermal energy source in a unit volume per unit time; $c$—specific heat; γ—specific weight.

*The initial conditions* of temperature distribution $T(x_i, 0) = f(x_i)$ should be added to equation (2.16). Besides, the conditions of heat exchange with the environment are to be satisfied on the boundary of a solid body (three types of heat exchange conditions; see, e.g., Moskvitin, 1972).

Under high enough temperatures, we should not ignore the temperature dependence of the elasticity constants. To describe the dependence, it is recommended to use *Bell's formula*[7] (Bell, 1984), which he has derived as a result of experimental investigations of more than 500 metals and alloys

$$\{G(T),\ E(T), K(T)\} = \{G(0),\ E(0), K(0)\}\, f(T)\ ,$$

where, $G(T)$, $E(T)$, and $K(T)$ are elasticity modules at temperature T (Kelvin); j($T$) is a linear function of temperature,

$$\phi(T) = \begin{cases} 1\ , & 0 < T/T_\mathrm{m} \le 0{,}06\,; \\ 1{,}03\,(1 - T/2T_\mathrm{m}), & 0{,}06 < T/T_\mathrm{m} \le 0{,}57; \end{cases}$$

$T_\mathrm{m}$ is melting temperature of the material; $G(0)$, $E(0)$, $K(0)$—values of elasticity parameters at so-called *zero-stress*. One can easily obtain these values if the value of one of the elasticity module at some, for instance, room temperature and melting temperature of the material are known. In particular, for duralumin:

$$T_m = 933\ K,\quad G(0) = 0{,}308 \times 10^5\, MPa,\quad E(0) = 0{,}829 \times 10^5\, MPa\,.$$

Generally speaking, $Q$ value in (2.16) for deformable bodies depends on the stress–strain state. Therefore, we speak about a *coupled thermoelasticity problem*.

Hooke's law (2.11) for anisotropic bodies with regard to temperature will be as follows:

$$\sigma_{ij} = E_{ijkl}(\varepsilon_{kl} - \alpha_{kl} T)\,.$$

where, $\alpha_{kl}$ is a *tensor of thermal expansion coefficients of the material*. Constants $E_{ijkl}$ shall be defined under isothermal conditions at $T = T_0$. If the temperature increments are not small, values $E_{ijkl}$ and $\alpha_{kl}$ shall be considered as temperature functions.

## SSS CASE STUDY

Let us presume that for a point of an elastic body the stress tensor is known (in MPa):

$$T_s = (\sigma_{ij}) = \begin{pmatrix} \sigma_{11} & \sigma_{12} & \sigma_{13} \\ \sigma_{21} & \sigma_{22} & \sigma_{23} \\ \sigma_{31} & \sigma_{32} & \sigma_{33} \end{pmatrix} = \begin{pmatrix} 2 & 1 & 3 \\ 1 & 9 & 6 \\ 3 & 6 & 4 \end{pmatrix}.$$

*It is required to*:

1. Plot the initial components of stress tensors on the faces of the elementary parallelepiped isolated in the vicinity of the point under consideration.
2. Determine the values of the normal and shear stresses on the area element with outer normal ν, in case we set cosines of the angles between the normal and coordinate axes $l_i = \cos(\nu, x_i)$: $l_1 = l_2 = 1/2$, $l_3 = \sqrt{2}/2$.
3. Expand a given stress tensor into spherical and deviatoric parts and show their components on the faces of the elementary parallelepiped.
4. Calculate components of the deformation tensor, volumetric deformation, stress, and strain intensity in this point, if $E = 2\times10^5\, MPa$; $\nu = 0.3$.
5. Calculate the principal stresses and maximum shear stresses assuming that in the initial stress tensor $\sigma_{33} = \sigma_{31} = \sigma_{32} = 0$.

*Solution*

1. The general location of the preset stress tensor components on the faces of the element isolated in the vicinity of a point is shown in Fig. 2.2. Since all components are positive, their direction on visible faces coincides with that of corresponding axes of the coordinates. On invisible faces, similar stresses are directed oppositely. For negative stresses, the directions should be substituted for opposite ones.
2. To determine normal $\sigma_n$ and shear $\tau_n$ stresses on the preset oblique planes, let us calculate first coordinate projections of stress vector $\sigma_\nu$ on this plane using formulas (1.8) $\sigma_{\nu i} = \sigma_{ij} l_j$. In the expanded form, we obtain the following projection of the stress vector onto axis $x_1$:

$$\sigma_{\nu 1} = \sigma_{1j} l_j = \sigma_{11} l_1 + \sigma_{12} l_2 + \sigma_{13} l_3 =$$
$$2\times\tfrac{1}{2} + 1\times\tfrac{1}{2} + 3\times\tfrac{\sqrt{2}}{2} = 3.62 \quad \text{MPa},$$

Along axes $x_2$, $x_3$, we have by analogy

$$\sigma_{\nu 2} = \sigma_{21} l_1 + \sigma_{22} l_2 + \sigma_{23} l_3 =$$
$$1\times\tfrac{1}{2} + 9\times\tfrac{1}{2} + 6\times\tfrac{\sqrt{2}}{2} = 9.24 \quad \text{MPa},$$

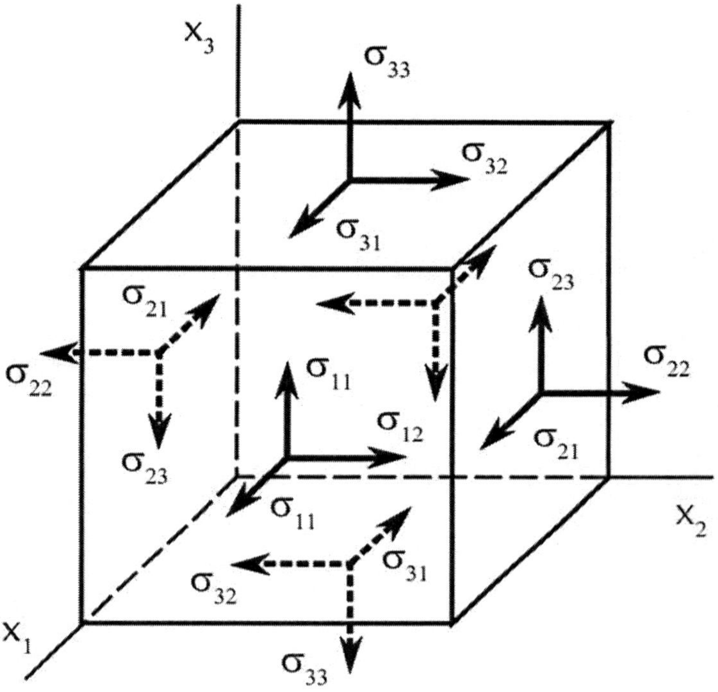


**Figure 2.2.**

$$\sigma_{\nu 3} = \sigma_{31} l_1 + \sigma_{32} l_2 + \sigma_{33} l_3 = 3 \cdot \tfrac{1}{2} + 6 \cdot \tfrac{1}{2} + 4 \cdot \tfrac{\sqrt{2}}{2} = 7.33 \text{ MPa.}$$

The modulus of the combined stress vector on this area element will be obtained through a sum of squares of coordinate stresses:

$$\sigma_\nu = \sqrt{\sigma_{\nu 1}^2 + \sigma_{\nu 2}^2 + \sigma_{\nu 3}^2} = \sqrt{3.62^2 + 9.24^2 + 7.33^2} = 12.34 \text{ MPa.}$$

The normal stress may be calculated as a sum of projections of coordinate stresses onto the normal to the area element:

$$\sigma_n = \sigma_{\nu i} l_i = 3.62 \cdot \tfrac{1}{2} + 9.24 \cdot \tfrac{1}{2} + 7.33 \cdot \tfrac{\sqrt{2}}{2} = 11.61 \text{ MPa.}$$

The desired shear stresses may be determined now using the *Pythagorean theorem*[8]:

$$\tau_n = \sqrt{\sigma_\nu^2 - \sigma_n^2} = \sqrt{12.34^2 - 11.61^2} = 4.18 \text{ MPa.}$$

3. Expansion of the stress tensor into the deviatoric (1.23) and spherical (1.22) parts $\sigma_{ij} = s_{ij} + \sigma\delta_{ij}$ looks in the matrix form:

$$\begin{pmatrix}\sigma_{11} & \sigma_{12} & \sigma_{13}\\ \sigma_{21} & \sigma_{22} & \sigma_{23}\\ \sigma_{31} & \sigma_{32} & \sigma_{33}\end{pmatrix}=\begin{pmatrix}\sigma_{11}-\sigma & \sigma_{12} & \sigma_{13}\\ \sigma_{21} & \sigma_{22}-\sigma & \sigma_{23}\\ \sigma_{31} & \sigma_{32} & \sigma_{33}-\sigma\end{pmatrix}+\begin{pmatrix}\sigma & 0 & 0\\ 0 & \sigma & 0\\ 0 & 0 & \sigma\end{pmatrix}=$$

$$\begin{pmatrix}\sigma_{11}-\sigma & \sigma_{12} & \sigma_{13}\\ \sigma_{21} & \sigma_{22}-\sigma & \sigma_{23}\\ \sigma_{31} & \sigma_{32} & \sigma_{33}-\sigma\end{pmatrix}+\begin{pmatrix}\sigma & 0 & 0\\ 0 & \sigma & 0\\ 0 & 0 & \sigma\end{pmatrix} \tag{2.17}$$

The spherical (mean) stress

$$\sigma=\tfrac{1}{3}(\sigma_{11}+\sigma_{22}+\sigma_{33})=\tfrac{1}{3}(2+9+4)=5\ \text{MPa}.$$

Let us substitute this value and components of the preset stress tensor in (2.17). Then,

$$\begin{pmatrix}2 & 1 & 3\\ 1 & 9 & 6\\ 3 & 6 & 4\end{pmatrix}=\begin{pmatrix}-3 & 1 & 3\\ 1 & 4 & 6\\ 3 & 6 & -1\end{pmatrix}+\begin{pmatrix}5 & 0 & 0\\ 0 & 5 & 0\\ 0 & 0 & 5\end{pmatrix}.$$

This expansion may be shown graphically on the faces of the elementary parallelepiped (Fig. 2.3).

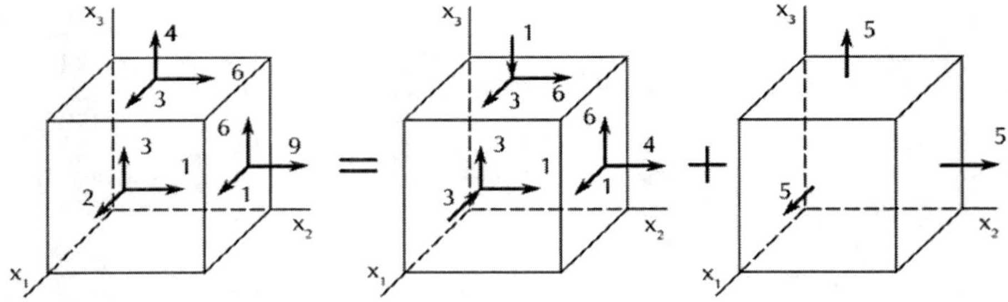


**Figure 2.3.**

4. The components of the deformation tensor shall be calculated using the relation of the generalized Hooke's law (2.5). The values of stresses and Young's modulus are substituted in megapascals.

$$\varepsilon_{11}=\frac{1}{E}\left[\sigma_{11}-\nu\left(\sigma_{22}+\sigma_{33}\right)\right]=\frac{1}{2\cdot10^{5}}\left[2-0.3(9+4)\right]=-9.5\cdot10^{-6},$$

$$\varepsilon_{22}=\frac{1}{E}\left[\sigma_{22}-\nu\left(\sigma_{33}+\sigma_{11}\right)\right]=\frac{1}{2\cdot10^{5}}\left[9-0.3(4+2)\right]=36\cdot10^{-6},$$

$$\varepsilon_{33}=\frac{1}{E}\left[\sigma_{33}-\nu\left(\sigma_{11}+\sigma_{22}\right)\right]=\frac{1}{2\cdot10^{5}}\left[4-0.3(9+2)\right]=3.5\cdot10^{-6},$$

$$G=\frac{E}{2(1+\nu)}=\frac{2\times10^5}{2(1+0.3)}=0.769\times10^5 \text{ MPa},$$

$$K=\frac{E}{3(1-2\nu)}=\frac{2\cdot10^5}{3(1-0.6)}=1.67\cdot10^5 \text{ MPa},$$

$$\varepsilon_{12}=\frac{\sigma_{12}}{2G}=\frac{1}{2\cdot0.769\cdot10^5}=0.65\cdot10^{-5},$$

$$\varepsilon_{23}=\frac{\sigma_{23}}{2G}=\frac{6}{2\cdot0.769\cdot10^5}=3.9\cdot10^{-5},$$

$$\varepsilon_{31}=\frac{\sigma_{31}}{2G}=\frac{3}{2\cdot0.769\cdot10^5}=1.95\cdot10^{-5}.$$

We shall determine the volumetric deformation by formula (1.28):

$$\theta=\varepsilon_{11}+\varepsilon_{22}+\varepsilon_{22}=(-9.5+36+3.5)\cdot10^{-6}=30\cdot10^{-6}=0.003\ \%.$$

The stress and strain rates are calculated by formulas (1.24) and (1.31):

$$\sigma_u=\tfrac{\sqrt{2}}{2}\sqrt{(\sigma_{11}-\sigma_{22})^2+(\sigma_{22}-\sigma_{33})^2+(\sigma_{33}-\sigma_{11})^2+6(\sigma_{12}^2+\sigma_{23}^2+\sigma_{31}^2)}=$$

$$\tfrac{\sqrt{2}}{2}\sqrt{(2-9)^2+(9-4)^2+(4-2)^2+6(1^2+6^2+3^2)}=13.3 \text{ MPa}.$$

$$\varepsilon_u=\tfrac{\sqrt{2}}{3}\sqrt{(\varepsilon_{11}-\varepsilon_{22})^2+(\varepsilon_{22}-\varepsilon_{33})^2+(\varepsilon_{33}-\varepsilon_{11})^2+6(\varepsilon_{12}^2+\varepsilon_{23}^2+\varepsilon_{31}^2)}$$

$$=\tfrac{\sqrt{2}}{3}\sqrt{(-9.5-36)^2+(36-3.5)^2+(3.5-(-9.5))^2+6(6.5^2+39^2+19.5^2)}\cdot10^{-6},$$

$\varepsilon_u == 0{,}576\cdot10^{-4}=0.00576\%$,

5. Let us consider the plane stress state. We shall assume that in the initial stress tensor $\sigma_{33}=\sigma_{31}=\sigma_{32}=0$. The tensor matrix looks like:

$$T_s=\begin{pmatrix}2&1&0\\1&9&0\\0&0&0\end{pmatrix}.$$

To calculate the principal stresses, let us write a secular equation

$$\sigma^3-J_1\sigma^2+J_2\sigma-J_3=0.$$

The coefficients here are the stress tensor invariants

$$J_1=\sigma_{11}+\sigma_{22}+\sigma_{33}=2+9+0=11 \text{ MPa},$$

$$J_2=\sigma_{11}\sigma_{22}+\sigma_{22}\sigma_{33}+\sigma_{33}\sigma_{11}-\sigma_{12}^2-\sigma_{23}^2-\sigma_{31}^2=$$

$$2\cdot9+9\cdot0+0\cdot2-1^2-0-0=17 \text{ (MPa)}^2,$$

$$J_3 = \begin{vmatrix} 2 & 1 & 0 \\ 1 & 9 & 0 \\ 0 & 0 & 0 \end{vmatrix} = 0.$$

The cubic equation for determining the principal stresses becomes:

$$\sigma^3 - 11\sigma^2 + 17\sigma = 0.$$

Hence, we obtain one root $\sigma = 0$; two other satisfies the quadratic equation:

$$\sigma^2 - 11\sigma + 17\sigma = 0.$$

Its solutions are

$$\sigma = \frac{11 \pm \sqrt{53}}{2}.$$

Consequently, the following values will be the principal stresses:

$$\sigma_1 = 9.14 \text{ MPa}, \ \sigma_2 = 1.86 \text{ MPa}, \ \sigma_3 = 0.$$

The maximum shear stresses we calculate by equation (1.21):

$$\tau_{max} = (\sigma_1 - \sigma_3)/2 \, \tau_{max} = (9{,}14 - 0)/2 = 4.57 \text{ MPa}.$$

Thus, we have calculated SSS for a solid body point according to a given stress tensor.

## NOTES

1. The same law was independently discovered in 1680 by a French physicist *Mariotte* (1620–1684).
2. *Young, Thomas* (1773–1829)—English scientist, in his honor the longitudinal elasticity modulus was named.
3. *Green, George* (1793–1841), English mathematician and physicist; worked on integral calculus, theory of elasticity.
4. *Castigliano* (1847–1884), Italian researcher in mechanics.
5. *Neumann, F.E.* (1798–1895), German mathematician.
6. *Laplace, Pierre Simon* (1749–1827), French mathematician and physicist; is known for works on celestial mechanics, heat, probability theory, mathematical physics.
7. *Bell, James Frederic* (Bell J. F.), Professor of Johns Hopkins University (USA); well-known for his experimental research in deformable body mechanics.
8. *Pythagor*, the Samian (Πιυαγόρας, VI century B.C.), ancient Greek thinker, mathematician; researched whole numbers, proved well-known theorems.

**KEYWORDS**

- **Electromagnetic**
- **Inhomogeneous**
- **Octahedral shear stress**
- **Orthotropic body**
- **Thermoelasticity**
- **Zero-stress**

# Chapter 3

## Statements and Problems Solving Procedurs of Elasticity Theory

### BOUNDARY PROBLEMS

To solve the *direct problems of the elasticity theory*, that is, to determine 15 unknown functions $u_i$, $\sigma_{ij}$, $\varepsilon_{ij}$ $(i, j = 1, 2, 3)$, we use the equilibrium equations, Cauchy relations and Hooke's law:

$$\sigma_{ij},_j + \rho F_i = 0, \tag{3.1}$$

$$\varepsilon_{ij} = \tfrac{1}{2}(u_i,_j + u_j,_i), \tag{3.2}$$

$$\sigma_{ij} = 2\mu\varepsilon_{ij} + \lambda\theta\delta_{ij}. \tag{3.3}$$

Consequently, we have 15 linear equations in partial derivatives. In the case the displacements are not included into the number of unknowns, equations (3.2) are substituted for the deformation compatibility conditions (without summation over the repeated indices)

$$\varepsilon_{\alpha\alpha},_{\beta\beta} + \varepsilon_{\beta\beta},_{\alpha\alpha} - 2\varepsilon_{\alpha\beta},_{\alpha\beta} = 0,$$

$$(\varepsilon_{\alpha\beta},_\gamma - \varepsilon_{\beta\gamma},_\alpha + \varepsilon_{\gamma\alpha},_\beta),_\alpha - \varepsilon_{\alpha\alpha},_{\beta\gamma} = 0. \tag{3.4}$$

Here and further on we assume the following boundary conditions as postulated. We shall take that the surface of a body is $S = S_\sigma + S_u$, that is, consists of two parts. Let us accept that the following conditions are given in each point of the surface:

$$u_i = u_{i0}(x) \text{ on } S_u, \quad R_{\nu i} = \sigma_{ij} l_j \text{ on } S_\sigma. \tag{3.5}$$

This simplest case does not exhaust all potentialities. For example, when a rigid punch is pressed into an elastic body, friction is often assumed to be absent. So, if to direct axis $x_3$ normally to the body surface, the boundary conditions under the punch will be:

$$u_i = u_{i0}, \sigma_{13} = \sigma_{23} = 0.$$

However, it makes no sense in the general theory to set the task like this on one surface for both displacements and stresses. The discourse presented below can be easily extended on above cases.

Hence, the elasticity theory problem consists in solving equations (3.1)–(3.4) under boundary conditions (3.5).

In a general case, we distinguish between *three types of boundary problems*. The first one relates to determining stresses and displacements inside the elastic body in the equilibrium state if displacements of the points on the surface ($S = S_u$) are known. The *second* type of the boundary problems presupposes that distribution of forces on the surface ($S = S_\sigma$) is known. Above formulated statement can be related to the *third* or mixed boundary problem. Other combinations of boundary conditions are also possible. For example, if a body contains an infinite point, we add a requirement to the boundary conditions on the solution regularity at infinity, which is commonly brought to a boundedness condition.

The forces of the body are assumed to be known in all these cases. Depending on the statement of the elasticity theory problem, the sought functions are either displacements or stresses.

*An inverse problem* of the *elasticity theory* is possible as well, in which case the stresses, strains or displacements are preset for all inner points as the functions of coordinates. We have to determine the conditions on the body boundaries fitting the preset SSS.

## STATEMENT OF THE ELASTICITY THEORY PROBLEM IN DISPLACEMENTS (LAME'S EQUATIONS)

If to substitute Hooke's law (3.3) in the equilibrium equation (3.1) and exclude deformations with the help of Cauchy relations (3.2), we obtain a system consisting of three differential equations with three unknown functions $u_i$ (*Lame's equations*[1]):

$$(\lambda+\mu)\theta,_i+\mu\Delta u_i+\rho F_i=0, \tag{3.6}$$

where $\theta = u_1,_1 + u_2,_2 + u_3,_3 = u_i,_i$—volume deformation, $\Delta$—Laplacian operator

$$\Delta u_i \equiv u_i,_{jj} = u_i,_{11} + u_i,_{22} + u_i,_{33}.$$

Thus obtained system of equations is *elliptical* according to *Petrovsky*[2] in the area $V$ occupied by the body at all Poisson's ratio values, except for $\nu = 0.5$ and $\nu = 1$.

The boundary conditions are applied to three equations (3.6). On one part ($S_u$) of the boundary surface, displacements $u_{0i}$ may be set as the known functions of the coordinates:

$$u_i = u_{i0}(x) \text{ on } S_u. \tag{3.7}$$

On the other part ($S_\sigma$) of the boundary surface the external surface forces $R_i = R_i(x_i)$ are preset. As far as the problem is solved in displacements, $R_i$ values shall be grouped with the sought $u_i$ values using Hooke's law and Cauchy relations. As a result, the following boundary conditions in displacements are met on $S_\sigma$:

$$\lambda\theta\delta_{ij}l_j+\mu(u_i,_j+u_j,_i)l_j=R_{\nu i}. \tag{3.8}$$

Despite the simplicity of Lame's equations, their solution presents a complex mathematical problem. Nevertheless, they are widely used in various calculation procedures of the elasticity theory problems since they avoid the deformation compatibility equations, which are met in this case identically.

The homogeneous problem in which the volumetric forces may be given equal to zero is of particular importance. Its solution at specified boundary conditions (3.7) and (3.8) presents the major difficulty. In the presence of volumetric forces (or body forces, e.g., gravity or centrifugal forces) the partial solutions can always be found. They may not meet the boundary conditions, and due to linearity of the basic equations, may be added to the homogeneous problem solution with the altered boundary conditions.

Here are some conclusions on the properties of sought displacements $u_i$ resulting from (3.6). By assuming that $F_i = 0$, we shall differentiate each of five equations (3.6) with respect to a corresponding coordinate $x_i$, convolve (summed over index *i*) and obtain:

$$(\lambda+\mu)\theta,_{ii} + \mu\Delta u_i,_i = 0. \quad (3.9)$$

Since

$$\theta,_{ii} = \theta,_{11} + \theta,_{22} + \theta,_{33} = \Delta\theta, \Delta u_i,_i = \Delta(u_1,_1 + u_2,_2 + u_3,_3) = \Delta\theta,$$

it follows from (3.9) that

$$\Delta\theta = 0. \quad (3.10)$$

Thus, the volumetric deformation in an elastic isotropic body in the absence of the body forces is a *harmonic function*, which is the function meeting the Laplacian equation (3.10). Now, let us take the Laplacian operator from the left-hand side of Lame's equation, assuming that $F_i = 0$:

$$(\lambda+\mu)\Delta\theta,_i + \mu\Delta\Delta u_i = 0.$$

From this equation and with respect to (3.10), we obtain $\Delta\theta,_i = (\Delta\theta),_i = 0$, therefore,

$$\Delta\Delta u_i = 0\cdot \quad (3.11)$$

Hence, each component of the displacement vector is a *biharmonic* (satisfying the double Laplacian equation) *function* of coordinates.

However, neither the elasticity theory problem can be reduced to integration of system (3.11), nor can $\theta$ be a value found by the known methods for the Laplacian equation. Value $\theta$ can never be set on the boundary. It is impossible to determine it by solving *Dirichlet's problem.*[3] System (3.11) is of the twelfth order while the initial system (3.6) is of the sixth order (the order of a system is a product of the maximal derivative order by the number of equations).

To determine the biharmonic function, it is necessary to preset two conditions on the area boundary, for example, $u_i$ and $\partial u_i / \partial n$, that is, a normal derivative of $u_i$. To solve equation (3.6), it is suffice to set only values $u_i$ in each point of the surface. It is easy enough to form three biharmonic functions that accept the preset values, although may not satisfy equations (3.6).

The boundary problem for incompressible materials when ν = 0.5 loses its ellipticity and its solution may be nonunique.

## STATEMENT OF THE ELASTICITY THEORY PROBLEM IN STRESSES (BELTRAMI–MICHELL EQUATIONS)

The elasticity theory problem may be set not only in displacements, but in stresses as well. It is convenient to set stresses on the boundaries. If to accept the components of the stress tensor as the sought unknown functions, the differential equilibrium equations at $F_i = 0$ will be:

$$\sigma_{ij},_j = 0. \tag{3.12}$$

Above three equations are insufficient for determining the stress state in an elastic body (six independent components of the stress tensor). We may obtain additional equations by expressing deformations through stresses with the help of Hooke's law (3.3) followed by substitution into the deformation compatibility equation (3.4). As a result, we obtain a *Beltrami–Michell system of equations*[4] also known as the *stress compatibility equations*:

$$(1+\nu)\Delta\sigma_{ij} + 3\sigma,_{ij} = 0\,, \tag{3.13}$$

where $\sigma = (\sigma_{11} + \sigma_{22} + \sigma_{33})/3$—mean hydrostatic stress.

It should be noted that stresses as well as the displacement vector components possess the properties of (3.10) and (3.11). In fact, by summing equations (3.13) at $i = j$, we come to $\Delta\sigma = 0$. Proceeding from obtained equations and using the Laplacian operator for equations (3.13), it follows that stresses in the static problem are biharmonic functions, providing the body forces are not available.

$$\Delta\Delta\sigma_{ij} = 0. \tag{3.14}$$

The system of equations (3.13) is of the twelfth order. By making differentiation at deducing it (in deformation compatibility equations), we have artificially raised the order of the initial system. As a result, possible solutions of Beltrami–Michell equations generate a class of functions wider than the solutions of the elasticity theory problems. These solutions may not satisfy the equilibrium equations.

To illustrate above-said, let us consider the following example. Let the stresses be arbitrarily linear functions of coordinates $\sigma_{ij} = \alpha_{ijk}x_k$. Since equations (3.13) contain only the second-order derivatives of $\sigma_{ij}$, they will be fulfilled identically at any $\alpha_{ijk}$ values. The substitution of stresses into equations (3.12) shows that in the absence of bulk forces these constants are bound by three conditions $\alpha_{ijj} = 0$. If some forces were set over the entire surface of a body, the constants shall be defined from the boundary conditions of the type (3.5). In the case the displacements of points on the surface were preset (3.7), it becomes impossible to state the general boundary conditions in stresses. Sometimes, for example, in plane problems of the elasticity theory, this can be successfully realized.

Pobedrya[5] and Georgievskii (1999) have proposed a new statement of the elasticity theory problem in stresses, which fits better the numerical methods. To find six independent components of the tress tensor, six generalized compatibility equations are to be solved. There are six boundary conditions for them: three equilibrium equations (3.1) brought to the boundary of area *S* are added to three conditions of the stress states (3.5).

Although the general concept of the elasticity theory problem solution in displacements or stresses is clear enough, its realization presents certain difficulties. It seems as yet impossible to solve these equations in a general form. Some solutions may be obtained only for the simplest cases. They are valuable, being in some way a standard for comparing approximate solutions deduced with the assistance of certain simplifying hypotheses.

## CLAPEYRON THEOREM

We assume that a linear elastic body experiencing volume ($F_i$) and surface ($R_i$) forces is in the equilibrium state. So, scalar multiplication of equilibrium equations by displacements $u_i$ and subsequent volume integration *V* of the body (with surface *S*) gives:

$$\int_V u_i\sigma_{ij},_j \,\mathrm{d}V + \int_V u_i\rho F_i \,\mathrm{d}V = 0 .$$

The first integral in the left-hand side is transformed into the surface integral by the formula

$$(u_i\sigma_{ij}),_j = u_i,_j\,\sigma_{ij} + u_i\sigma_{ij},_j ,$$

therefore,

$$\int_V (u_i\sigma_{ij}),_j \,dV + \int_V u_i\rho F_i dV = \int_V u_i,_j\,\sigma_{ij} dV .$$

The first integral in the left-hand side transfers into the surface integral using known *Gauss*[6]*–Ostrogradsky's* equation (see 14.1). Owing to a symmetry of the stress tensor, we satisfy in the right-hand side the equality $u_i,_j\,\sigma_{ij} = \varepsilon_{ij}\sigma_{ij}$. As a result, we obtain the following relation:

$$\int_S u_i R_i \,\mathrm{d}S + \int_V u_i\rho F_i \,\mathrm{d}V = 2\int_V U \,\mathrm{d}V .$$

According to Clapeyron's formula (2.12) the integrand in the right-hand side is equal to double specific potential strain energy. Consequently,

$$\int_S u_i R_i \,\mathrm{d}S + \int_V u_i\rho F_i \,\mathrm{d}V = 2\int_V U\mathrm{d}V . \qquad (3.15)$$

Proceeding from above, the work on the elastic deformation equals to a half work done by the external surface and volume forces at displacements from the initial equilibrium to the final state (*Clapeyron's theorem*).

## THE EXISTENCE AND UNIQUENESS OF THE ELASTICITY THEORY PROBLEM SOLUTION

From the viewpoint of physical notions, each elastic body subjected to en external loading and backed up properly is found at least in one equilibrium state. As far as the mathematical statement of the elastic theory problem is based on the fundamental physical principles, we may anticipate that the derived relations cannot bring us to any absurd results. This speaks in favor of the existence of the elastic boundary problem solution. Moreover, this is a most complex mathematical problem solved by now only under the general conditions. We are not going to give rather bulky and intricate substantiation, but just offer instead a corresponding solution that meets both differential equations and boundary conditions of the problem.

The proof to the uniqueness of the boundary problem solution for a static elastic body is based on the hypothesis that the assumption on nonuniqueness will lead to a contradiction. The uniqueness was first proved by Kirchhoff (1859). It was based on the positive definiteness of the potential strain energy (see 2.1). We may assume that there exist two solutions: $u_i^{(1)}, \sigma_{ij}^{(1)}, \varepsilon_{ij}^{(1)}$ and $u_i^{(2)}, \sigma_{ij}^{(2)}, \varepsilon_{ij}^{(2)}$, which meet the same boundary conditions and the basic equations (3.1)–(3.3) at identical volume forces. Then, from the equilibrium equations and boundary conditions for stresses we can obtain:

$$\sigma_{ij}^{(1)},_j = \sigma_{ij}^{(2)},_j = -\rho F_i, \; \sigma_{ij}^{(1)} l_j = \sigma_{ij}^{(2)} l_j = R_i. \tag{3.16}$$

In view of linearity of the problems, the following differences

$$\sigma_{ij}^* = \sigma_{ij}^{(1)} - \sigma_{ij}^{(2)}, u_i^* = u_i^{(1)} - u_i^{(2)}$$

are also the solutions of the basic equations of the elasticity theory for the cases of zero volume and surface forces $F_i^* = 0$, $R_i^* = 0$ because

$$\sigma_{ij}^*,_j = \sigma_{ij}^{(1)},_j - \sigma_{ij}^{(2)},_j = -\rho F_i + \rho F_i = 0,$$

$$\sigma_{ij}^* l_j = \sigma_{ij}^{(1)} l_j - \sigma_{ij}^{(2)} l_j = R_i - R_i = 0.$$

Proceeding from (3.16), Clapyeron's theorem (3.15) for the values with asterisks takes the form:

$$\int_V U \mathrm{d}V = 0.$$

Due to the positive definiteness of the specific potential energy of the elastic deformation, this equality may be satisfied only for $\varepsilon_{ij}^* = \varepsilon_{ij}^{(1)} - \varepsilon_{ij}^{(2)} = 0$. This implies that

$$\varepsilon_{ij}^{(1)} = \varepsilon_{ij}^{(2)}, \; \sigma_{ij}^{(1)} = \sigma_{ij}^{(2)},$$

which proves the uniqueness of the boundary problem solution.

For the first boundary problem the displacements are reduced to zero on the boundary. In the case of a mixed problem the same argumentation is valid. We suppose that, if necessary, the displacements become a single-valued function of coordinates of the point.

It should be noted, that we have proved the uniqueness theorem for the geometrically linear elasticity theory. For the nonlinear theory and large deformations, above method of proving is invalid, because the positive definiteness of the strain energy may be violated in this case. The latter fact means one of two things: either the accepted model of continuum is incorrect, or the material is unstable. The unstable material may be exemplified by a material with a declining tensile stress–strain curve, when two different strain values (Rzhanicyn, 1968) correspond to one and the same stress value. For the geometrically nonlinear systems the uniqueness theorem is invalid: violation of the uniqueness corresponds to the loss of stability by the elastic body. We are not going to touch upon the problems of the kind in this book. For detailed information on stability problems refer please to the well-known publications on the elasticity theory (Trefftz, 1934) or the monograph by N.A. Alfutov (1991).

## SEMI-INVERSE SAINT-VENANT'S METHOD

The stress and strain tensors and displacement vectors induced by the external forces on a deformable solid body are considered in the direct problems of the elasticity theory.

The inverse problems preset either displacements or the strain tensor components in the considered body, and specify all other values, including external forces. The inverse problems do not present any special difficulties, although their solution is of little practical interest.

A so-called *semi-inverse method* was proposed by Saint-Venant in 1853. Its essence consists in that only a part of displacement and stress components is involved in solution of the elasticity theory problems. The missing components are found by the equations of the elasticity theory so as to satisfy all these equations and boundary conditions. As a result, we obtain an accurate solution of the elasticity theory problem. Using this method, Saint-Venant has solved the problems for torsion and bending of a noncircular in section bar.

It is evident that the semi-inverse method is not a general approach. It requires certain intuition in order to guess the part of displacement and stress components needed. Nevertheless, this method may turn useful for solving some elasticity theory problems. Let us take as an example a problem of torsion of an arbitrarily shaped, uniform in section bar.

We shall accept here designations $x_1 = x$, $x_2 = y$, $x_3 = z$, and so forth. instead of arbitrary coordinates $x_i$ $(i = 1, 2, 3)$. The axial displacements will be $u$, $v$ and $w$, respectively.

In contrast to a circular bar, we may observe *deplanation* of the sections (bending in Eigen plane) when torsion of an arbitrary in section bar takes place, having in mind that the *hypothesis on plane sections* introduced in the course of strength of materials becomes invalid. Saint-Venant has solved the problem of a bar torsion in the assumption that nothing prevents deplanation of the sections, meaning that axial displacements $w$ of the bar are independent of coordinate $z$ and normal stresses are $\sigma_{zz} = 0$, while stress tensor components $\sigma_{xx}$, $\sigma_{yy}$, $\sigma_{xy}$ are considered equal to zero.

Then, by neglecting volume forces, the following equilibrium equations may be derived from (1.10):

$$\frac{\partial \sigma_{zx}}{\partial z} = 0, \frac{\partial \sigma_{zy}}{\partial z} = 0, \frac{\partial \sigma_{xz}}{\partial x} + \frac{\partial \sigma_{yz}}{\partial y} = 0 \cdot \qquad (3.17)$$

It follows from the first two equations (3.17) that stresses $\sigma_{xz}$, $\sigma_{yz}$ are independent of coordinate $z$, that is the distribution of the tangential stresses in all cross-sections of the bar is identical.

Let us introduce a stress function $\varphi$, based on which stresses $\sigma_{xz}$, $\sigma_{yz}$ take the form:

$$\sigma_{xz} = \frac{\partial \phi}{\partial y}, \sigma_{yz} = -\frac{\partial \phi}{\partial x}. \qquad (3.18)$$

So, the third equilibrium equation (3.17) is met identically:

$$\frac{\partial^2 \varphi}{\partial x \partial y} - \frac{\partial^2 \varphi}{\partial y \partial x} = 0 .$$

Let us now refer to the deformation compatibility equations in the form of Beltrami–Michell ones (3.13) based on the assumptions accepted earlier for the stress tensor components

$$\sigma_{zz} = \sigma_{xx} = \sigma_{yy} = \sigma_{xy} = 0$$

the first four equations of system (3.13) are satisfied under any dependence of function $\varphi$, and the last two equations will become:

$$\Delta \sigma_{xz} = 0, \Delta \sigma_{yz} = 0 \cdot \qquad (3.19)$$

Having substituted (3.18) into (3.19), we obtain:

$$\frac{\partial}{\partial y}\left(\frac{\partial^2 \phi}{\partial x^2} + \frac{\partial^2 \phi}{\partial y^2}\right) = 0, \frac{\partial}{\partial x}\left(\frac{\partial^2 \phi}{\partial x^2} + \frac{\partial^2 \phi}{\partial y^2}\right) = 0 \qquad (3.20)$$

It follows from (3.20) that the bracketed expression $\partial^2 \varphi / \partial x^2 + \partial^2 \varphi / \partial y^2$ is a constant value:

$$\frac{\partial^2 \varphi}{\partial x^2} + \frac{\partial^2 \varphi}{\partial y^2} = C \cdot \qquad (3.21)$$

When the lateral surface of the bar is free from external loads (the torsion moment is applied on the bar ends), the projection of the tangential stresses on the normal to the boundary should be equal to zero. In other words, the shear stresses about the boundary should be directed tangentially to it. This condition may be written as follows:

$$\frac{\partial \varphi}{\partial y}\frac{\mathrm{d}y}{\mathrm{d}s}+\frac{\partial \varphi}{\partial x}\frac{\mathrm{d}x}{\mathrm{d}s}=0, \text{ or } \frac{\mathrm{d}\varphi}{\mathrm{d}s}=0.$$

This means that stress function φ over the whole contour has one and the same value (for simply connected domains). In particular, for a bar of a solid cross-section it may be accepted equal to zero. Thus, the problem of the bar torsion is brought to finding such a stress function φ, which is able to meet equation (3.21) and be equal to zero all over the boundary.

Torsion moments $M_z$ of the bar in the form of shear stresses $p_x$, $p_y$ distributed over the bar end surface are applied to the end sections. Hence, the static boundary conditions on the bar ends may be written as follows:

$$\sigma_{xz}=\pm p_x, \sigma_{yz}=\pm p_y.$$

Taking into account equation (3.18) for $\sigma_{xz}$, $\sigma_{yz}$, and using a formula known from the course of strength of materials relating the torque to shear stresses, we may obtain after transformations:

$$M_z=\int_F (x\sigma_{yz} - y\sigma_{xz})dF=2\int_F \phi dxdy=2\int_F \phi\,\mathrm{d}x\,\mathrm{d}y. \tag{3.22}$$

Let us consider a case where the bar section represents an ellipse, and its equation is as follows:

$$\frac{x^2}{a^2}+\frac{y^2}{b^2}=1,$$

where *a*—size of the major semiaxis, and *b*—size of the minor semiaxis of the ellipse.

We shall assume for φ the following expression:

$$\varphi=m\left(\frac{x^2}{a^2}+\frac{y^2}{b^2}-1\right). \tag{3.23}$$

All over the boundary φ = 0. After substitution (3.23) into (3.21), we obtain:

$$2m\left(\frac{1}{a^2}+\frac{1}{b^2}\right)=C, \quad m=C\frac{a^2b^2}{2(a^2+b^2)}.$$

Therefore, the expression for φ takes the form:

$$\varphi=C\frac{a^2b^2}{2(a^2+b^2)}\left(\frac{x^2}{a^2}+\frac{y^2}{b^2}-1\right). \tag{3.24}$$

By substituting (3.24) in (3.22), we find an expression for *C* through the torsion moment $M_z$:

$$C = -\frac{2M_z(a^2+b^2)}{\pi a^3 b^3}.$$

Finally, the formula for φ will be:

$$\phi = -\frac{M_z}{\pi ab}\left(\frac{x^2}{a^2}+\frac{y^2}{b^2}-1\right). \qquad (3.25)$$

The shear stresses are found from (3.18), (3.25):

$$\sigma_{xz} = -\frac{M_z}{\pi ab^3}y, \sigma_{yz} = \frac{M_z}{\pi a^3 b}x. \qquad (3.26)$$

The maximum shear stress takes place at the end of the minor semiaxis section at $y = b$:

$$\sigma_{\max} = \frac{M_z}{\pi ab^2}.$$

To determine the relative torsion angle θ and deplanation $w$ of the section it is necessary to consider the stress-displacement relation. Since $\sigma_{zz} = \sigma_{xx} = \sigma_{yy} = 0$, it follows from Hooke's law that

$$\frac{\partial u}{\partial x} = \frac{\partial v}{\partial y} = \frac{\partial w}{\partial z} = 0.$$

The shear strain will be:

$$2\varepsilon_{xy} = \frac{\partial u}{\partial y}+\frac{\partial v}{\partial x} = 0, 2\varepsilon_{xz} = \frac{\partial u}{\partial z}+\frac{\partial w}{\partial x} = \frac{\sigma_{xz}}{G}, 2\varepsilon_{yz} = \frac{\partial v}{\partial z}+\frac{\partial w}{\partial y} = \frac{\sigma_{yz}}{G}. \qquad (3.27)$$

Axis $z$ of the bar remains intact at torsion. The sections are rotated about the axis so as to satisfy equations (3.27). A straight line whose points remain immovable at rotating of the bar is called the *axis of torsion*.

If we accept the expressions for displacements in the form $u = -\theta yz$ and $v = \theta xz$, the first equation (3.27) will be satisfied, and from the latter two we will have:

$$-\theta y+\frac{\partial w}{\partial x} = \frac{\sigma_{xz}}{G}, \theta x+\frac{\partial w}{\partial y} = \frac{\sigma_{yz}}{G},$$

or

$$\frac{\partial w}{\partial x} = \frac{\sigma_{xz}}{G}+\theta y, \frac{\partial w}{\partial y} = \frac{\sigma_{yz}}{G}+\theta x, \frac{\partial w}{\partial y} = \frac{\sigma_{yz}}{G}+\theta x. \qquad (3.28)$$

If to differentiate the second equation (3.27) with respect to $y$ and the third one with respect to $x$, and then subtract the second result from the first one, we obtain

$$\frac{\partial^2 u}{\partial y \partial z}-\frac{\partial^2 v}{\partial x \partial z} = \frac{1}{G}\left(\frac{\partial \sigma_{xz}}{\partial y}-\frac{\partial \sigma_{yz}}{\partial x}\right). \qquad (3.29)$$

Having the expressions assumed for $u$ and $v$, the left-hand side member of equation (3.29) will be -2θ, and the right-hand one with account of (3.18) is:

$$\frac{1}{G}\left(\frac{\partial^2\varphi}{\partial y^2}+\frac{\partial^2\varphi}{\partial x^2}\right).$$

Thus, it follows from (3.29) and (3.21) that $-2G\theta = C$, and the relative angle of torsion of the section is $\theta = -C/2G$.

For the problem of torsion of the elliptic in section bar, the relation for θ becomes:

$$\theta=\frac{M_z(a^2+b^2)}{\pi a^3b^3G}. \tag{3.30}$$

By substitution of (3.26) and (3.30) in (3.28) and integration, we may obtain, the following expression for deplanation $w$:

$$w=\frac{M_z(b^2-a^2)}{\pi a^3b^3G}xy.$$

The problem for torsion of the elliptical bar is solved in the assumption that external stresses $p_x$, $p_y$ are distributed over the bar ends by obeying the law like that for $\sigma_{xz}$ and $\sigma_{yz}$, that is on the grounds of equations (3.26), in which case deplanation of the sections is not constraint. However, according to *Saint-Venant's principle,* SSS of a body distant from the load application is conditioned mainly by the static load equivalent. Therefore, the solution obtained is valid for the sections, distant enough from the ends, under any distribution of external loads over the bar ends.

Nevertheless, we should bear in mind that Saint-Venant's principle is inapplicable in some cases. Statically equivalent changes in external loads across the ends of thin-slab structures (plates, shells, thin-walled rods) may rather lead to variations in SSS of the whole thin-walled element than to local stresses and deformations.

## STATEMENT OF THE ELASTICITY THEORY PROBLEM IN CYLINDRICAL AND SPHERICAL COORDINATES

Let us consider the basic relations of the elasticity theory in cylindrical and spherical coordinates. The *cylindrical coordinates* $r$, φ, $z$ are related to Cartesian coordinates in the following way (Fig. 3.1):

$$x_1 = x = r\cos\varphi,\ x_2 = y = r\sin\varphi,$$

$$x_3 = z = z.$$

The linear element is preset in a quadratic form:

$$(\mathrm{d}s)^2 = (\mathrm{d}r)^2 + r^2(\mathrm{d}\varphi)^2 + (\mathrm{d}z)^2$$

The components of the displacement vector are: $u_1 = u_r$, $u_2 = u_\varphi$, $u_3 = u_z$.

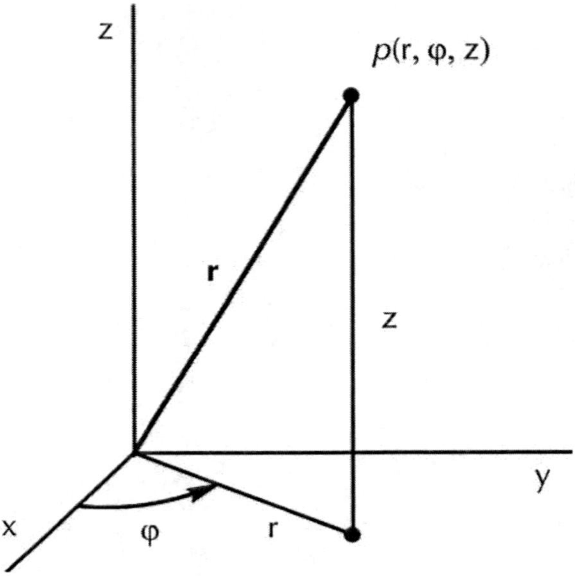


**Figure 3.1.**

While the components of the strain tensor:

$$\varepsilon_{11} = \varepsilon_{rr}, \varepsilon_{22} = \varepsilon_{\varphi\varphi} \varepsilon_{33} = \varepsilon_{zz},$$

$$\varepsilon_{12} = \varepsilon_{r\phi}, \varepsilon_{23} = \varepsilon_{\phi z}, \varepsilon_{31} = \varepsilon_{zr}.$$

The following kinematic relations are valid:

$$\varepsilon_{rr} = \frac{\partial u_r}{\partial r}, \varepsilon_{\varphi\varphi} = \frac{1}{r}\frac{\partial u_\varphi}{\partial \varphi} + \frac{u_r}{r}, \varepsilon_{zz} = \frac{\partial u_z}{\partial z}$$

$$\varepsilon_{r\varphi} = \frac{1}{2}\left(\frac{1}{r}\frac{\partial u_r}{\partial \varphi} + \frac{\partial u_\varphi}{\partial r} - \frac{u_\varphi}{r}\right), \quad e_{jz} = \frac{1}{2}\left(\frac{\partial u_j}{\partial z} + \frac{1}{r}\frac{\partial u_z}{\partial \varphi}\right),$$

$$\varepsilon_{rz} = \frac{1}{2}\left(\frac{\partial u_z}{\partial r} + \frac{\partial u_r}{\partial z}\right).$$

The components of the stress tensor $\sigma_{rr}$, $\sigma_{\varphi\varphi}$, $\sigma_{zz}$, $\tau_{r\varphi}$, $\tau_{\varphi z}$, $\tau_{zr}$ are related to strains via Hooke's law similarly to the cylindrical coordinates:

$$\sigma_{rr} = 2\mu\varepsilon_{rr} + \lambda\theta, ..., \tau_{r\varphi} = 2\mu\varepsilon_{r\varphi}, ...,,$$

where the relative volume deformation is:

$$\theta = \varepsilon_{rr} + \varepsilon_{\varphi\varphi} + \varepsilon_{zz}.$$

The equilibrium equations in stresses take the form:

$$\frac{1}{r}\frac{\partial}{\partial r}(r\sigma_{rr})+\frac{1}{r}\frac{\partial}{\partial \varphi}+\frac{\partial \tau_{rz}}{\partial z}-\frac{\sigma_{\varphi\varphi}}{r}+\rho F_r=0,$$

$$\frac{1}{r^2}\frac{\partial}{\partial r}(r^2\tau_{r\varphi})+\frac{1}{r}\frac{\partial \sigma_{\varphi\varphi}}{\partial \varphi}+\frac{\partial \tau_{\varphi z}}{\partial z}+\rho F_\varphi=0,$$

$$\frac{1}{r}\frac{\partial}{\partial r}(r\tau_{rz})+\frac{1}{r}\frac{\partial \tau_{\varphi z}}{\partial \varphi}+\frac{\partial \sigma_{zz}}{\partial z}+\rho F_z=0,$$

where $\rho F_r$, $\rho F_\varphi$, $\rho F_z$ are the components of the vector of volume forces.

*In spherical coordinates* (Fig. 3.2) the following transformation formulas are valid

$$x_1=x=R\sin\vartheta\cos\varphi,\quad x_2=y=R\sin\vartheta\sin\varphi,$$

$$x_3=z=R\cos\vartheta$$

with a squared linear element

$$(\mathrm{d}s)^2=(\mathrm{d}R)^2+R^2\sin^2\vartheta(\mathrm{d}\varphi)^2+R^2(\mathrm{d}\vartheta)^2.$$

The displacements are of the kind

$$u_1=u_R,\, u_2=u_\varphi,\, u_3=u_\vartheta$$

$$\varepsilon_{11}=\varepsilon_{RR},\, \varepsilon_{22}=\varepsilon_{\varphi\varphi},\, \varepsilon_{33}=\varepsilon_{\vartheta\vartheta},$$

$$\varepsilon_{12}=\varepsilon_{R\varphi},\, \varepsilon_{23}=\varepsilon_{\varphi\vartheta},\, \varepsilon_{31}=\varepsilon_{R\vartheta}.$$

The following kinematic relations are valid:

$$\varepsilon_{RR}=\frac{\partial u_R}{\partial r},\, \varepsilon_{\varphi\varphi}=\frac{1}{R\sin\vartheta}\frac{\partial u_\varphi}{\partial \varphi}+\frac{u_R}{R}+\mathrm{ctg}\vartheta\frac{u_\varphi}{R},$$

$$\varepsilon_{\vartheta\vartheta}=\frac{1}{R}\frac{\partial u_\varphi}{\partial \vartheta}+\frac{u_R}{R},\, \varepsilon_{R\varphi}=\frac{1}{2}\left(\frac{1}{R\sin\vartheta}\frac{\partial u_R}{\partial \varphi}+\frac{\partial u_\varphi}{\partial R}-\frac{u_\varphi}{R}\right),$$

$$\varepsilon_{R\vartheta}=\frac{1}{2}\left(\frac{1}{R}\frac{\partial u_R}{\partial \vartheta}+\frac{\partial u_\varphi}{\partial R}-\frac{u_\varphi}{R}\right),$$

$$\varepsilon_{\phi\vartheta}=\frac{1}{2}\left(\frac{1}{R\sin\vartheta}\frac{\partial u_\vartheta}{\partial \phi}+\frac{1}{R}\frac{\partial u_\phi}{\partial \vartheta}-\frac{u_\phi}{R}\mathrm{ctg}\vartheta\right).$$

The components of the stress tensor $\sigma_{RR}$, $\sigma_{\varphi\varphi}$, $\sigma_{\vartheta\vartheta}$, $\tau_{R\varphi}$, $\tau_{\varphi\vartheta}$, $\tau_{\vartheta R}$ are related to strains by Hooke's law similarly to the cylindrical system of coordinates:

$$\sigma_{RR} = 2\mu\varepsilon_{RR} + \lambda\theta\,,\ \dots\,,\ \tau_{R\varphi} = 2\mu\varepsilon_{R\varphi}\,,\ \dots\,,$$

where

$$\theta = \varepsilon_{RR} + \varepsilon_{\varphi\varphi} + \varepsilon_{\vartheta\vartheta}\,.$$

The equilibrium equations look like:

$$\frac{1}{R}\frac{\partial(R^2\sigma_{RR})}{\partial R} + \frac{1}{R\sin\varphi}\frac{\partial(\tau_{R\phi}\sin\varphi)}{\partial\varphi} + \frac{1}{R\sin\varphi}\frac{\partial\tau_{R\vartheta}}{\partial\vartheta} - \frac{\sigma_{\varphi\varphi} + \sigma_{\vartheta\vartheta}}{R} + \rho F_R = 0\,,$$

$$\frac{1}{R^3}\frac{\partial(R^3\tau_{R\varphi})}{\partial R} + \frac{1}{R\sin^2\varphi}\frac{\partial(\tau_{\varphi\vartheta}\sin^2\vartheta)}{\partial\varphi} + \frac{1}{R\sin\varphi}\frac{\partial\sigma_{\varphi\varphi}}{\partial\varphi} + \rho F_\varphi = 0\,,$$

$$\frac{1}{R^3}\frac{\partial(R^3\tau_{R\vartheta})}{\partial R} + \frac{1}{R\sin\varphi}\frac{\partial(\sigma_{\vartheta\vartheta}\sin\vartheta)}{\partial\vartheta} + \frac{1}{R\sin\vartheta}\frac{\partial\tau_{\varphi\vartheta}}{\partial\varphi} - \frac{\mathrm{ctg}\vartheta}{R}\sigma_{\varphi\varphi} + \rho F_\vartheta = 0\,.$$

Above-considered relations may be simplified essentially for the case of axial symmetry (when derivatives $\partial/\partial\varphi$ and all mixed derivatives containing $\varphi$ vanish).

## NOTES

1. The equations in displacements were first derived by Navier (1821) in the assumption that elastic properties of isotropic materials are described by a single constant of elasticity ("rariconstant" theory supposed that $\nu = 0.25$). Based on the experimental data, Lame (1852) has come to the two-constant ("multiconstant") equations which were generally accepted later.
2. *Petrovsky, I.G.* (1901–1973), a Russian mathematician, rector of Moscow State University (1951), a member of the USSR Academy of Sciences.
3. *Dirichlet, P.G.* (1805–1859), a German mathematician. *Dirichlet's problem* consists in finding a function that meets the Laplacian equation inside the area and set on its boundary. In *Neumann's problem* the derivative of the sought function is set on the boundary.
4. *Beltrami, E.* (1835–1900), Italian mathematician, formulated stress compatibility equations in 1892; *Michell*, John (1863–1940), Australian mechanical engineer, derived them in nineteenth century with volumetric forces different from zero.
5. *Pobedrya, B.E.,* Russian mechanical engineer, Professor of Moscow State University; works in the theories of plasticity and viscoelasticity.
6. *Gauss, C.F.* (1777–1855), German mathematician and physicist; is famous for works in algebra, theory of numbers and differential geometry.

## KEYWORDS

- **Biharmonic functions**
- **Boundedness condition**
- **Clapyeron's theorem**
- **Components**
- **Deplanation**
- **Elasticity theory**
- **Ellipticity**

# Chapter 4

## Variational Methods

We have already mentioned the concepts of work and energy in the preceding chapters, which are critical for general mechanics. These values are interrelated in a certain manner. The forces in a mechanical system may perform a work and the system may possess energy. A number of methods of continuum mechanics are based on the energy concept. The expediency of their application proceeds from the fact that energy is a well-defined value and it is independent of the coordinate system. Various energy principles are interconnected since they stem from the fundamental laws of continuum mechanics.

### LAGRANGIAN PRINCIPLE OF VIRTUAL DISPLACEMENTS

The *principles of virtual displacements* with reference to solids have been formulated by Lagrange in his analytical mechanics (1788). This principle has been applied for the first time to elastic bodies (a system of rods) by *Poisson* in 1833. Similarly to the principle of virtual displacements, which allows to obtain equilibrium equations for solids, the geometric derivation of equilibrium equations for elastic bodies can be replaced by the analytical one (see 6.14).

Let us consider a body of volume $V$ and surface $S = S_\sigma + S_u$ found in the equilibrium state under the applied external surface ($R_i$) and volume ($F_i$) forces. Stresses $\sigma_{ij}$, strains $\varepsilon_{ij}$ and displacements $u_i$ occur inside the body (Fig. 4.1), which are the functions of external loads.

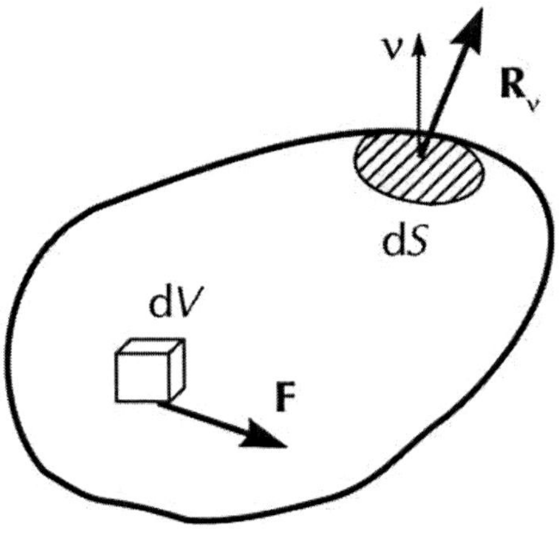


Figure 4.1.

Let us change the external forces for infinitesimal values $dR_j$, $dF_j$. Then, actual displacement $u_i$ will be imparted in infinitesimal increment $du_i$. This increment of the displacement function $u_i = u_i(R_j, F_j)$ will take place due to the varying arguments representing the loads. Let us now consider a series of displacements of some arbitrary points, which might be imparted to them at a given moment $t$ in response to external constraints imposed on the body, but not realized actually due to invariable external forces. We shall call any infinitesimal imaginary displacement performed a point may at a given fixed moment according to the constraints imposed, a probable or *virtual displacement.*

The virtual displacement in contrast to the actual one $du_i$, will be denoted as $\delta u_i$, where δ is a *variation* obeying the laws similar to the ones accepted for the operator-differential d. Remember that these laws do not cover arguments $R_j$, $F_j$ of function $u_i$. In other words, variation of a function ($u_i$, in this case) is a change of the type of the function itself at fixed coordinates of the point in question.

Let us impart virtual increments $\delta u_i$ to displacements $u_i$ leading to *virtual deformations* $\delta\varepsilon_{ij}$. In the assumption that variations $\delta u_i$ are sufficiently small and do not affect the equilibrium of the external forces and internal stresses, they are compatible with the conditions restricting the body within the boundary and continuity conditions inside the body. This means that $\delta u_i$ are *admissible kinematic functions*, that is, $\delta u_i = 0$ on $S_u$ (Fig. 4.2). In all other respects, virtual displacements may be arbitrary continuous functions.

For virtual deformations the following equation is true:

$$\delta\varepsilon_{ij} = \tfrac{1}{2}\delta(u_{i,j} + u_{j,i}) = \tfrac{1}{2}[(\delta u_i)_{,j} + (\delta u_j)_{,i}]. \tag{4.1}$$

We have used the following rule of permutation here:

$$\delta(u_{i,j}) = (\delta u_i)_{,j}\,. \tag{4.2}$$

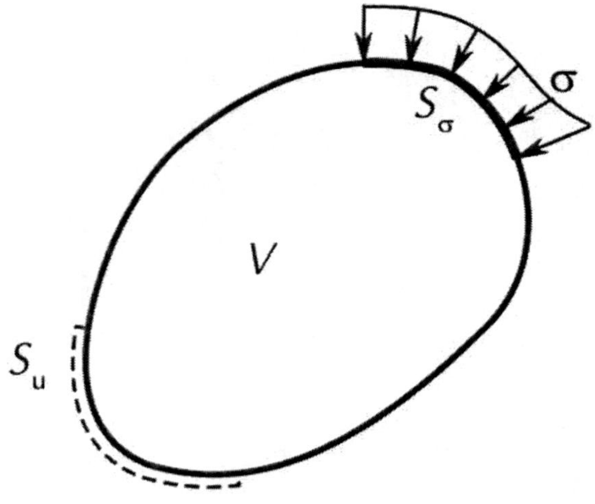


**Figure 4.2.**

According to the hypothesis of continuity, a body may be treated as a system of material points. So far, we may apply the *Lagrangian principle of virtual displacements* to it, which states: to ensure the equilibrium in a system of material points with stationary nonliberating and ideal links, it is necessary and sufficient to have the sum of elementary works of all active forces affecting the system at any possible displacement of the system equal to zero.

The work of surface and mass forces at virtual displacements we call the *virtual (probable) work of external forces* definable by a sum of integrals:

$$\delta A = \int_{S_\sigma} R_i \delta u_i \, \mathrm{d}S + \int_V \rho F_i \delta u_i \, \mathrm{d}V . \tag{4.3}$$

In this connection, we must remember that the actual forces have been applied fully to the body prior to the occurrence of the virtual displacements, and are not interrelated.

In a similar way, the *virtual work of the internal forces* is defined as a work of the actual internal stresses at virtual deformations:

$$\delta W = \int_V \sigma_{ij} \delta \varepsilon_{ij} \, \mathrm{d}V . \tag{4.4}$$

We can prove that, in the equilibrium state of the body these works are equal. With this aim, the first integral in (4.3) is transformed into the volume one using the divergence theorem (Gauss-Ostrogradski's theorem) with account that $R_i = \sigma_{ij} l_j$:

$$\int_{S_\sigma} R_i \delta u_i \, \mathrm{d}S = \int_{S_\sigma} \sigma_{ij} l_j \delta u_i \, \mathrm{d}S = \int_V (\sigma_{ij} \delta u_i),_j \, \mathrm{d}V .$$

Hence,

$$\delta A = \int_V \left[ (\sigma_{ij} \delta u_i),_j + \rho F_i \delta u_i \right] \mathrm{d}V .$$

For the first summand of the integrand with regard to the rule of permutation (4.2) it follows

$$(\sigma_{ij} \delta u_i),_j = \sigma_{ij},_j \, \delta u_i + \sigma_{ij} \delta u_i,_j .$$

This implies that

$$\delta A = \int_V (\sigma_{ij},_j + \rho F_i) \delta u_i \, \mathrm{d}V + \int_V \sigma_{ij} \delta u_i,_j \, \mathrm{d}V .$$

The first of the integrals vanishes since the equations of the body equilibrium are fulfilled

$$\sigma_{ij},_j + \rho F_i = 0 .$$

On the other hand, owing to symmetry of the stress tensor and the expression for the deformation variation (4.1), we have

$$\sigma_{ij} \delta(u_i,_j) = \sigma_{ij} \tfrac{1}{2} (\delta u_i,_j + \delta u_j,_i) = \sigma_{ij} \delta \varepsilon_{ij} .$$

Consequently,

$$\delta A = \int_V \sigma_{ij} \delta\varepsilon_{ij} \, \mathrm{d}V ,$$

where the integral represents a virtual work of the internal forces (4.4). By substitution of the virtual work of the external forces, we obtain

$$\int_{S_\sigma} R_i \delta u_i \, \mathrm{d}S + \int_V \rho F_i \delta u_i \, \mathrm{d}V = \int_V \sigma_{ij} \delta\varepsilon_{ij} \, \mathrm{d}V, \tag{4.5}$$

or

$$\delta A = \delta W \cdot$$

Thus, the virtual work of the external forces acting on the body in the equilibrium state is equal to the work of the internal stresses for the corresponding virtual deformations.

Equation (4.5) is valid for both elastic and inelastic bodies since we have not yet used Hooke's law at derivation.

Let us show now that the equilibrium condition for the body in question follows from the equality of works (4.5). The left-hand side integrand is rewritten as follows:

$$\sigma_{ij} \delta\varepsilon_{ij} = \sigma_{ij} \delta(u_i,_j) = (\sigma_{ij} \delta u_i),_j - \sigma_{ij},_j \delta u_i \, .$$

Then, equation (4.5) will be in the form

$$\int_V (\sigma_{ij} \delta u_i),_j \, \mathrm{d}V = \int_V (\sigma_{ij},_j + \rho F_i) \delta u_i \, \mathrm{d}V + \int_{S_\sigma} R_i \delta u_i \, \mathrm{d}S \, .$$

Using the known Gauss-Ostrogradski's formula for transformation of the volume integral into the surface one (see 3.4), we obtain

$$\int_V (\sigma_{ij},_j + \rho F_i) \delta u_i \, \mathrm{d}V + \int_{S_\sigma} (R_i - \sigma_{ij} l_j) \delta u_i \, \mathrm{d}S = 0 \, . \tag{4.6}$$

Since variations $\delta u_i$ are arbitrary, the bracketed integrand is equated to zero to satisfy equality (4.6). As a result, known differential equilibrium equations and static boundary conditions in stresses are obtained due to the Lagrangian variational equation (4.5) in this case. Consequently, when solving the problems by the Lagrangian variational principle, we must not satisfy the static boundary conditions because they are fulfilled automatically.

For the elastic behavior of materials the virtual work of deformation corresponds to the virtual energy of deformation of the body $\delta U_0$, that is

$$\delta W = \delta U_0 = \int_V \delta U \, \mathrm{d}V = \int_V \sigma_{ij} \delta\varepsilon_{ij} \, \mathrm{d}V ,$$

where $U$ is a specific elastic deformation energy of a body with elastic potential $\partial U/\partial\varepsilon_{ij} = \sigma_{ij}$.

As a result, remembering that the external forces and internal stresses do not vary, and following relations (4.5), we have

$$\int_V \delta_\varepsilon(\sigma_{ij}\varepsilon_{ij})\,\mathrm{d}V - \int_{S_\sigma} \delta_\varepsilon(R_i u_i)\,\mathrm{d}S - \int_V \delta_\varepsilon(\rho F_i u_i)\,\mathrm{d}V = 0 .$$

Symbol $\delta_\varepsilon$, indicating that only deformations and displacements vary, may be taken outside the integral sign. Then,

$$\delta_\varepsilon\left[\int_V \sigma_{ij}\varepsilon_{ij}\,\mathrm{d}V - \int_{S_\sigma} R_i u_i\,\mathrm{d}S - \int_V \rho F_i u_i\,\mathrm{d}V\right] = 0 .$$

The first bracketed integral presents a potential energy of deformations, and the rest—the potential of the external forces. Let us denote the total quantity by Π and call it a *potential energy of the body*,

$$\Pi = \int_V \sigma_{ij}\varepsilon_{ij}\,\mathrm{d}V - \int_{S_\sigma} R_i u_i\,\mathrm{d}S - \int_V \rho F_i u_i\,\mathrm{d}V . \qquad (4.7)$$

By means of (4.7) the preceding equation may be brought to

$$\delta_\varepsilon \Pi = 0 . \qquad (4.8)$$

Equation (4.8) expresses a so-called *principle of potential energy* meaning that at specified external forces and boundary conditions actual displacements $u_i$ are such that for any virtual displacement the first variation of the total potential energy is equal to zero, that is, the total potential energy Π has a stationary value.

We may prove that the potential energy of the system in a stable equilibrium is minimal, so the second variation is $\delta^2\Pi \geq 0$ (*theorem of Lagrange-Dirichlet*). Consequently, among the displacements meeting specified boundary conditions only the actual displacements satisfying equilibrium may give a minimal value of the potential energy.

It follows that for a specified loading on a body one should find such $u_i$ functions at which condition (4.8) is met, wherefrom we can define true displacements of the body and the problem of the elasticity theory in displacements will be solved. This is the variational formulation of the elasticity theory problem based on the Lagrangian principle.

The inverse theorem is also true, which says that in case the potential energy reaches a bare minimum for some field of displacements satisfying boundary conditions in displacements, then this field should also obey the equilibrium equations inside the body and the boundary conditions in stresses (Hahn, 1985).

## CASTIGLIANO'S PRINCIPLE OF VIRTUAL FORCES

Castigliano supposed that instead of considering virtual displacements from the equilibrium state, we may vary the stresses. As opposed to Lagrange's principle, according to which the state of a body is characterized by *displacement functions*, Castigliano's principle accepts *stresses*, as the determining factors. This principle involves calculations of the virtual work being done by *virtual forces* $\delta F_i$, $\delta R_i$, and *stresses* $\delta\sigma_{ij}$ on actual displacements $u_i$, irrespective of actual forces and stresses meeting the equilibrium of the body. The displacements and the corresponding stress state are set in each point of the body to satisfy the equilibrium equations inside the body and the boundary conditions on surface $S_\sigma$.

Virtual forces $\delta F_i$, $\delta R_i$ and stresses $\delta\sigma_{ij}$ are statically admissible functions, for which the equilibrium equations inside the body and the boundary conditions on its surface are valid:

$$\delta\sigma_{ij},_j + \rho\delta F_i = 0 \text{ in volume } V,$$

$$\delta\sigma_{ij} l_j = \delta R_i = 0 \text{ on } S. \tag{4.9}$$

Values $\delta\sigma_{ij}$ and $\delta R_i$ on $S_u$ are arbitrary ones. There exists an infinite number of such stress systems in each problem of the elasticity theory since this problem is indeterminable statically. Notice that three equilibrium equations (4.9) include six unknown stress functions. Castigliano's principle isolates the stresses out of all statically possible, which ensure not only equilibrium, but also compatibility of deformations in a body, being therefore, a single sought solution of the elasticity theory problem.

Let us assume additionally those variations $\delta R_i$ vanish on the surface area, where superficial forces $R_i$ have been set, that is, on $S_\sigma$.

To formulate the principle, we shall define the value of the *virtual extra work $\delta A^*$ of the external virtual forces and the virtual extra work of the virtual internal stresses $\delta W^*$:*

$$\delta A^* = \int_{S_u} u_i \delta R_i \, \mathrm{d}S + \int_V u_i \rho \delta F_i \, \mathrm{d}V ,$$

$$\delta W^* = \int_V \varepsilon_{ij} \delta\sigma_{ij} \, \mathrm{d}V . \tag{4.10}$$

We can prove the equality of works (4.10) for a body in equilibrium, which follows, for example, from variational equilibrium equations (4.9). If we multiply them by $u_i$, sum up over $i$ and integrate over the body volume, we shall have

$$\int_V u_i (\delta\sigma_{ij},_j + \rho\delta F_i) \mathrm{d}V = 0 . \tag{4.11}$$

Let us transform the first term in the integrand

$$u_i \delta\sigma_{ij},_j = (u_i \delta\sigma_{ij}),_j - u_i,_j \delta\sigma_{ij} .$$

On the basis of the divergence theorem and variational boundary conditions (4.9), it follows from (4.11) that

$$\int_S u_i \delta R_i \,\mathrm{d}S - \int_V u_{i},_j \delta\sigma_{ij} \,\mathrm{d}V + \int_V u_i \rho \delta F_i \,\mathrm{d}V = 0 . \tag{4.12}$$

When the integrating over the whole surface of the body ($S$) for the first term, we substitute by integration over a part of the surface ($S_u$), since the virtual surface forces on boundary $S_\sigma$ tend to zero. Owing to symmetry $\delta\sigma_{ij}$ from the second integral in (4.12), the following equality is derived:

$$\int_V u_{i},_j \delta\sigma_{ij} \,\mathrm{d}V = \int_V \varepsilon_{ij} \delta\sigma_{ij} \,\mathrm{d}V .$$

We can finally obtain an expression from (4.12):

$$\int_{S_u} u_i \delta R_i \,\mathrm{d}S + \int_V u_i \rho \delta F_i \,\mathrm{d}V = \int_V \varepsilon_{ij} \delta\sigma_{ij} \,\mathrm{d}V , \tag{4.13}$$

or

$$\delta A^* = \delta W^* .$$

It presents a mathematical formulation of the *principle of virtual forces* for any deformed body:

> The sum of works of increments of all external forces $\delta F_i$, $\delta R_i$ spent on actual displacements of the body at any virtual variation of its stress state equals to the increment of the extra work of the virtual internal stresses.

As far as the volume forces are set constant, $\delta F_i = 0$ as a rule, equality (4.13) is valid for any elastic body irrespective of the material properties, and the body is found in the equilibrium state if this equation is met.

*Result.* Since the displacements and deformations remain invariable, we obtain an expression from the principle of virtual forces (4.13) by factorizing the variation sign:

$$\delta_\sigma \left[ \int_V \varepsilon_{ij} \sigma_{ij} \,\mathrm{d}V - \int_{S_u} u_i R_i \,\mathrm{d}S - \int_V u_i \rho F_i \,\mathrm{d}V \right] = 0 \cdot \tag{4.14}$$

The index at $\delta_\sigma$ indicates that only stresses and forces vary. The first bracketed integral presents a complementary potential energy of deformation, and the rest—the complementary potential of external forces. Let us denote the total quantity by $\Pi^*$ and call it the *complementary potential energy of the body*:

$$\Pi^* = \int_V \sigma_{ij} \varepsilon_{ij} \,\mathrm{d}V - \int_{S_u} R_i u_i \,\mathrm{d}F - \int_V \rho F_i u_i \,\mathrm{d}V . \tag{4.15}$$

By means of (4.15), equation (4.14) may be presented as

$$\delta_\sigma \Pi^* = 0.$$

The zeroth first variation of the complementary potential energy of the system means that it acquires a stationary value for the forces and stresses meeting the equilibrium equations, which correspond to a true deformed state.

Based on this principle, we may conclude that in case of a stable equilibrium the extreme value of $\Pi^*$ corresponds to a minimum (*Castigliano's principle*). Consequently, among all stress states that satisfy the equilibrium equations, the one that for which the complementary energy reaches the minimal value is the actual.

## CASTIGLIANO'S THEOREMS

A number of theorems for the work valuable for practical applications, especially for the strength of materials theory and structural mechanics used to calculate deformations in statically determinable and statically indeterminable systems are based on the principle of the virtual work, or the minimum principles.

Let us consider an elastic body in a stable equilibrium resting immovably on two small areas of its surface (Fig. 4.3). The volume forces may be ignored, while the load is set by the surface forces in the form of concentrated forces $P_k$ and moments $M_n$ realizable, for example, by a couple of forces. Let us denote the displacements corresponding to forces $P_k$, by $u_k$; rotations $\varphi_n$ correspond to torques $M_n$.

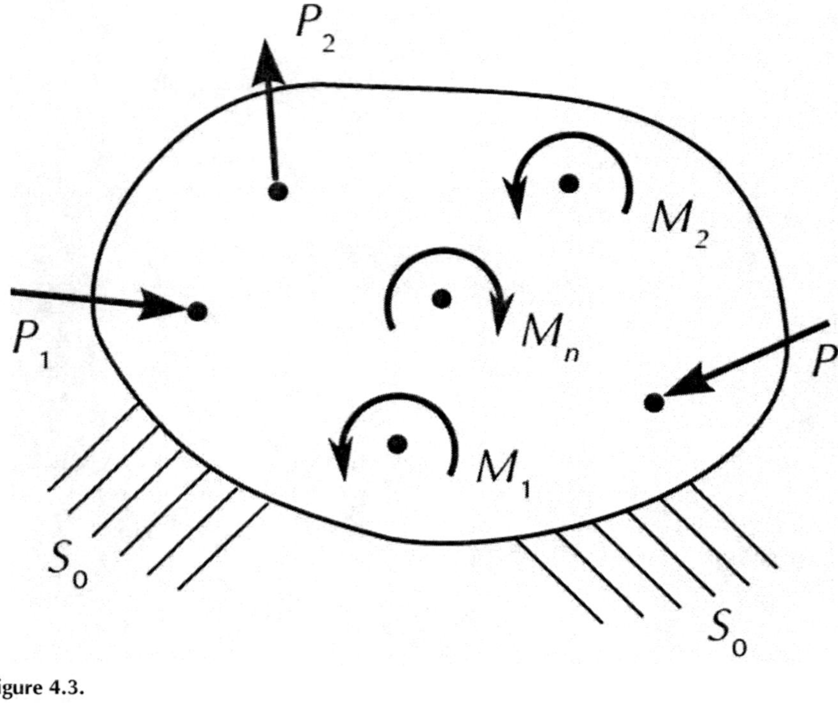


**Figure 4.3.**

Then, stresses and strains as well as the strain energy are determined as functions $u_k$ and $\varphi_n$: $U = U(u, \varphi)$.

*The first Castigliano's theorem states*:

$$\frac{\partial U}{\partial u_k} = P_k,$$

$$\frac{\partial U}{\partial \phi_n} = M_n. \tag{4.16}$$

This conclusion follows from the principle of virtual displacements when only one virtual displacement $\delta u_k$ in direction of force $P_k$ is considered. So, relation $\delta U = P_k \delta u_k$ is valid, wherefrom we obtain the first relation (4.16) by passage to the limit. The second relation is obtained in a similar way.

The first Castigliano's theorem may be used, for example, to calculate indeterminable statically systems, but its significance for practical applications is rather low.

*The second Castigliano's* theorem is of great importance, which for linearly elastic materials is of the form:

$$\frac{\partial U^*}{\partial P_k} = u_k, \frac{\partial U^*}{\partial M_n} = \varphi_n,$$

where $U^* = U^*(P, M)$ means the strain energy expressed by the external forces and moments (assuming validity of Hooke's law).

It is easy to calculate deformations in the point of force application for statically determinable and statically indeterminable structures using this theorem. To calculate deformations in arbitrary points we introduce fictitious auxiliary forces (moments), which are then assumed equal to zero.

## BETTI'S RECIPROCAL THEOREM

*Betti's*[1] *reciprocal theorem*, inter relating various states of equilibrium of a linearly elastic body under various loads turns to be very useful for practice. We shall consider two equilibrium states of an elastic body (of volume $V$ and surface $S$) known as, respectively, I and II and characterized by values $u_i^{\mathrm{I}}$, $\varepsilon_{ij}^{\mathrm{I}}$, $\sigma_{ij}^{\mathrm{I}}$, generated by forces $\rho F_i^{\mathrm{I}}$ and $R_i^{\mathrm{I}}$, along with values $u_i^{\mathrm{II}}$, $\varepsilon_{ij}^{\mathrm{II}}$, $\sigma_{ij}^{\mathrm{II}}$ generated by forces $\rho F_i^{\mathrm{II}}$ and $R_i^{\mathrm{II}}$.

Then the following identity is valid:

$$\sigma_{ij}^{\mathrm{I}} \varepsilon_{ij}^{\mathrm{II}} = \sigma_{ij}^{\mathrm{II}} \varepsilon_{ij}^{\mathrm{I}}, \tag{4.17}$$

which is easy to check under the linear stress to strain relations (2.11):

$$\sigma_{ij}^{\mathrm{I}} = E_{ijkl} \varepsilon_{kl}^{\mathrm{I}}, \sigma_{ij}^{\mathrm{II}} = E_{ijkl} \varepsilon_{kl}^{\mathrm{II}}.$$

By substitution of these relations into (4.17) when elasticity module $E_{ijkl} = E_{klij}$ are symmetrical, we obtain

$$\sigma_{ij}^{\rm I}\varepsilon_{ij}^{\rm II} = E_{ijkl}\varepsilon_{kl}^{\rm I}\varepsilon_{ij}^{\rm II} = E_{klij}\varepsilon_{ij}^{\rm II}\varepsilon_{kl}^{\rm I} = \sigma_{kl}^{\rm II}\varepsilon_{kl}^{\rm I} = \sigma_{ij}^{\rm II}\varepsilon_{ij}^{\rm I}.$$

The next equality can be derived from (4.17):

$$\int_V \sigma_{ij}^{\rm I}\varepsilon_{ij}^{\rm II}\,{\rm d}V = \int_V \sigma_{ij}^{\rm II}\varepsilon_{ij}^{\rm I}\,{\rm d}V, \tag{4.18}$$

which presents *one of the kinds of Betti's reciprocal theorem*. Its left-hand side is transformed in the following way:

$$\int_V \sigma_{ij}^{\rm I}\tfrac{1}{2}(u_i^{\rm II},_j + u_j^{\rm II},_i)\,{\rm d}V = \int_V \sigma_{ij}^{\rm I}u_i^{\rm II},_j\,{\rm d}V = \int_V (\sigma_{ij}^{\rm I}u_i^{\rm II}),_j\,{\rm d}V - \int_V \sigma_{ij}^{\rm I},_j\,u_i^{\rm II}\,{\rm d}V.$$

Based on the Gauss-Ostrogradski formula and equilibrium equations ($\sigma_{ij}^{\rm I},_j = -\rho F_i^{\rm I}$), we have

$$\int_V \sigma_{ij}^{\rm I}\varepsilon_{ij}^{\rm II}\,{\rm d}V = \int_S R_i^{\rm I}u_i^{\rm II}\,{\rm d}S + \int_V \rho F_i^{\rm I}u_i^{\rm II}\,{\rm d}V.$$

The right-hand side of (4.18) may be transformed in a similar way. As a result, we come to *Betti's reciprocal theorem* stating that the work of the system of external forces I on displacements caused by system II equals to the work of the system of external forces II on displacements caused by system I:

$$\int_S R_i^{\rm I}u_i^{\rm II}\,{\rm d}S + \int_V \rho F_i^{\rm I}u_i^{\rm II}\,{\rm d}V = \int_S R_i^{\rm II}u_i^{\rm I}\,{\rm d}S + \int_V \rho F_i^{\rm II}u_i^{\rm I}\,{\rm d}V. \tag{4.19}$$

The significance of Betti's theorem consists in the possibility to obtain a relation between the forces applied and displacements of system I using an arbitrarily selected system II. System II is an auxiliary one and may be very simple, for example, a homogeneous stress state. The theorem may then offer an explanation for various properties of solution I. It may also serve as the initial one for the so-called integral boundary equations.

Note. Betti's theorem is also valid for an elastic body loaded by concentrated forces (Fig. 4.4). In the absence of the mass forces, for example, in case only two concentrated forces in specified points are acting, then proceeding from (4.19), equality $P^{\rm I}u^{\rm I} = P^{\rm I}u^{\rm I}$ is valid where $u^{\rm II}$ is a displacement of the point of force application $P^{\rm I}$ (in direction $P^{\rm I}$) due to the action of force $P^{\rm II}$. Correspondingly, $u^{\rm I}$ is a displacement of the point of force application $P^{\rm II}$ (in direction $P^{\rm II}$) due to the action of force $P^{\rm I}$.

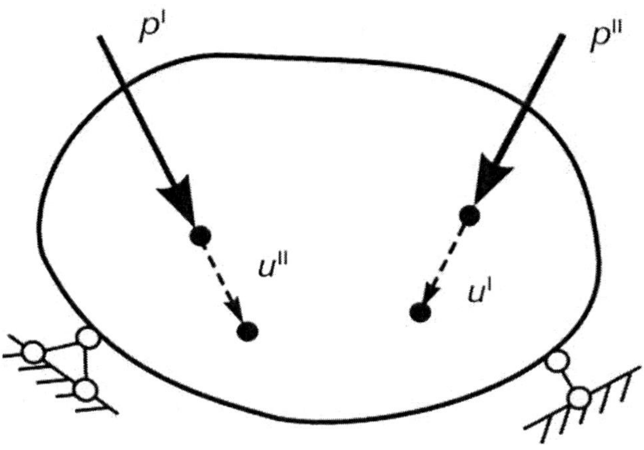


**Figure 4.4.**

If a series of forces are acting, then

$$\sum_s P_s^I u_s^I = \sum_s P_s^I u_s^I$$

The corresponding relations are obtained also for the concentrated torques and rotations.

## THE VARIATIONAL RAYLEIGH–RITZ METHOD

Most problems in the elasticity theory are brought to solution of differential equations with preset boundary conditions. The calculations are often complicated. The direct variational methods allow us to avoid some mathematical difficulties. Instead of the fundamental differential equations of the elasticity theory, we may state a problem on finding functions $u_i$, $\varepsilon_{ij}$, $\sigma_{ij}$ to meet boundary conditions and minimize functional[3] $\Phi(u_i, \varepsilon_{ij}, \sigma_{ij})$, for example, the total potential energy $\Pi$ or complementary energy $\Pi^*$.

According to the *method* of *Rayleigh–Ritz,* displacements $u_i$ present a series of functions, each meeting the geometrical boundary conditions. For instance,

$$u_i = A_k \varphi_k^i(x_1, x_2, x_3) \ (i = 1, 2, 3; k = 1, 2, 3, \ldots), \qquad (4.20)$$

where $A_{ik}$—indeterminate coefficients to be found from condition (4.8); $\varphi_k^i$—known functions meeting the boundary conditions. By substituting their expressions (4.20) into (4.7), (4.8) instead of $u_i$, and integrating, we obtain

$$\Pi = \Pi(A_{ik}) \cdot \qquad (4.21)$$

Based on condition (4.7) written as

$$\delta\Pi = \frac{\partial \Pi}{\partial A_{ik}} \delta A_{ik} = 0,$$

in view of arbitrariness of variations $\delta A_{ik}$ it follows

$$\frac{\partial \Pi}{\partial A_{ik}} = 0. \tag{4.22}$$

If we leave $n$ members in (4.20), then conditions (4.22) give a $3n$ system of linear algebraic equations for the case of the linear elasticity

$$\delta_{jk} A_{ik} + \Delta_j = 0 \ (k, j = 1, 2, \ldots, n; i = 1, 2, 3), \tag{4.23}$$

able to find parameters $A_{ik}$. Equations (4.22) or (4.23) are known as *Rayleigh–Ritz equations*. Substitution of parameters $A_{ik}$, obtained from these equations in (4.20), gives us an approximate solution of the problem.

## BUBNOV–GALERKIN'S METHOD

This method has been first proposed by I.G. Bubnov in 1913 as applied to the elasticity theory problems. Later on, it was developed by B.G. Galerkin in 1915. In case function $\varphi_k^i$ in displacement equations (4.20) is chosen beforehand so as to meet both geometrical and static boundary conditions for stresses, then the surface integral in equation (4.6) vanishes, and becomes:

$$\int_V (\sigma_{ij},_j + \rho F_i)\delta u_i \, \mathrm{d}V = 0$$

If we introduce here the expression

$$\delta u_i = \delta A_{ik} \varphi_k^i$$

and take into account arbitrariness of variations $\delta A_{ik}$, we can obtain *Bubnov-Galerkin's equations* to determine parameters $A_{ik}$

$$\int_V (\sigma_{ij},_j + \rho F_i)\varphi_k^i \, \mathrm{d}V = 0 .$$

In the cases when the static boundary conditions may be satisfied, Bubnov-Galerkin's method simplifies considerably the calculations.

## RITZ–LAGRANGE'S METHOD

This method combines Ritz method and the method of Lagrangian multiplier. According to Ritz's approach, functions $\varphi_k^i$ are selected so that each of them meets the geometrical boundary condition. In some cases it is difficult to satisfy this requirement. Instead we can use the *Lagrangian undeterminate multipliers* so that the boundary conditions are met not by each function but by the relation for displacements $u_i$ as a whole. In this case, coefficients $A_{ik}$ will satisfy some auxiliary relations of the form:

$$f_k(A_k) = 0 . \tag{4.24}$$

Let us assume that the number of such relations is $m$. Then, according to Lagrange, we should write a function

$$Э = \Pi - \lambda_k f_k \ (k = 1, 2, \ldots, m)$$

and solve the problem on a constrained extremum of this function, that is, set the extreme conditions

$$\frac{\partial Э}{\partial A_{ik}} = 0 \ (i = 1, 2, \ldots, n). \tag{4.25}$$

Equations (4.24), (4.25) compose a system of $m + n$ equations with $m + n$ unknowns $A_{ik}$ and $\lambda_k$. Above-described method will turn especially convenient if functions $\varphi_k^{\,i}$ become orthogonal, that is:

$$\int_0^l \varphi_k^i \varphi_k^i \, \mathrm{d}x = 0 \, .$$

In this case, each equation (4.25) will contain only one unknown $A_{ik}$, and the problem will be reduced to a system of $m$ equations (4.24).

As an example of application of the principle of Lagrangian virtual displacements we may consider the solution of equilibrium equations and boundary conditions for an elastic circular three-layer plate (see 6.14).

## NOTES

1. Betti, E. (1823–1892), Italian mathematician.
2. The main idea of this method was used in 1877 by Lord Rayleigh (1842–1919) for approximate determination of natural frequencies of mechanical oscillating system. It is also called the Rayleigh–Ritz method.
3. A functional is a variable, which attaches a certain number to each sought function (curve, surface) of some class. We mean that the functional is a composite function.

## KEYWORDS

- **Admissible kinematic functions**
- **Infinitesimal**
- **Statically indeterminable systems**
- **Virtual displacement**

# Chapter 5

## The Plane Elastic Problem

The problems in which the sought function depends only on two coordinates are of great practical significance. To these belong also the plane elastic problems. We distinguish two cases with coinciding mathematical description, namely the *plane strain state* and *the plane stress state*.

### PLANE STRAIN STATE

To consider this case, we take a long prismatic body (with longitudinal axis $z$) loaded by the surface forces independent of $z$ without any constituents along this axis. The example presented in Fig. 5.1 shows a cylinder loaded by forces $p_1, p_2, p_3$ linearly distributed over its generatrix. The elastic body may be either infinitely long or of a finite length with appropriately fixed edges.

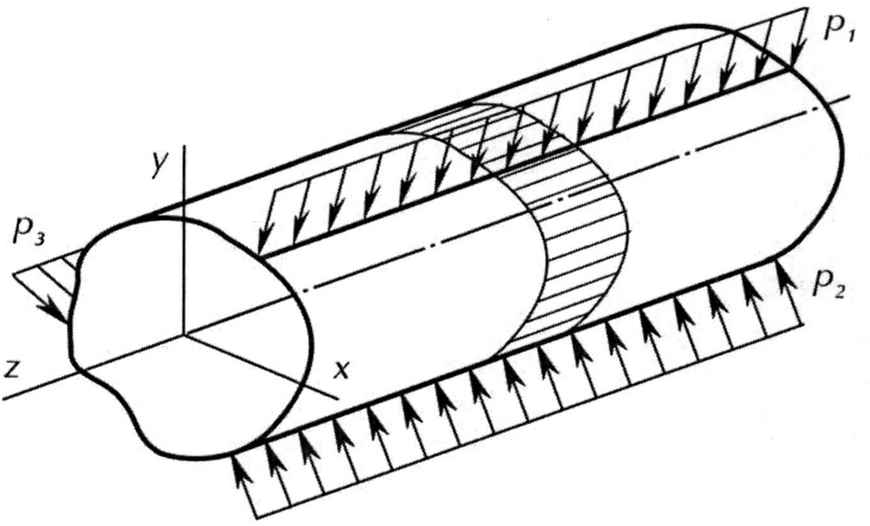


**Figure 5.1.**

Then, the determinant in each cross section of the body is the plane strain state.

We shall further use the simplified relations in this text, so the indices (tensors) will appear only to ensure compactness of the formulas. The displacements in Cartesian coordinates $x$, $y$, $z$ will be denoted by, respectively, $u$, $v$, $w$. For the plane strain state they are given by

$$u = u(x, y), \quad v = v(x, y), \quad w = \text{const} \quad or \quad w = 0; \tag{5.1}$$

Besides, all the derivatives with respect to $z$ (to $x_3$) become zero. The strains, including volume one $\theta$, will be:

$$\varepsilon_{xx} = \frac{\partial u}{\partial x}, \varepsilon_{yy} = \frac{\partial v}{\partial y}, \varepsilon_{xy} = \frac{1}{2}\left(\frac{\partial u}{\partial y} + \frac{\partial v}{\partial x}\right),$$
$$\varepsilon_{xz} = \varepsilon_{yz} = \varepsilon_{zz} = 0, \theta = \frac{\partial u}{\partial x} + \frac{\partial v}{\partial y} \tag{5.2}$$

Only one of deformation compatibility conditions (1.33), (1.34) is left

$$\frac{\partial^2 \varepsilon_{xx}}{\partial y^2} + \frac{\partial^2 \varepsilon_{yy}}{\partial x^2} = 2\frac{\partial^2 \varepsilon_{xy}}{\partial x \partial y}. \tag{5.3}$$

Nonzero components of the stress tensor are $\sigma_{xx}$, $\sigma_{xy}$, $\sigma_{yy}$, $\sigma_{zz}$. Hooke's formulas of the generalized law will take the form:

$$\sigma_{xx} = 2\mu\varepsilon_{xx} + \lambda\theta, \sigma_{yy} = 2\mu\varepsilon_{yy} + \lambda\theta, \sigma_{xy} = 2\mu\varepsilon_{xy}, \sigma_{zz} = \lambda\theta. \tag{5.4}$$

Deformations are expressed in terms of stresses by the formulas from (2.8):

$$\varepsilon_{xx} = \frac{1}{E}\left[\sigma_{xx} - \nu(\sigma_{yy} + \sigma_{zz})\right], \varepsilon_{yy} = \frac{1}{E}\left[\sigma_{yy} - \nu(\sigma_{zz} + \sigma_{xx})\right], \varepsilon_{xy} = \frac{1+\nu}{E}\sigma_{xy}.$$

We can easily show that in the case of a body plane strain the number of independent components of the stress tensor is three. From Hooke's law, in provision that $\varepsilon_{zz} = 0$, it follows

$$\sigma_{zz} = \nu(\sigma_{xx} + \sigma_{yy}). \tag{5.5}$$

From the equilibrium equations, we leave the following:

$$\frac{\partial \sigma_{xx}}{\partial x} + \frac{\partial \sigma_{xy}}{\partial y} = 0, \frac{\partial \sigma_{yy}}{\partial y} + \frac{\partial \sigma_{xy}}{\partial x} = 0, \tag{5.6}$$

wherefore, the volume forces are not taken into account here.

The boundary conditions for stresses, with $R_x$ and $R_y$ stresses set on the cylinder edge (the *first boundary problem*), are

$$\sigma_{xx}\cos(\nu, x) + \sigma_{xy}\cos(\nu, y) = R_x, \sigma_{xy}\cos(\nu, x) + \sigma_{yy}\cos(\nu, y) = R_y. \tag{5.7}$$

If we take a prismatic body of length $l$ with supported end sections, the following boundary conditions will be valid:

$$w(x, y, 0) = w(x, y, l) = 0.$$

Then the forces acting in the end cross-sections are:

$$P_z = \int_S \sigma_{zz} dS \,,$$

Hence, the stresses are integrated over the cross-section of the prismatic body. So, equilibrium equations in Lame's displacements (3.6) have the form

$$\Delta u + \frac{1}{1-2\nu}\frac{\partial}{\partial x}\left(\frac{\partial u}{\partial x}+\frac{\partial v}{\partial y}\right)=0,\ \Delta v + \frac{1}{1-2\nu}\frac{\partial}{\partial y}\left(\frac{\partial u}{\partial x}+\frac{\partial v}{\partial y}\right)=0\,.$$

Correspondingly, the equations for Beltrami-Michell's stresses (3.13) become:

$$\Delta\sigma_{xx} + \frac{3}{1+\nu}\frac{\partial^2\sigma}{\partial x^2}=0,\ \Delta\sigma_{yy} + \frac{3}{1+\nu}\frac{\partial^2\sigma}{\partial y^2}=0,\ \Delta\sigma_{xy} + \frac{3}{1+\nu}\frac{\partial^2\sigma}{\partial x\partial y}=0\,.$$

Here, the mean stress with regard to equation (5.5) for $\sigma_z$ is given by

$$\sigma = \frac{1}{3}(\sigma_{xx}+\sigma_{yy}+\sigma_{zz}) = \frac{1+\nu}{3}(\sigma_{xx}+\sigma_{yy})\,.$$

If the end faces of a finite-length prismatic body are not free from stresses, the plane deformation conditions are violated. However, again, for the sections, sufficiently distant from the end faces, it may be assumed according to Saint-Venant's principle that SSS corresponds to the conditions in plane deformation.

## PLANE STRESS STATE

In this case, we take a thin plane elastic body (a plate) loaded only by the in-plane forces, so far the through-thickness normal stresses are absent (Fig. 5.2). The forces applied are either uniformly distributed through the thickness, being therefore independent of *z* (always satisfied for thin plates with good approximation), or distributed symmetrically about so-called *median plane* (an imaginary surface bisecting the plate thickness). Then, we may introduce their average (across plate thickness) value.

In the case of the plane stress state, there are the following nonzero components of the stress tensor in plane *x*, *y*:

$$\sigma_{xx} = \sigma_{xx}(x, y),\ \ \sigma_{yy} = \sigma_{yy}(x, y),$$
$$\sigma_{xy} = \sigma_{xy}(x, y)\,.$$

The rest being

$$\sigma_{zz} = \sigma_{xz} = \sigma_{yz} = 0\,.$$

Displacement components *u*, *v* and *w* are here fully independent of coordinate *z*. Notice that, the kinematic equations correspond to equations (5.2) for the plane strain state, while compatibility conditions to equation (5.3), equilibrium equations to equations (5.6), and the boundary conditions to conditions (5.7).

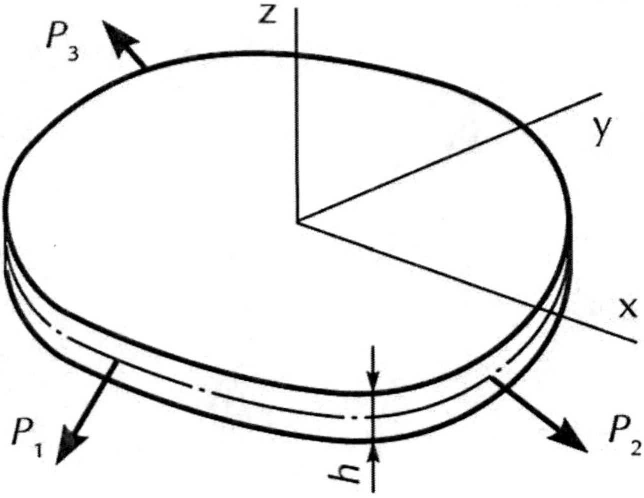


**Figure 5.2.**

The differences between the plane strain and plane stress states become apparent when considering the deformations like in Hooke's law. Since $\sigma_{zz} = 0$, it follows that

$$\varepsilon_{xx} = \frac{1}{E}\left[\sigma_{xx} - \nu\sigma_{yy}\right],\ \varepsilon_{yy} = \frac{1}{E}\left[\sigma_{yy} - \nu\sigma_{xx}\right],$$

$$\varepsilon_{zz} = -\frac{\nu}{E}(\sigma_{xx} + \sigma_{yy}),\ \varepsilon_{xy} = \frac{\sigma_{xy}}{2G}. \quad (5.8)$$

The inversion of (5.8) gives the expressions for stresses in terms of strains

$$\sigma_{xx} = \frac{E}{1-\nu^2}(\varepsilon_{xx} + \nu\varepsilon_{yy}),\ \sigma_{yy} = \frac{E}{1-\nu^2}(\varepsilon_{yy} + \nu\varepsilon_{xx}),\ \sigma_{xy} = \frac{E}{1+\nu}\varepsilon_{xy},$$

that differ from the corresponding formulas (5.4), (5.5).

Proceeding from above, the plane stress state is realized not accurately, but approximately in sufficiently thin plates.

L.N.G. Filon[1] has proved that a generalized case may be considered and the reduced relations may be modified. It is assumed therefore, that the applied forces are distributed symmetrically about the plate middle plane. If to presume that stresses $\sigma_{zz}$ are not available in the plate, while $\sigma_{zx}$ and $\sigma_{zy}$ are equal to zero on the external surfaces of the plate, then the average values are calculated for displacements and stresses through thickness of the plate, for instance

$$\sigma_{xx}^{*} = \frac{1}{2h}\int_{-h}^{h} \sigma_{xx}\,\mathrm{d}z\,,$$

where $2h$ is the plate thickness. As a result, the problem is brought to the previous one.

The stress state, arising in such plates, often occurs in practice and is, therefore, rather important for consideration. Based on Love's formulation, we call it a *generalized plane stress state*.

## COMPATIBILITY EQUATIONS FOR STRESSES

To solve the plane problem in stresses, we are to present the deformation compatibility equation (5.3) in stresses using Hooke's law (5.8), which gives

$$\frac{1}{E}\frac{\partial^2}{\partial y^2}\left[\sigma_{xx} - \nu\sigma_{yy}\right] + \frac{1}{E}\frac{\partial^2}{\partial x^2}\left[\sigma_{yy} - \nu\sigma_{xx}\right] - 2\frac{1}{2G}\frac{\partial^2\sigma_{xy}}{\partial x\partial y} = 0\,.$$

Let us now express the derivatives of tangential stresses by equilibrium equations (5.6)

$$\frac{\partial^2\sigma_{xy}}{\partial x\partial y} = -\frac{\partial^2\sigma_{xx}}{\partial x^2},\ \frac{\partial^2\sigma_{xy}}{\partial x\partial y} = -\frac{\partial^2\sigma_{yy}}{\partial y^2}$$

and substitute them into the preceding relation. By changing shear modulus $2G = E/(1+\nu)$, we have

$$\frac{\partial^2}{\partial y^2}\left[\sigma_{xx} - \nu\sigma_{yy}\right] + \frac{\partial^2}{\partial x^2}\left[\sigma_{yy} - \nu\sigma_{xx}\right] + (1+\nu)\left(\frac{\partial^2\sigma_{xx}}{\partial x^2} + \frac{\partial^2\sigma_{yy}}{\partial y^2}\right) = 0\,.$$

After cancellations

$$\left(\frac{\partial^2}{\partial x^2} + \frac{\partial^2}{\partial y^2}\right)(\sigma_{xx} + \sigma_{yy}) = 0\,,$$

or, using the Laplacian operator in plane coordinates $\Delta \equiv \partial^2/\partial x^2 + \partial^2/\partial y^2$, we come to

$$\Delta(\sigma_{xx} + \sigma_{yy}) = 0\,. \tag{5.9}$$

Consequently, the sum of stresses is a harmonic function.

## AIRY'S[2] STRESS FUNCTION

In the absence of body forces, equilibrium equations (5.6) are satisfied identically by introduction of stress functions

$$\sigma_{xx} = \frac{\partial^2\phi}{\partial y^2},\ \sigma_{yy} = \frac{\partial^2\phi}{\partial x^2},\ \sigma_{xy} = -\frac{\partial^2\phi}{\partial x\partial y}\,. \tag{5.10}$$

The sought function $\Phi(x, y)$ is known as Airy's stress function. If we sum up the normal stresses:

$$\sigma_{xx} + \sigma_{yy} = \frac{\partial^2 \phi}{\partial x^2} + \frac{\partial^2 \phi}{\partial y^2} = \Delta\phi .$$

Let us take the Laplacian operator from both sides of above equality. Since the deformation compatibility equations in stresses should be satisfied (5.9), the basic differential equation for Airy's stress function takes the form

$$\Delta\Delta\phi = \frac{\partial^4 \phi}{\partial x^4} + 2\frac{\partial^4 \phi}{\partial x^2\,\partial y^2} + \frac{\partial^4 \phi}{\partial y^4} = 0 \cdot \tag{5.11}$$

This equation (5.11) was first derived by Maxwell[3] and is known as a *biharmonic equation*, while its solutions—as *biharmonic functions*.

The biharmonic equation is typically applied in the elasticity theory. Its numerous partial solutions are also well-known, each corresponding to a certain stress state meeting the equilibrium and compatibility equations. For example,

$$x^2,\ y^2,\ xy,\ x^3,\ y^3,\ x^2y,\ xy^2,\ x^4 - y^4,\ \cos(\lambda x)\,\mathrm{ch}(\lambda y),\ \cos(\lambda y)\,\mathrm{ch}(\lambda x),$$

The main difficulty in deriving solutions lies in selection of the functions that satisfy the boundary conditions. A number of important elasticity theory problems were solved by superposition. There is no whatever general solution for the biharmonic equation, nor there are any general approaches to its solution.

In order to close the mathematical boundary problem, the boundary conditions and displacements should be expressed in terms of Airy's stress function. Corresponding calculations are described in paragraph 11.2.

*Note.* To obtaining a correct solution of the elasticity theory problem we are to find the functions able to satisfy the differential equations of the problem, for example biharmonic equation (5.11), and meet equally strictly the boundary conditions in each point of the solid surface. This postulate does not work very often. Therefore, there is a tendency to soften the boundary conditions by obeying above conditions rather integrally than in each point of the surface. For instance, in the case we know that there are no stresses on a given side of the plate, instead of equating these stresses to zero in each point, we may formulate an easier attainable condition when their principal vector and principal moment are equal to zero. As a result, the true system of stresses on the considered side of the body is substituted by another system, which is statically equivalent to the first one. According to Saint-Venant's principle, this substitution will be perceptible only in the vicinity of the studied surface area, while at sufficient distance from it the solution will in fact be independent of above-mentioned substitution of surface loads.

## EXAMPLES OF SOLUTIONS OF THE PLANE ELASTIC PROBLEM

***Example 5.1.*** We are to study, what external force on the plate edges (Fig. 5.3) corresponds to Airy's stress function

$$\phi(x, y) = 2x^3,\ L = 20h \cdot$$

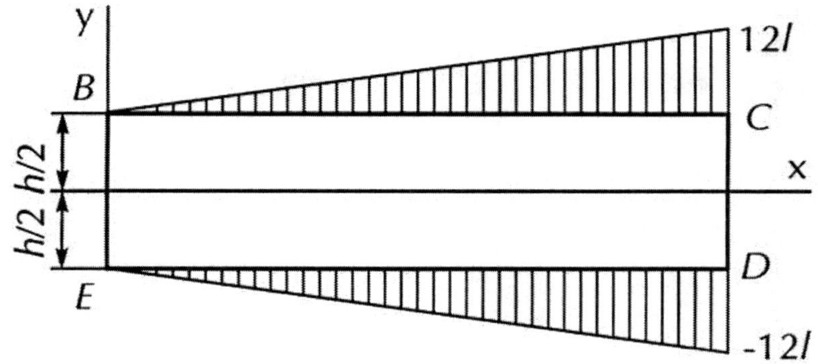


**Figure 5.3.**

Let us check suitability of the given function $\Phi(x, y) = 2x^3$ for the proposed problem. In this connection, we substitute it into the deformation compatibility equation of the plane elastic problem (5.11)

$$\Delta\Delta\phi = \frac{\partial^4\phi}{\partial x^4} + 2\frac{\partial^4\phi}{\partial x^2\,\partial y^2} + \frac{\partial^4\phi}{\partial y^4} = 0.$$

We obtain the identity 0 = 0, that is the function at issue meets the biharmonic equation.

Let us calculate the stresses corresponding to the given function $\phi(x, y) = 2x^3$ by formulas (5.10):

$$\sigma_{xx} = \frac{\partial^2\phi}{\partial y^2} = \frac{\partial^2(2x^3)}{\partial y^2} = 0,\ \sigma_{yy} = \frac{\partial^2\phi}{\partial x^2} = \frac{\partial^2(2x^3)}{\partial x^2} = 12x,$$

$$\sigma_{xy} = -\frac{\partial^2\phi}{\partial x\partial y} = -\frac{\partial^2(2x^3)}{\partial x\partial y} = 0.$$

By using the boundary conditions (5.7),

$$\sigma_{xx}l_x + \sigma_{xy}l_y = R_x,\ \sigma_{xy}l_x + \sigma_{yy}l_y = R_y,$$

$$l_x = \cos(\nu, x),\quad l_y = \cos(\nu, y),$$

and calculating the directional cosines of the normal to the faces, we define the external stress on the plate faces.

*Face BC*. Normal $\nu_1$ is parallel to axis *y*;

$$l_{1x} = \cos(\nu_1, x) = \cos 90^\circ = 0,\quad l_{1y} = \cos(\nu_1, y) = \cos 0^\circ = 1.$$

Condition $y = h/2$ is met on the face. Then,

$$R_{1x} = (\sigma_{xx}l_{1x} + \sigma_{xy}l_{1y})\Big|_{y=h/2} = 0,\ R_{1y} = (\sigma_{xy}l_{1x} + \sigma_{yy}l_{1y})\Big|_{y=h/2} = 12x$$

*Face CD*. Normal $\nu_2$ is parallel to axis $x$;

$$l_{2x} = \cos(\nu_2, x) = \cos 0^\circ = 1, \quad l_{2y} = \cos(\nu_2, y) = \cos 90^\circ = 0\,.$$

On the face, $x = 20h$. Consequently,

$$R_{2x} = (\sigma_{xx} l_{2x} + \sigma_{xy} l_{2y})\Big|_{x=20h} = 0,\ R_{2y} = (\sigma_{xy} l_{2x} + \sigma_{yy} l_{2y})\Big|_{x=20h} = 0\,.$$

*Face DE*. Normal $\nu_3$ is parallel to axis $y$, but is directed to the opposite side;

$$l_{3x} = \cos(\nu_3, x) = 0, \quad l_{3y} = \cos(\nu_3, y) = \cos 180^\circ = -1\,.$$

On this face, $y = -h/2$. That is why

$$R_{3x} = (\sigma_{xx} l_{3x} + \sigma_{xy} l_{3y})\Big|_{y=-h/2} = 0,\ R_{3y} = (\sigma_{xy} l_{3x} + \sigma_{yy} l_{3y})\Big|_{y=-h/2} = \text{-}12x\,.$$

*Face EB*. Normal $\nu_4$ is parallel to axis $x$, but directed to the opposite side;

$$l_{4x} = -1, \quad l_{4y} = 0\,.$$

On this face, $x = 0$. This implies that

$$R_{2x} = (\sigma_{xx} l_{2x} + \sigma_{xy} l_{2y})\Big|_{x=20h} = 0,\ R_{2y} = (\sigma_{xy} l_{2x} + \sigma_{yy} l_{2y})\Big|_{x=20h} = 0\,.$$

As a result, load $R_x$ is absent on all faces. The diagrams for the calculated external distributed load $R_y$ are shown on the plate faces (Fig. 5.3). In the presence of both nonzero stresses, diagrams $R_x$ and $R_y$ should be plotted separately.

***Example 5.2.*** Let us check suitability of the stress function

$$\Phi(x, y) = ax^3 + bx^2y^2 + cy^4$$

for the plane elastic problem we are to make appropriate corrections if the function turns to be unfit, and rewrite it.

The given function $\Phi(x, y)$ should meet biharmonic equation (5.11)

$$\Delta\Delta\phi = \frac{\partial^4\phi}{\partial x^4} + 2\frac{\partial^4\phi}{\partial x^2\,\partial y^2} + \frac{\partial^4\phi}{\partial y^4} = 0\,.$$

The derivatives included into this equation, will be as follows:

$$\frac{\partial^4\phi}{\partial x^4} = \frac{\partial^4}{\partial x^4}(ax^3 + bx^2y^2 + cy^4) = 0\,,$$

$$\frac{\partial^4\phi}{\partial y^4} = \frac{\partial^4}{\partial y^4}(ax^3 + bx^2y^2 + cy^4) = 24c\,,$$

$$\frac{\partial^4\phi}{\partial x^2\partial y^2} = \frac{\partial^4}{\partial x^2\partial y^2}(ax^3 + bx^2y^2 + cy^4) = 4b\,.$$

where from,

$$\Delta\Delta\phi = 24c + 8b = 0\,.$$

Hence, to ensure suitability of the stress function for solving the plane elastic problem, the following condition is to be met:

$$b = -3c\,.$$

Airy's stress function shall be finally of the kind

$$\phi(x, y) = ax^3 - 3cx^2y^2 + cy^4\,.$$

***Example 5.3.*** Let us check the possibility of the plane strain state (5.1) with the following components of the strain tensor:

$$\varepsilon_{xx} = 5(x^2 + y^2),\ \varepsilon_{yy} = 5y^2,\ \varepsilon_{xy} = 5xy\,.$$

We shall check, whether the deformation compatibility equation (5.3) is met

$$\frac{\partial^2\varepsilon_{xx}}{\partial y^2} + \frac{\partial^2\varepsilon_{yy}}{\partial x^2} - 2\frac{\partial^2\varepsilon_{xy}}{\partial x\,\partial y} = 0\,.$$

For this purpose, let us calculate the derivatives of the given deformations and substitute them into this equation: 10 + 0 – 2.5 = 0. The equation is satisfied, so the given functions obey the plane strain state.

## NOTES

1. *Filon, L.N.G.* (1875–1937) English mathematician and physicist.
2. *Airy, George Biddell* (1801–1892), English mathematician and astronomer. In 1863 he published his famous work devoted to stresses in rectangular beams, where he introduced a stress function satisfying boundary conditions.
3. *Maxwell, James Clark* (1831–1879) obtained differential compatibility equations for plane problems with respect to Airy stress function.

## KEYWORDS

- **Biharmonic functions**
- **Median plan**
- **Shear modulus**

# Chapter 6

## Plate Bending

The analysis of plates at bending as a constituent part of the applied theory of elasticity is currently widespread in various engineering branches, such as construction, aviation, shipbuilding, mechanical engineering, and so forth. This is attributed to the lightweight and rationality of the profile peculiar to a thin-walled structure in combination with its high load-bearing capacity, efficiency, and perfect workability.

### BASIC CONCEPTS AND HYPOTHESES

A locus that bisects a plate through its thickness is called a *median plane* of the plate (Fig. 6.1a, b). It is as much important for the theory of plate bending as the neutral layer at beam bending is critical for the theory of strength of materials. The line bounding the median plane of the plate is called the *plate contour*.

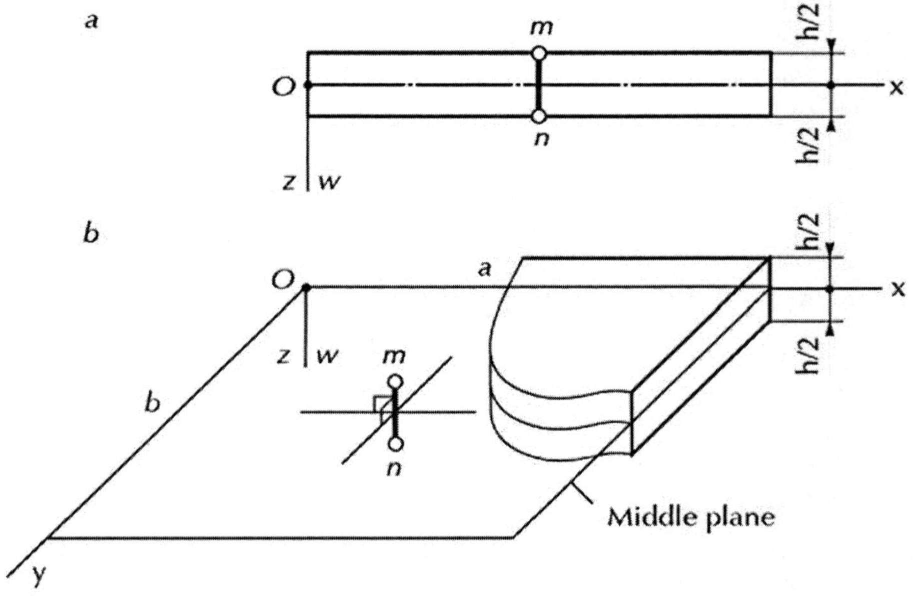


**Figure 6.1.**

Let us assume that axes $x$ and $y$ are located in the plate median plane, while axis $z$ is directed downwards. A displacement of the median plane along $z$ axis is called a deflection of the plate, denoted by $w$. The median plane transforms at bending into a slightly curved surface of deflections $w = w(x, y)$. It is called a *median surface* of a *bent plate*.

Thickness of a plate significantly affects its properties of bending. We distinguish three types of the plates depending on relation $a/h$, which is a characteristic size in plane $a$ versus plate thickness $h$.

*For a thick plate* this relation is $a/h \leq 8 \ldots 10$. It is calculated with regard to all three components of the stress state similarly to a solid 3D body.

*Membranes* are referred to as plates at $a/h \geq 80 \ldots 100$. They are operable only if their edges are securely fixed over the contour. Their resistance to bending is negligibly small, so the tensile and shear forces in the median surface play a main part in perceiving the transverse load. These forces create a projection on the vertical axis $z$, balancing thereby the load applied on each element of the membrane.

*Thin plates*, for which

$$8 \ldots 10 \leq a/h \leq 80 \ldots 100$$

present the broadest type of plates. Depending on the value of relation $w/h$ of the plate maximum deflection to its thickness, the role of bending and membrane forces may be different. In this connection, this type of plates is subdivided into two more classes: rigid and flexible plates.

If deflections are small and

$$w/h \leq 0.2 \ldots 0.5,$$

then the bending force factors play the leading role, while deformations in the median plane and membrane forces may be ignored. Such plates are called *rigid*. When $w/h$ exceeds the given approximate limits, the plate is simultaneously bending and operates like a membrane. These factors are similarly important, and the role of stretching of the median surface increases with augmenting deflection growth. In this case, the plate is *elastic*. For example, the reinforced concrete slabs are usually rigid plates, while steel plates, depending on the load, may work as both rigid and elements.

The analysis of bending of thin rigid plates is given below. Thanks to introduction of certain hypotheses, the theory of these plates is simple enough and is brought to the linear differential equations. Deformations of elastic plates (as well as membranes and shells) are described by a series of linear equations, which complicate the problem essentially.

Let us formulate the assumptions and limitations used in the theory of thin rigid plates:

- the intercept $mn$ of the normal to the median plane (see Fig. 6.1) remains straight and normal to the deformed median surface $m_1n_1$ at bending, that is the *hypothesis of direct normal's* is valid;
- the stresses of the horizontal layers of the plate occurred at pressing each other ($\sigma_{zz}$) are ignored as compared to stresses $\sigma_{xx}$ and $\sigma_{yy}$ acting in the plane of the layers;
- the deflections are assumed to be so small ($w/h \leq 0.2 \ldots 0.5$) that the membrane forces in the median surface may be neglected.

The first two assumptions are known in literature as *Kirhhoff's hypothesis* (*the hypothesis of a thin plate*). As applied to the shells, this is *Kirhhoff-Love's hypothesis.*

It will be proved further that the definition of stresses and forces across the plate sections is a statically indeterminate problem. It can be better solved in displacements, for which purpose the deflection $w = w(x, y)$ is accepted as a principal sought function. Then, all the rest unknown values are expressed through it and the resulting equation is solved with respect to $w$. After determination of deflections, we calculate all the rest parameters. This is a general way of solving the problem plate bending.

## DISPLACEMENTS AND DEFORMATIONS PLATES

Under the influence of a transverse load $q = q(x, y)$, the plate undergoes bending, and its middle layer, after being distorted, forms a flexure surface $w = w(x, y)$. Let us isolate a small element from the plate with the sizes in plane $\Delta x$ and $\Delta y$, and height $h$ (Fig. 6.2a). We shall further consider the displacements of an arbitrary point $O$ of the middle layer and the normal $mn$ passing through it. According to above the assumptions, the middle layer is not stretched, so point $O$ will shift only vertically by a deflection value $w$, and the normal $mn$ will make a turn in space.

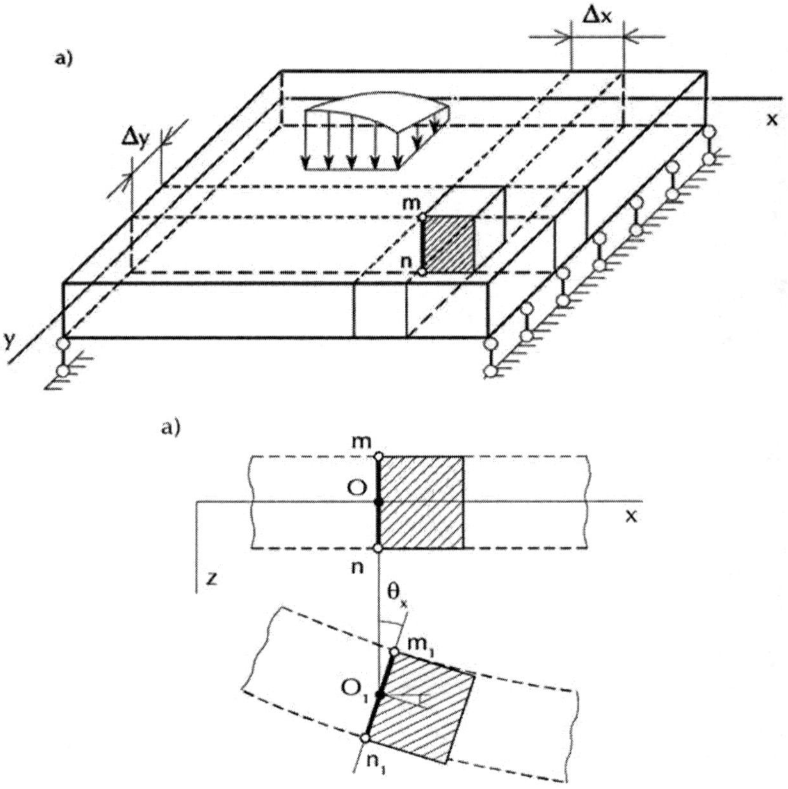


**Figure 6.2.**

The isolated element is a part of crossing «bars» singled out imaginary from the plate and shown by a dashed line in Fig. 6.2a. The edges of the element are cross-sections of these bars deformed together with the plate. Figure 6.2b shows a projection of the bending bar parallel to axis $x$ onto plane $xz$. It is evident that normal $mn$ in this plane is turned by an angle $\theta_x = \partial w/\partial x$. A similar picture is observed in plane $yz$, where the angle of rotation of the normal is equal to $\theta_y = \partial w/\partial y$.

Hence, the characteristic displacements of the median plane of an arbitrary point is a deflection $w = w(x, y)$, and angles of rotation of the normal $\theta_x = \partial w/\partial x$ and $\theta_y = \partial w/\partial y$. This leads to distortion of the isolated element in perpendicular direction and to its twisting.

Let us denote the tangential displacements of the points of the considered arbitrary normal, located at a distance $z$ from the median surface, by $u$ and $v$. Based on Fig. 6.3, we can find a displacement $u$ (by analogy with $v$):

$$u = -z\theta_x = -z\frac{\partial w}{\partial x},$$

$$v = -z\frac{\partial w}{\partial y}. \qquad (6.1)$$

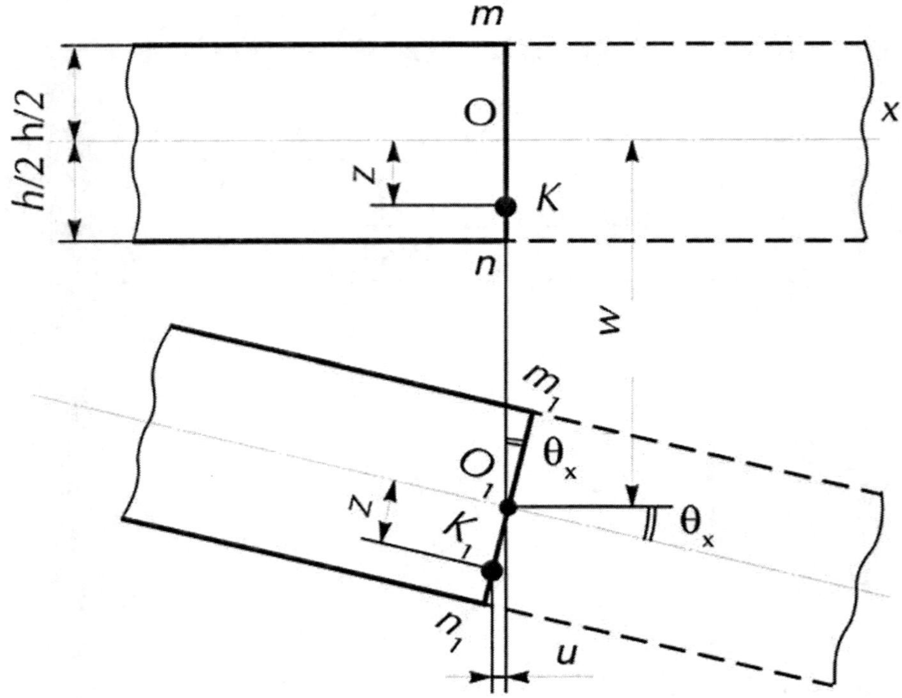


**Figure 6.3.**

The minus means that if the values $\theta_x$ and $\theta_y$ are positive, the displacements of the point, where $z > 0$, are directed to the opposite side with respect to axes $x$ or $y$.

Using the displacement (6.1) and the accepted assumptions we obtain by Cauchy formulas the following strains in the plate:

$$\varepsilon_{xx} = \frac{\partial u}{\partial x} = -z\frac{\partial^2 w}{\partial x^2}, \varepsilon_{yy} = \frac{\partial v}{\partial y} = -z\frac{\partial^2 w}{\partial y^2}, \varepsilon_{xy} = \frac{1}{2}\left(\frac{\partial u}{\partial y} + \frac{\partial v}{\partial x}\right) = -z\frac{\partial^2 w}{\partial x \partial y},$$

$$\varepsilon_{xz} = \frac{\partial u}{\partial z} + \frac{\partial w}{\partial x} = 0, \varepsilon_{yz} = \frac{\partial v}{\partial z} + \frac{\partial w}{\partial y} = 0, \varepsilon_{zz} = \frac{\partial w}{\partial z} = 0. \tag{6.2}$$

It is evident that the strains of an arbitrary horizontal layer of the plate change linearly across its thickness and independently of three characteristic values:

$$\kappa_x = \frac{1}{\rho_x} = -\frac{\partial^2 w}{\partial x^2}, \kappa_y = \frac{1}{\rho_y} = -\frac{\partial^2 w}{\partial y^2}, \kappa = -\frac{\partial^2 w}{\partial x \partial y}.$$

When the deflections are insignificant, the values $\kappa_x$ and $\kappa_y$ make up the curvatures of the median surface of the isolated element; $\kappa$—its twisting; $\rho_x$, $\rho_y$—corresponding radii of the curvature.

## STRESSES AND INTERNAL FORCES IN PLATES

Stresses in a plate may be calculated by formulas (5.8) for the plane stress state according to the accepted hypotheses, $\sigma_{zz} = 0$. Then,

$$\sigma_{xx} = \frac{E}{1-\nu^2}(\varepsilon_{xx} + \nu\varepsilon_{yy}), \sigma_{yy} = \frac{E}{1-\nu^2}(\varepsilon_{yy} + \nu\varepsilon_{xx}), \sigma_{xy} = \frac{E}{1+\nu}\varepsilon_{xy}.$$

By substitution of strains (6.2), we obtain

$$\sigma_{xx} = -\frac{Ez}{1-\nu^2}\left(\frac{\partial^2 w}{\partial x^2} + \nu\frac{\partial^2 w}{\partial y^2}\right), \sigma_{yy} = -\frac{Ez}{1-\nu^2}\left(\frac{\partial^2 w}{\partial y^2} + \nu\frac{\partial^2 w}{\partial x^2}\right),$$

$$\sigma_{xy} = -\frac{Ez}{1+\nu}\frac{\partial^2 w}{\partial x \partial y}. \tag{6.3}$$

Let us introduce the intensities of the internal force factors with the help of the relations:

$$M_x = \int_{-h/2}^{h/2} \sigma_{xx} z \, dz, M_y = \int_{-h/2}^{h/2} \sigma_{yy} z \, dz, M_{xy} = \int_{-h/2}^{h/2} \sigma_{xy} z \, dz,$$

$$Q_x = \int_{-h/2}^{h/2} \sigma_{zx} \, dz, Q_y = \int_{-h/2}^{h/2} \sigma_{zy} \, dz. \tag{6.4}$$

Here, $M_x$, $M_y$—bending moments, $M_{xy} = M_{yx}$—torques and $Q_x$, $Q_y$—shearing forces related to a unit length of the sections parallel to planes *xz* and *yz* (Fig. 6.4).

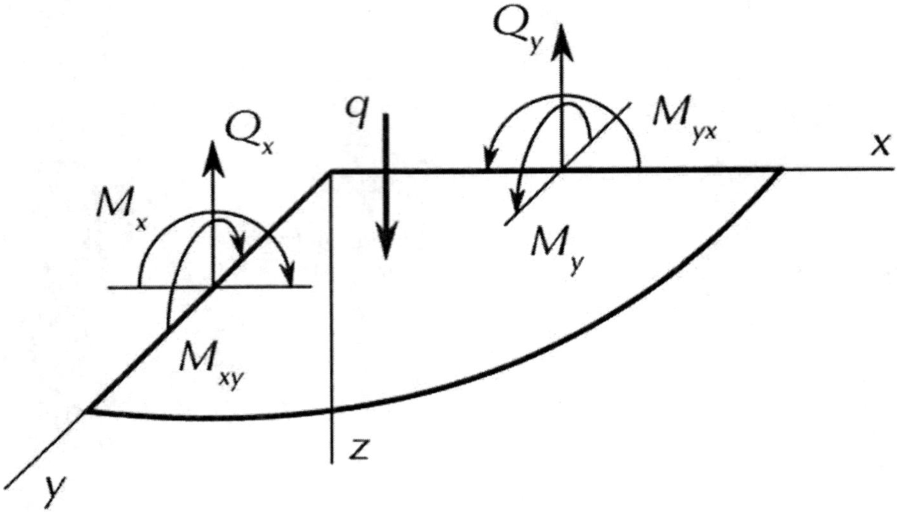


**Figure 6.4.**

In contrast to bending of the beams, the indices of bending moments for the plates correspond to directions of the stresses they were created by. We shall take bending moments positive if they tend to bend a plate element with a convexity directed downwards. According to strains (6.2), the conditions

$$\sigma_{zx} = 2\mu\varepsilon_{zx} = 0, \sigma_{zy} = 2\mu\varepsilon_{zy} = 0$$

should be obeyed in the equilibrium equations.

Despite this fact, we should account for the resulting forces $Q_x$ and $Q_y$ dependent on tangential stresses $\sigma_{zx}$ and $\sigma_{zx}$ as the values of the same order with the intensity of shearing force $q$, and moments $M_x$, $M_y$, $M_{xy}$.

By substitution of expressions for stresses (6.3) into the first three relations (6.4), integration over a homogeneous plate thickness and introduction of *cylindrical stiffness D*, we come to a the following formula for bending moments and the torque:

$$M_x = -D\left(\frac{\partial^2 w}{\partial x^2} + \nu\frac{\partial^2 w}{\partial y^2}\right), M_y = -D\left(\frac{\partial^2 w}{\partial y^2} + \nu\frac{\partial^2 w}{\partial x^2}\right),$$

$$M_{xy} = -D(1-\nu)\frac{\partial^2 w}{\partial x \partial y}, D = \frac{Eh^3}{12(1-\nu^2)}. \qquad (6.5)$$

Thus, relations (6.5) express the internal force in the plate in terms of its deflection.

## DIFFERENTIAL EQUATION OF PLATE BENDING

Let us consider an element cut out from a plate by two pairs of planes parallel to coordinate planes *xz* and *yz* (Fig. 6.5). To ensure an equilibrium of the element, the sums of the forces acting on it as well as their moments about *x* and *y* axes should be equal to zero. By assuming that the body forces are absent and ignoring the third-order values of smallness (without account of the moments produced by the increments in shearing forces), we have

$$-Q_x\,dy+\left(Q_x+\frac{\partial Q_x}{\partial x}dx\right)dy-Q_y\,dx+\left(Q_y+\frac{\partial Q_y}{\partial y}dy\right)dx+q\,dx\,dy=0,$$

$$M_y\,dx-\left(M_y+\frac{\partial M_y}{\partial y}dy\right)dx+M_{xy}\,dy-\left(M_{xy}+\frac{\partial M_{xy}}{\partial x}dx\right)dy+Q_y\,dx\,dy=0,$$

$$-M_{yx}\,dx+\left(M_{yx}+\frac{\partial M_{yx}}{\partial y}dy\right)dx-M_x\,dy+\left(M_x+\frac{\partial M_x}{\partial x}dx\right)dy-Q_x\,dx\,dy=0\;.$$

By reducing the terms and canceling d*x*d*y*, we shall have a series of equations:

$$\frac{\partial Q_x}{\partial x}+\frac{\partial Q_y}{\partial y}+q=0,\;-\frac{\partial M_{xy}}{\partial x}-\frac{\partial M_y}{\partial y}+Q_y=0,\;\frac{\partial M_{yx}}{\partial y}+\frac{\partial M_x}{\partial x}-Q_x=0. \tag{6.6}$$

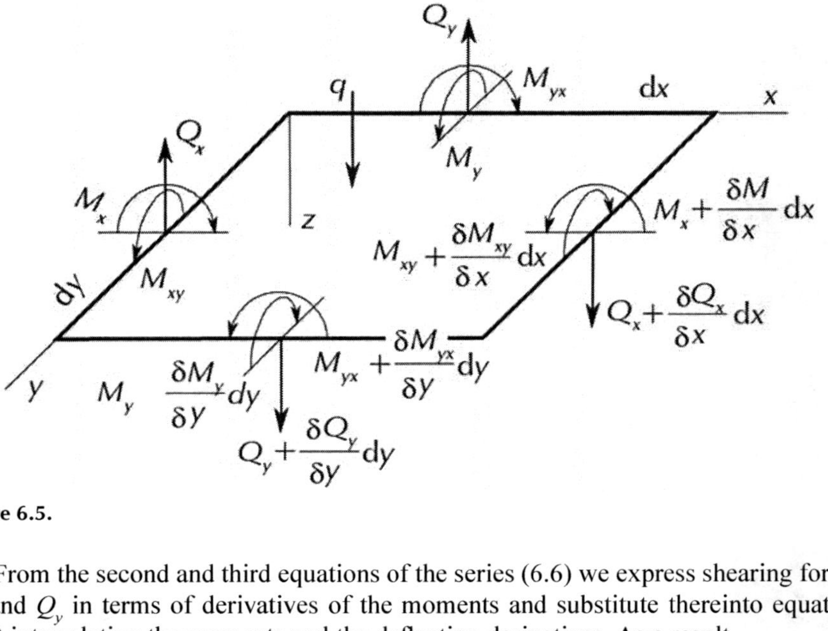


**Figure 6.5.**

From the second and third equations of the series (6.6) we express shearing forces $Q_x$ and $Q_y$ in terms of derivatives of the moments and substitute thereinto equation (6.5) interrelating the moments and the deflection derivatives. As a result,

$$Q_x = -D\frac{\partial}{\partial x}\left(\frac{\partial^2 w}{\partial x^2}+\frac{\partial^2 w}{\partial y^2}\right) = -D\frac{\partial}{\partial x}\Delta w\ ,\ Q_y = -D\frac{\partial}{\partial y}\Delta w\ . \qquad (6.7)$$

If we substitute the obtained equation (6.7) into the first equation (6.6), we obtain Sophie Germain's equation:

$$\frac{\partial^4 w}{\partial x^4}+2\frac{\partial^4 w}{\partial x^2 \partial y^2}+\frac{\partial^4 w}{\partial y^4}=\frac{q}{D}. \qquad (6.8)$$

Using the Laplacian operator in the plane system of coordinates

$$\Delta\Delta \equiv \Delta^2 \equiv \left(\frac{\partial^2}{\partial x^2}+\frac{\partial^2}{\partial y^2}\right)\left(\frac{\partial^2}{\partial x^2}+\frac{\partial^2}{\partial y^2}\right) \equiv \frac{\partial^4}{\partial x^4}+2\frac{\partial^4}{\partial x^2\partial y^2}+\frac{\partial^4}{\partial y^4},$$

equation (6.8) may be rewritten as:

$$\Delta^2 w = \frac{q}{D}.$$

Thus, the problem of the plate bending by a distributed stress $q$ is reduced to integration of equation (6.8).

## BOUNDARY CONDITIONS

Let us consider the boundary conditions for a rectangular plate with the edges fixed by some method, while axes $x$ and $y$ we shall direct parallel to the edges of the plate.

**Fixed edge.** If the plate edge $x = 0$ is fixed, the deflection in the points of this edge is equal to zero, whereas, the fixed section of the plate is non-rotational (the plane tangent to the bent median surface coincides with the median plane of the plate prior to bending):

$$w\big|_{x=0} = 0,\ \frac{\partial w}{\partial x}\bigg|_{x=0} = 0. \qquad (6.9)$$

**Hinged edge.** If the plate edge $x = 0$ is supported and may rotate freely, the deflection and bending moments on this edge should be zero:

$$w\big|_{x=0} = 0,\ M_x\big|_{x=0} = \left(\frac{\partial^2 w}{\partial x^2}+\nu\frac{\partial^2 w}{\partial y^2}\right)\bigg|_{x=0} = 0.$$

In this case, the hinged supports are assumed to be rigid, $w = 0$ and line $x = 0$ remains straight. Hence, the derivatives will be:

$$\frac{\partial w}{\partial y}\bigg|_{x=0} = \frac{\partial^2 w}{\partial y^2}\bigg|_{x=0} = 0.$$

Therefore, the boundary conditions for a hinged edge are

$$w\big|_{x=0} = 0, \left.\frac{\partial^2 w}{\partial x^2}\right|_{x=0} = 0 \cdot \tag{6.10}$$

**Free edge.** If the edge $x = 0$ is free then, from the first sight, we should set the bending moment $M_x$, torque $M_{xy}$ and shearing force $Q_x$ equal to zero along this edge:

$$M_x\big|_{x=0} = 0,\ M_{xy}\big|_{x=0} = 0,\ Q_x\big|_{x=0} = 0\,. \tag{6.11}$$

Thus, we have three boundary conditions here, while in the other cases there were only two of them. It was Poisson who has derived the conditions (6.11). Later on (in 1850), Kirhhoff has shown proceeding from Saint-Venant's principle, that two boundary conditions are suffice to define fully the deflection $w$ and meet the equation (6.8) since the latter two in (6.11) may be combined into one:

$$\left.\left(Q_x + \frac{\partial M_{xy}}{\partial y}\right)\right|_{x=0} = 0 \cdot \tag{6.12}$$

As a result, by substitution of expressions of the internal forces in terms of deflection (6.5), (6.7) into the first relation of (6.11) and in (6.12), we obtain the following boundary conditions for the free edge:

$$\left.\left(\frac{\partial^2 w}{\partial x^2} + \nu\frac{\partial^2 w}{\partial y^2}\right)\right|_{x=0} = 0, \left.\left(\frac{\partial^3 w}{\partial x^3} + (2-\nu)\frac{\partial^3 w}{\partial x \partial y^2}\right)\right|_{x=0} = 0\,. \tag{6.13}$$

In the case the plates are with curvilinear edges the system of coordinates shall be introduced normally and tangentially to the boundary. The corresponding internal forces on this boundary are expressed in terms of the former coordinate forces for the element equilibrium, and the boundary conditions are given similarly to already considered ones.

## CYLINDRICAL BENDING OF A RECTANGULAR PLATE

Let us consider a plate, infinitely long in direction of $y$ axis and bearing a transverse constant along this axis load (Fig. 6.6). This load may vary arbitrary along axis $x$: $q = q(x)$. In this case, the deflections $w = w(x)$ create a cylindrical surface. Assuming that the derivatives of $y$ are equal to zero in (6.8), we obtain the following equation for $w$:

$$\frac{\mathrm{d}^4 w}{\mathrm{d}x^4} = \frac{q}{D} \cdot \tag{6.14}$$

This definition is used here rather for a common derivative than for the partial one, since the deflection depends on one argument $w = w(x)$. Equation (6.14) describing cylindrical bending coincides with that for the beam bending, where the bending section stiffness factor is $EJ = D$. Wherefrom, value $D$ is termed as *bending stiffness*.

Integration of this equation is straightforward. Its solution consists of a sum of a general solution of the corresponding homogeneous equation ($q = 0$) and a partial solution $w_0$ of equation (6.14):

$$w = C_1 + C_2 x + C_3 x^2 + C_4 x^3 + w_0 . \tag{6.15}$$

Integration constants $C_1$, …, $C_4$ are found from the boundary conditions of a concrete problem.

***Example 6.1.*** Let us assume that a linearly varying load $q = q_0 x/a$, where $q_0$ = const and $a$ plate width, is acting on the plate. The conditions of fixing will be those shown in Fig. 6.6: on the left—fixing, on the right—hinged support. Then, the boundary conditions will be like (6.9) if $x = 0$, and (6.10) if $x = a$:

$$w\big|_{x=0} = 0,\ \frac{\partial w}{\partial x}\bigg|_{x=0} = 0,\ w\big|_{x=a} = 0,\ \frac{\partial^2 w}{\partial x^2}\bigg|_{x=a} = 0 .$$

The partial solution in this case may be

$$w = \frac{q_0 x^5}{120 a D} .$$

If we substitute it into the complete solution of (6.13) and use four accepted boundary conditions, we obtain:

$$C_1 = C_2 = 0 ,\ C_3 = \frac{7 q_0 a^2}{240 D},\ C_4 = -\frac{9 q_0 a}{240 D},$$

whereupon, the expression for the deflection will be

$$w = \frac{q_0 a^4}{240 D}\left(7\frac{x^2}{a^2} - 9\frac{x^3}{a^3} + 2\frac{x^5}{a^5}\right).$$

At cylindrical bending, when the derivatives of $y$ are equal to zero, expressions for the moments (6.5) are simplified:

$$M_x = -D\frac{\partial^2 w}{\partial x^2},\ M_y = -\nu D\frac{\partial^2 w}{\partial x^2} = -\nu M_x,\ M_y = -\nu D\frac{\partial^2 w}{\partial x^2} = -\nu M_x .$$

By substituting thereinto the expression found for the deflection, we obtain

$$M_y = -\nu D\frac{\partial^2 w}{\partial x^2} = -\nu M_x .$$

If a plate is of a finite size along $y$ axis, we should perform the same manipulations to create a cylindrical bending.

*Note.* In the case of a pure bending of a rectangular plate, the transverse load is absent $q = 0$, and the external distributed bending moments are acting over the unfixed contour. In this case, the deflection follows from (6.15) at $w_0 = 0$, where the integration constants are defined by the value and condition of the contour bending moments.

## A RECTANGULAR PLATE AT SINUSOIDAL LOAD

We shall consider a particular case of bending of a hinged rectangular plate (Fig. 6.7), when the external load varies sinusoidally along each coordinate axis, has a maximum $q_0$ and is given as

$$q(x,y) = q_0 \sin\frac{\pi x}{a}\sin\frac{\pi y}{b}. \tag{6.16}$$

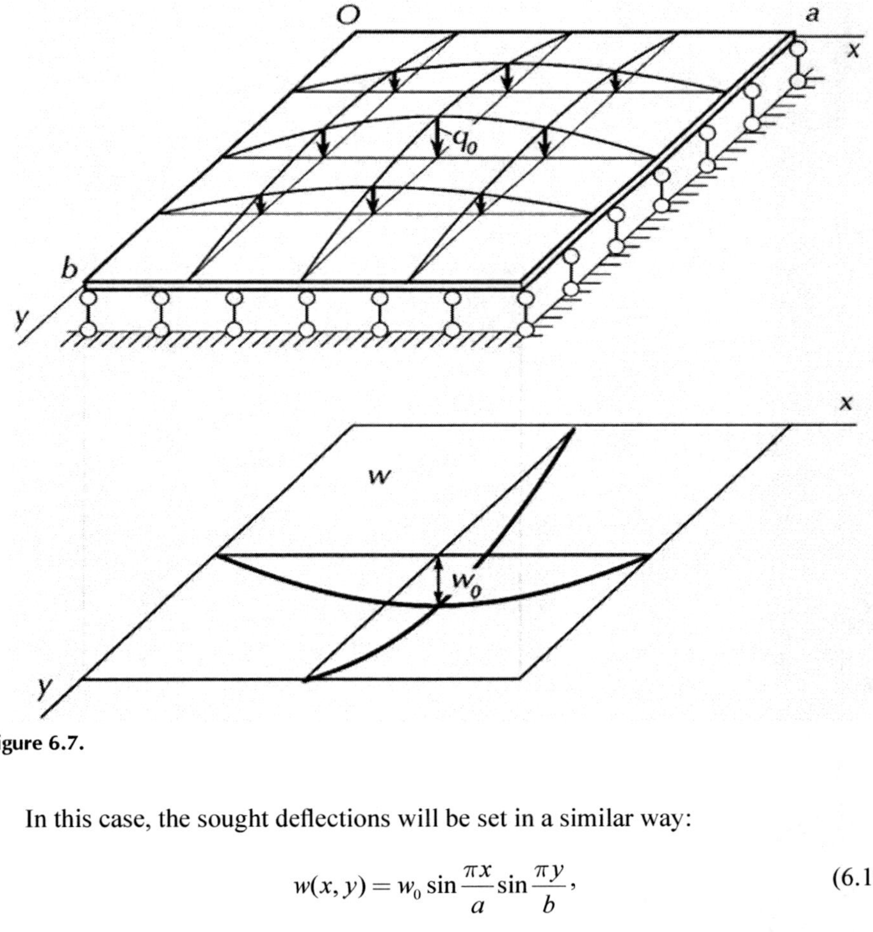


**Figure 6.7.**

In this case, the sought deflections will be set in a similar way:

$$w(x,y) = w_0 \sin\frac{\pi x}{a}\sin\frac{\pi y}{b}, \tag{6.17}$$

where $w_0$ is a multiplier to be defined. The bending moments are calculated in terms of the deflection by formulas (6.5)

$$M_x = w_0\pi^2 D\left(\frac{1}{a^2}+\nu\frac{1}{b^2}\right)\sin\frac{\pi x}{a}\sin\frac{\pi y}{b},$$

$$M_y = w_0\pi^2 D\left(\frac{1}{b^2}+\nu\frac{1}{a^2}\right)\sin\frac{\pi x}{a}\sin\frac{\pi y}{b}.$$

Consequently, the bending moments are distributed in the plate according to the same sinusoidal law like the deflections and the load.

The boundary conditions at hinged supporting impose that deflections and bending moments tend to zero on the contour. Since the product of sines on the contour gives zero, these conditions are met.

To define amplitude $w_0$ of the deflection, let us substitute (6.17) and (6.16) into Sophie Germain's differential equation (6.8):

$$\frac{\partial^4 w}{\partial x^4} + 2\frac{\partial^4 w}{\partial x^2 \partial y^2} + \frac{\partial^4 w}{\partial y^4} = \frac{q(x,y)}{D}.$$

Since even derivatives of the sine give a sine again, we obtain in the left-hand and right-hand sides of this equation common multipliers in the form of a product of two sines. By equating their coefficients, we have

$$w_0\left[\left(\frac{\pi}{a}\right)^4 + 2\left(\frac{\pi}{a}\right)^2\left(\frac{\pi}{b}\right)^2 + \left(\frac{\pi}{b}\right)^4\right] = \frac{q_0}{D}.$$

There is a squared sum of squares in square brackets, wherefrom

$$w_0 = \frac{q_0}{\pi^4 D\left(\frac{1}{a^2} + \frac{1}{b^2}\right)^2}. \tag{6.18}$$

Thus, the expression for deflections (6.17) meets the differential equation of the plate bending and the accepted boundary conditions. The deflection amplitude is defined by expression (6.18).

## SOLUTION IN DOUBLE TRIGONOMETRIC SERIES

The solution of this problem proposed by Navier is valid under the action of an arbitrary transverse load if all sides of the rectangular plate are hinged. In this case, a specified external stress $q(x, y)$ is expanded into a double trigonometric series

$$q(x,y) = \sum_{n=1}^{\infty}\sum_{m=1}^{\infty} q_{mn} \sin\frac{m\pi x}{a} \sin\frac{n\pi y}{b}, \tag{6.19}$$

where $q_{mn}$—load decomposition coefficients; $m, n = 1, 2, 3, \ldots$—whole numbers.

To calculate $q_{mn}$, let us use a property of fundamentality and orthogonality of the employed sinusoidal function system known from the mathematical analysis. Consequently,

$$\int_0^a \sin\frac{m\pi x}{a} \sin\frac{n\pi x}{a} \mathrm{d}x = \begin{cases} 0, & m \neq n \\ a/2, & m = n \end{cases},$$

$$\int_0^b \sin\frac{m\pi y}{b}\sin\frac{n\pi y}{b}\mathrm{d}\,y=\begin{cases}0, & m\neq n\\ b/2, & m=n\end{cases}. \tag{6.20}$$

The validity of formulas (6.20) may be confirmed by direct integration.

Then we multiply both sides of equality (6.19) by

$$\sin\frac{l\pi x}{a}\sin\frac{p\pi y}{b}\quad (l, p - \text{whole numbers})$$

and integrate over the plate area. If to attach specific values $m$, $n$ to $l, p$, then, in virtue of (6.20) all the items in the right-hand side will turn into zero, except for one possessing $m$, $n$. values. Hence, following this equality, we find

$$q_{mn}=\frac{4}{ab}\int_0^a\int_0^b q(x,y)\sin\frac{m\pi x}{a}\sin\frac{n\pi y}{b}\mathrm{d}\,x\,\mathrm{d}\,y\cdot \tag{6.21}$$

The given load is presented thereby as a set of separate sinusoidal components with coefficients $q_{mn}$ (6.21). This allows us to use the procedure considered earlier. Now, the problem reduces to defining deflections from separate members of the series (6.19) and to summation of the results obtained:

$$w(x,y)=\sum_{n=1}^{\infty}\sum_{m=1}^{\infty} w_{mn}\sin\frac{m\pi x}{a}\sin\frac{n\pi y}{b}, \tag{6.22}$$

where $w_{mn}$—the sought coefficients of deflection decomposition into a double trigonometric series. To find them, the term of the series (6.22) with terms $m$, $n$ and the term of the load series (6.19) with the same numbers shall be substituted into the differential equation (6.8) for the plate bending. As a result, after equating the coefficients and solution of the equality, we derive an expression for the coefficient similar to (6.18):

$$w_{mn}=\frac{q_{mn}}{\pi^4 D\left(\frac{m^2}{a^2}+\frac{n^2}{b^2}\right)^2}. \tag{6.23}$$

If the plate deflection is known and the coefficients from (6.23), it is easy to find the internal forces (6.3), (6.5).

## APPLICATION OF A SINGLE TRIGONOMETRIC SERIES

Above-stated Navier's solution in a double trigonometric series is restricted since all four edges of the plate must be hinged. Levy's proposal to apply a single trigonometric series for the plate bending expands the class of solvable problems essentially. The sought solution takes the form

$$w(x,y)=\sum_{m=1}^{\infty} Y_m\sin\frac{m\pi x}{a}, \tag{6.24}$$

where $Y_m = Y_m(y)$—functions of argument $y$ to be defined.

It is evident that deflection (6.24) meets the conditions (6.10) of the plate resting on the rigid hinged-supports of two parallel edges $x = 0$ and $x = a$, as the deflection itself and its second derivative turn to zero.

Functions $Y_m(y)$ are selected so that expression $w(x, y)$ meets Sophie Germain's equation (6.8) and boundary conditions on the longitudinal edges $y = \pm b/2$. These edges may be fixed arbitrarily.

Let us decompose the predetermined external load $q$ into a single series

$$q(x,y)=\sum_{m=1}^{\infty} q_m(y)\sin\frac{m\pi x}{a}. \tag{6.25}$$

Using the property of orthogonality (6.20) of the sines at various numbers $m$, let us find the multiplier $q_m$ in the form

$$q_m=\frac{2}{a}\int_0^a q(x,y)\sin\frac{m\pi x}{a}\mathrm{d}x.$$

Let us substitute the $m$th term of the series from (6.24) and (6.25) into the plate bending equation (6.17). After equating the coefficients, we obtain a heterogeneous ordinary differential equation with respect to function $Y_m(y)$:

$$Y_m^{\mathrm{IV}}-2\lambda^2 Y_m''+\lambda^4 Y_m=\frac{q_m(y)}{D},\quad \lambda=\frac{m\pi}{a}, \tag{6.26}$$

The primed values are related to $y$ derivatives.

A general integral of this equation $Y_m$ is made up of a general solution of a corresponding homogeneous equation $Y_{m0}$ (at $q_m = 0$) and partial solution $Y_{m1}$ of equation (6.26). A general solution of the homogeneous equation was obtained in two forms:

- the exponential form

$$Y_{m0}=C_1e^{-\lambda y}+C_2\lambda ye^{-\lambda y}+C_3e^{\lambda y}+C_4\lambda ye^{\lambda y},$$

- the hyperbolic-trigonometric form

$$Y_{m0}=C_1\mathrm{ch}\lambda y+C_2\lambda y\mathrm{sh}\lambda y+C_3\mathrm{sh}\lambda y+C_4\lambda y\mathrm{ch}\lambda y. \tag{6.27}$$

Notice that, in the second case the first two summands make up a symmetrical part of the solution about axis $x$, and the others—antisymmetrical. The integration constants $C_1$, …, $C_4$ are found from boundary conditions on longitudinal edges $y = \pm b/2$ of the plate.

The partial solution of equation (6.26) depends on the type of function $q_m(y)$. For example, if $q_m$ is a linear function of $y$: $q_m = A + By$, then by substitution it is easy to verify that

$$Y_{m1} = \frac{1}{D\lambda^4}(A + By).$$

Now the internal force in the plate may be obtained using formulas (6.3), (6.5) and solution (6.27).

## A RECTANGULAR PLATE ON ELASTIC FOUNDATION

If a plate rests on a solid deformable foundation, the differential equation for bending should take into account the bearing reaction of the foundation (repulse) distributed over the plate area (Fig. 6.8). By denoting the reaction intensity as $q_R = q_R(x, y)$, the equation of bend may become

$$\frac{\partial^4 w}{\partial x^4} + 2\frac{\partial^4 w}{\partial x^2 \partial y^2} + \frac{\partial^4 w}{\partial y^4} = \frac{q + q_R}{D}, \tag{6.28}$$

where $q$ is the intensity of the external distributed load.

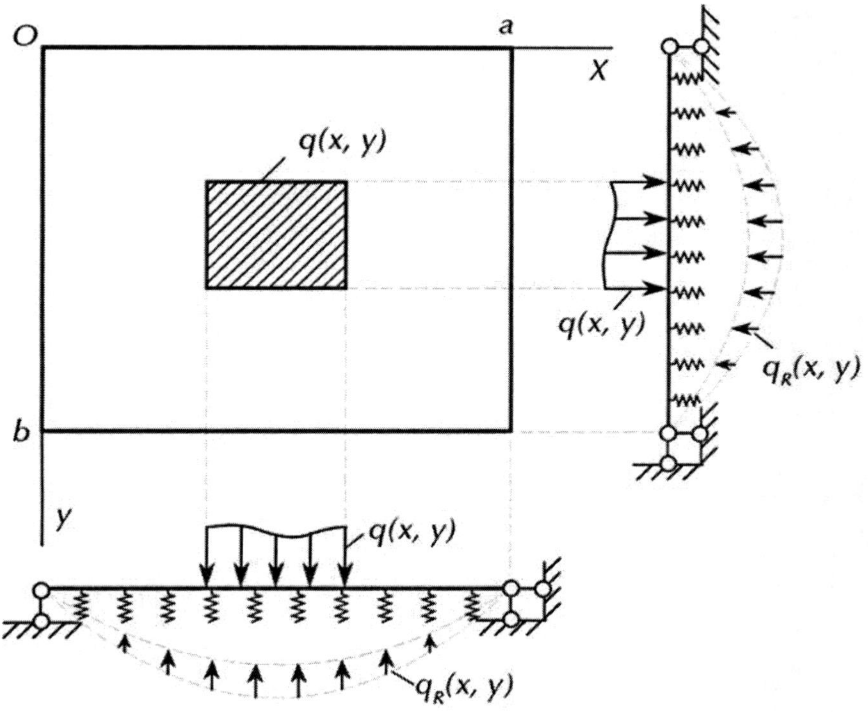


**Figure 6.8.**

Depending on the properties of the deformed foundation, the relation between the reaction and deflection may vary. In practice, we often use *Winkler's*[1] *mode*, known from the strength of materials, according to which

$$q_R = -\kappa w\,. \tag{6.29}$$

Here, κ is the *factor of stiffness of the elastic foundation* (modulus of a subgrade reaction); the minus indicates that the reaction is directed opposite to bending.

If we substitute (6.29) into (6.28) and transpose this term to the left-hand side, we obtain the bending equation for the plate lying on Winkler's foundation:

$$\frac{\partial^4 w}{\partial x^4} + 2\frac{\partial^4 w}{\partial x^2 \partial y^2} + \frac{\partial^4 w}{\partial y^4} + \frac{\kappa}{D} w = \frac{q}{D}\,. \tag{6.30}$$

To integrate equation (6.30) any of above considered methods may be used. Thus, in the case of a hinge-supported plate, the stresses and deflections (according to 6.8) may be presented as double trigonometric series:

$$q(x,y) = \sum_{n=1}^{\infty}\sum_{m=1}^{\infty} q_{mn} \sin\frac{m\pi x}{a} \sin\frac{n\pi y}{b}\,,$$

$$w(x,y) = \sum_{n=1}^{\infty}\sum_{m=1}^{\infty} w_{mn} \sin\frac{m\pi x}{a} \sin\frac{n\pi y}{b}\,.$$

Reiteration of the calculations described in 6.8, will give us the following formula for $w_{mn}$:

$$w_{mn} = \frac{q_{mn}}{\pi^4 D\left(\frac{m^2}{a^2} + \frac{n^2}{b^2}\right)^2 + \kappa}\,. \tag{6.31}$$

In a particular case, when κ = 0 the formula for deflection (6.31) coincides with (6.23).

This solution presumes that the relations between the plate and the elastic foundation are true for both compression and separation (bilateral constraints). This may be valid only if the total reaction in all supporting points of the plate will be compressive with account of all loads (e.g., dead weight of the plate). Otherwise, it will be necessary to define the area of separation of the plate from the foundation by means of sequential approximations, and equation (6.30) will be solved jointly for two areas—the indentation area of the plate into the foundation, where $\kappa_1 = \kappa$; and separation at $\kappa_2 = 0$.

It should be noted, that there is another widely applicable model for a deformed foundation, namely the model of the elastic semi-infinite space.

## CIRCULAR PLATE BENDING

When studying bending of circular plates it is convenient to use the polar coordinate system ($r$, $\varphi$) based on the formulas expressing the relationship between the polar and Cartesian coordinates,

$$r^2 = x^2 + y^2, \quad \varphi = \operatorname{arctg}\frac{y}{x}. \tag{6.32}$$

The harmonic operator takes the form

$$\Delta \equiv \frac{\partial^2}{\partial x^2} + \frac{\partial^2}{\partial y^2} \equiv \frac{\partial^2}{\partial r^2} + \frac{1}{r}\frac{\partial}{\partial r} + \frac{1}{r^2}\frac{\partial^2}{\partial \varphi^2}. \tag{6.33}$$

Consequently, the equation (6.8) for the plate bending in polar coordinates becomes:

$$\left(\frac{\partial^2}{\partial r^2} + \frac{1}{r}\frac{\partial}{\partial r} + \frac{1}{r^2}\frac{\partial^2}{\partial \varphi^2}\right)\left(\frac{\partial^2}{\partial r^2} + \frac{1}{r}\frac{\partial}{\partial r} + \frac{1}{r^2}\frac{\partial^2}{\partial \varphi^2}\right) w = \frac{q}{D}. \tag{6.34}$$

Let us denote bending moments acting in the sections with normals $\boldsymbol{r}$ and $\boldsymbol{\varphi}$, by $M_r$, $M_\varphi$, respectively, and the torque—by $M_{r\varphi}$. These moments are, as usual, related to a unit length. Let us presume that *ox* axis coincides with the polar radius $r$, then moments $M_r$, $M_\varphi$, and $M_{r\varphi}$ will have the same values as moments $M_x$, $M_y$, and $M_{xy}$ (Fig. 6.9).

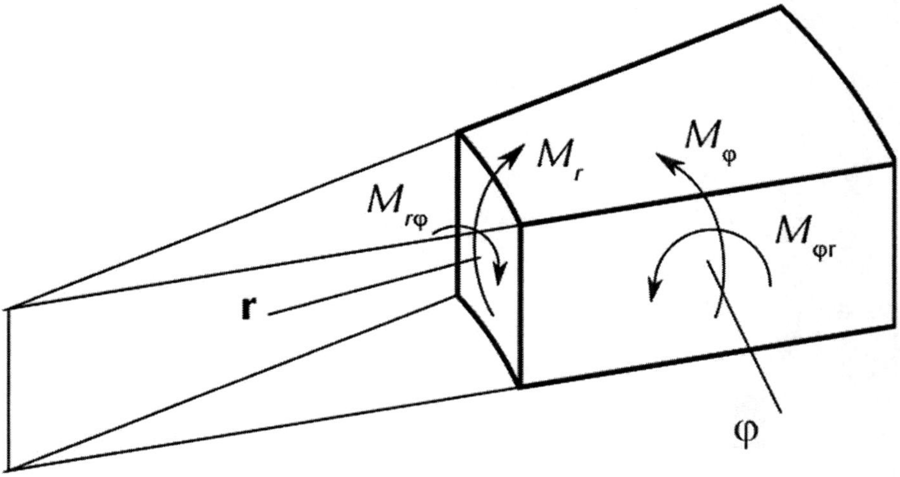


**Figure 6.9.**

Thus, at transfer from the Cartesian to polar coordinates with the help of (6.32), (6.33) and assuming that $\varphi = 0$, using formulas (6.3) and (6.5) we will finally have

$$M_r = -D\left(\frac{\partial^2 w}{\partial x^2} + \nu\frac{\partial^2 w}{\partial y^2}\right)_{\varphi=0} = -D\left(\frac{\partial^2 w}{\partial r^2} + \nu\left(\frac{1}{r}\frac{\partial w}{\partial r} + \frac{1}{r^2}\frac{\partial^2 w}{\partial \varphi^2}\right)\right),$$

$$M_\varphi = -D\left(\frac{1}{r}\frac{\partial w}{\partial r} + \frac{1}{r^2}\frac{\partial^2 w}{\partial \varphi^2} + \nu\frac{\partial^2 w}{\partial r^2}\right),$$

$$M_{\varphi r} = (1-\nu)D\left(\frac{1}{r}\frac{\partial^2 w}{\partial r\partial\varphi} - \frac{1}{r^2}\frac{\partial^2 w}{\partial \varphi^2}\right),$$

$$Q_r = -D\frac{\partial}{\partial r}\left(\frac{\partial^2 w}{\partial r^2} + \frac{1}{r}\frac{\partial w}{\partial r} + \frac{1}{r^2}\frac{\partial^2 w}{\partial \varphi^2}\right),$$

$$Q_\varphi = -D\frac{1}{r}\frac{\partial}{\partial \varphi}\left(\frac{\partial^2 w}{\partial r^2} + \frac{1}{r}\frac{\partial w}{\partial r} + \frac{1}{r^2}\frac{\partial^2 w}{\partial \varphi^2}\right). \tag{6.35}$$

If the edge of a circular plate of radius $a$ is pinched, then

$$w\big|_{r=a} = 0, \quad \frac{\partial w}{\partial r}\bigg|_{r=a} = 0. \tag{6.36}$$

For hingedly supported

$$w\big|_{r=a} = 0, \quad M_r\big|_{r=a} = 0. \tag{6.37}$$

For simply supported edge

$$M_r\big|_{r=a} = 0, \quad \left(Q_r - \frac{1}{r}\frac{\partial M_\varphi}{\partial \varphi}\right)_{r=a} = 0.$$

The general solution of equation (6.34) may be

$$w = w_0 + w_1,$$

where $w_0$ is its partial solution of some kind, $w_1$ is a general solution of the homogeneous equation

$$\left(\frac{\partial^2}{\partial r^2} + \frac{1}{r}\frac{\partial}{\partial r} + \frac{1}{r^2}\frac{\partial^2}{\partial \varphi^2}\right)\left(\frac{\partial^2}{\partial r^2} + \frac{1}{r}\frac{\partial}{\partial r} + \frac{1}{r^2}\frac{\partial^2}{\partial \varphi^2}\right)w_1 = 0. \tag{6.38}$$

The general solution of this equation was presented by Klebsch in the form

$$w_1 = R_0^{(0)}(r) + \sum_{n=1}^{\infty} R_n^{(1)}(r)\cos n\varphi + \sum_{n=1}^{\infty} R_n^{(2)}(r)\sin n\varphi.$$

Term $R_0^{(0)}(r)$ independent of angle φ is a symmetrical bending of a circular plate.

Let us substitute Klebsch's solution into the homogeneous equation (6.38), then for the $n^{\text{th}}$ member of the series, we obtain

$$\left(\frac{\partial^2}{\partial r^2}+\frac{1}{r}\frac{\partial}{\partial r}-\frac{n^2}{r^2}\right)\left(\frac{\partial^2 R_n^{(k)}}{\partial r^2}+\frac{1}{r}\frac{\partial R_n^{(k)}}{\partial r}-\frac{n^2}{r^2}R_n^{(k)}\right)=0,$$

where $k = 0, 1, 2$. So, a general solution of this equation is:

at $n = 0$ – $R_0^{(0)} = A_0 + B_0 r^2 + C_0 \ln r + D_0 r^2 \ln r$,

at $n = 1$ – $R_1^{(k)} = A_1^{(k)} + B_1^{(k)} r^3 + C_1^{(k)} r^{-1} + D_1^{(k)} r \ln r$,

at $n \geq 2$ – $R_n^{(k)} = A_n^{(k)} r^n + B_n^{(k)} r^{-n} + C_n^{(k)} r^{n+2} + D_n^{(k)} r^{-n+2}$.

Integration constants $A_n^{(k)}$, $B_n^{(k)}$, $C_n^{(k)}$ and $D_n^{(k)}$ are defined in terms of the boundary conditions.

## SYMMETRICAL BENDING OF A CIRCULAR PLATE

If stress $q$ is distributed symmetrically with respect to the center of the plate, deflection $w$ does not depend on coordinate φ. In this case, it follows from equation (6.34) that

$$\left(\frac{\mathrm{d}^2}{\mathrm{d}r^2}+\frac{1}{r}\frac{\mathrm{d}}{\mathrm{d}r}\right)\left(\frac{\mathrm{d}^2}{\mathrm{d}r^2}+\frac{1}{r}\frac{\mathrm{d}}{\mathrm{d}r}\right)w=\frac{q(r)}{D}.$$

Let us present this equation in a short-cut form suitable for direct integration,

$$\frac{1}{r}\frac{\mathrm{d}}{\mathrm{d}r}\left(r\frac{\mathrm{d}}{\mathrm{d}r}\left(\frac{1}{r}\frac{\mathrm{d}}{\mathrm{d}r}\left(r\frac{\mathrm{d}w}{\mathrm{d}r}\right)\right)\right)=\frac{q(r)}{D}. \quad (6.39)$$

By a quadruple integration of equation (6.39) in a series, we obtain the deflection

$$w = C_1 + C_2 r^2 + C_3 \ln r + C_4 r^2 \ln r + w_0. \quad (6.40)$$

Here,

$$w_0 = \frac{1}{D}\int\left(\int\left(\int\left(\int qr\,\mathrm{d}r\right)\frac{1}{r}\mathrm{d}r\right)r\,\mathrm{d}r\right)\frac{1}{r}\mathrm{d}r. \quad (6.41)$$

Integration constants $C_1$, ..., $C_4$ are defined in terms of the plate fixing conditions.

***Example 6.2.*** Let us consider the case, when $q$ = const. From (6.41) it follows that

$$w_0 = \frac{qr^4}{64D}.$$

In the case of continuous plates for the boundedness of solution (6.40) in the origin of coordinates we should set that $C_3 = 0$, then the deflection is

$$w = C_1 + C_2 r^2 + C_4 r^2 \ln r + \frac{qr^4}{64D}. \tag{6.42}$$

On the basis of the formula for a shearing force from (6.35) and expression (6.40), we obtain

$$Q_r = -D\left(\frac{q}{2D}r + 4C_4\frac{1}{r}\right). \tag{6.43}$$

On the other hand, when cutting the circle of radius $r$ out of the plate by a circular section, we may find $Q_r$ from the sum of projections of all forces onto axis $z$:

$$Q_r 2\pi r + \int_0^r \int_0^{2\pi} qr \, \mathrm{d}r \, \mathrm{d}\phi = 0, Q_r = -\frac{qr}{2}. \tag{6.44}$$

By equating (6.43) and (6.44), we obtain that $C_4 = 0$. As a result of solving (6.42), we have

$$w = C_1 + C_2 r^2 + \frac{qr^4}{64D}. \tag{6.45}$$

If the plate edge is fixed, then by substituting (6.45) into the boundary conditions (6.36), we obtain a system of the linear algebraic equations

$$C_1 + C_2 a^2 + \frac{qa^4}{64D} = 0,$$

$$2C_2 a^2 + \frac{3qa^2}{16D} + \nu\left(2C_2 + \frac{qa^2}{16D}\right) = 0.$$

Based on the constants $C_1$, $C_2$, the expression of bending the plate with a fixed contour will be as follows:

$$w = \frac{q}{64D}\left(a^2 - r^2\right)\left(\frac{5+\nu}{1+\nu}a^2 - r^2\right).$$

In the case of a hinged contour of the plate, after substitution of deflection (6.45) into the boundary conditions (6.37), we come to:

$$C_1 + C_2 a^2 + \frac{qa^4}{64D} = 0, 2C_2 a + \frac{qa^3}{16D} = 0.$$

Once we have found integration constants $C_1$, $C_2$, we may obtain the deflection under these boundary conditions:

$$w = \frac{q}{64D}\left(a^2 - r^2\right)^2.$$

Finally, stresses in the plate may be calculated by the formulas like (6.3).

## ELLIPTIC PLATE

Let us consider a uniformly loaded pinched elliptic plate (Fig. 6.10). The equation of the plate contour is

$$\frac{x^2}{a^2}+\frac{y^2}{b^2}-1=0.$$

The differential equation of the plate bending

$$\Delta^2 w=\frac{q}{D}$$

and the boundary conditions for the contour

$$w=0,\quad \frac{\partial w}{\partial n}=0$$

will be met if we accept for the deflection the expression proposed by Bryan[2]

$$w=w_0\left(1-\frac{x^2}{a^2}-\frac{y^2}{b^2}\right)^2. \tag{6.46}$$

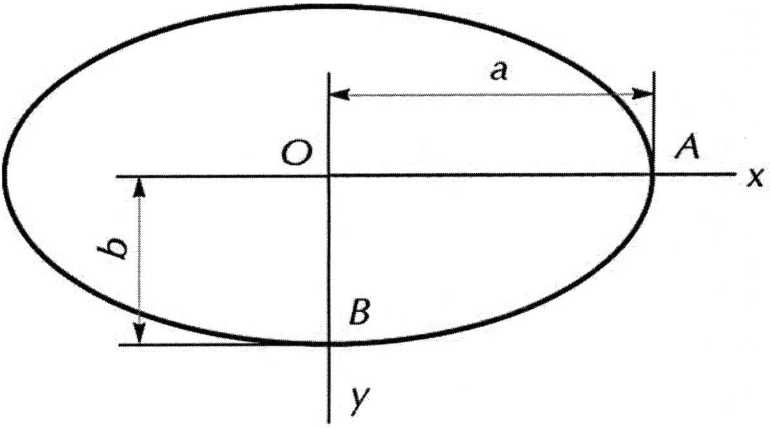


**Figure 6.10.**

This expression and its first derivatives of $x$ and $y$ become zero on the contour. After substitution of (6.46) in the equation for the plate bending, it becomes evident that it would be met if,

$$w_0=\frac{q}{D\left(\frac{24}{a^4}+\frac{24}{b^4}+\frac{16}{a^2b^2}\right)}. \tag{6.47}$$

Thus, the solution of (6.46) is exact for a uniformly loaded pinched elliptic plate. After substitution of $x = y = 0$ in expression (6.46), we obtain that $w_0$ defined by formula (6.47) is the deflection plate in center. When $a = b$, we have a solution for a circular restrained plate. At $a = \infty$ we have the deflection of a uniformly loaded strip of spacing $2b$ with restrained ends.

The internal moments in the plate may be calculated by substitution of the expression for deflection (6.46) into formulas. It is apparent that the maximum bending stress is observed at the ends of the shorter axis of the ellipse.

When a linearly varying pressure $q = q_0 x$ affects the restrained elliptical plate, the solution will be as follows:

$$w_0 = \frac{q_0 x}{24D} \frac{\left(1 - \frac{x^2}{a^2} - \frac{y^2}{b^2}\right)^2}{\frac{5}{a^4} + \frac{1}{b^4} + \frac{2}{a^2 b^2}}.$$

Similarly to the previous case and proceeding from above expression, one can calculate the internal bending moments and torques, and analyze the stress state of the plate.

In all previously considered cases of the plate bending problems, the equilibrium equations have been derived statically using throughout the article Kirhhoff's kinematic hypotheses. If to reject these hypotheses, we may find the equilibrium equations and boundary conditions of the forces using the Variational-Lagrangian principle.

## A CIRCULAR ELASTIC THREE-LAYER PLATE

Multilayered structural elements are widely applicable at present in extensively developing branches of industry and construction. This is explained by such advantages as high specific rigidity, perfect heat-insulating and sound-proofing properties, high aerodynamic characteristics. As an example of calculations let us consider a symmetric bending of an elastic sandwich circular plate being asymmetric across thickness (Fig. 6.11).

Let us state and solve the problem in cylindrical coordinates $r$, $\varphi$, $z$. We shall assume that the median plane of the filler is the coordinate one and the direct axis $z$ perpendicular to layer 1. Kirhhoff's hypotheses are accepted for the thin outer bearing layers of $h_1 \neq h_2$ thickness. The hypothesis on rectilinearity and incompressibility of the deformed normal is valid for a thick rigid filler ($h_3 = 2c$), which perceives the load in tangential direction. The projections of the external load onto the vertical and radial axes of coordinates are: $q = q(r)$, $p = p(r)$. A rigid membrane preventing relative shear of the layers ($\psi = 0$ at $r = 1$) is assumed to be available on the plate contour.

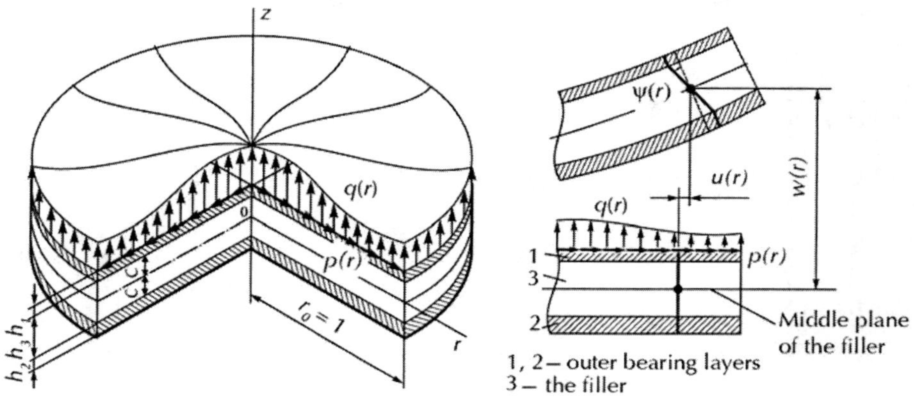


**Figure 6.11.**

Owing to symmetry of the load there are no tangential displacements in the layers: $u_\varphi^{(k)} = 0$ ($k$—layer number), and the plate deflection $w$, relative shear in the filler $\psi$ and radial displacement of the coordinate surface $u$ are independent of coordinate $\varphi$, that is, $w = w(r)$, $u = u(r)$, $\psi = \psi(r)$. Hereinafter, these functions are assumed to be the sought ones. All the displacements and linear dimensions of the plate are related to its radius $r_0$, force characteristics—to 1 Pa; thickness of the $k^{\text{th}}$ layer is denoted by $h_k$. Using the hypothesis of the normal straightness of the filler,

$$2\varepsilon_{rz}^{(3)} = u_r^{(3)},_z + w,_r = \psi ,$$

after integration we obtain the expressions for radial displacements in the layers $u_r^{(k)}$ in terms of the sought functions:

$$\begin{aligned} u_r^{(1)} &= u + c\psi - zw,_r , & c \le z \le c + h_1 , \\ u_r^{(3)} &= u + z\psi - zw,_r , & -c \le z \le c , \\ u_r^{(2)} &= u - c\psi - zw,_r , & -c - h_2 \le z \le -c , \end{aligned} \tag{6.48}$$

where $z$ is a distance from the considered fiber to the median plane of the filler, $u + c\psi$ is a displacement value of the outer bearing layer caused by the filler deformation. For the second bearing layer this displacement will be $u - c\psi$. The comma in the subscript means differentiation with respect to a subsequent coordinate.

Deformations in the layers are found from (6.48) and Cauchy's relations

$$\varepsilon_r^{(1)} = u,_r + c\psi,_r - zw,_{rr} , \varepsilon_\phi^{(1)} = \frac{1}{r}(u + c\psi - zw,_r), \varepsilon_{rz}^{(1)} = 0 ,$$

$$\varepsilon_r^{(3)} = u,_r + z\psi,_r - zw,_{rr} , \varepsilon_\phi^{(3)} = \frac{1}{r}(u + z\psi - zw,_r), \quad \varepsilon_{rz}^{(3)} = \tfrac{1}{2}\psi ,$$

$$\varepsilon_r^{(2)} = u,_r - c\psi,_r - zw,_{rr}, \varepsilon_\phi^{(2)} = \frac{1}{r}(u - c\psi - zw,_r), \varepsilon_{rz}^{(2)} = 0 \tag{6.49}$$

Let us introduce the generalized internal forces and the moments in the plate:

$$T_\alpha \equiv \sum_{k=1}^{3} T_\alpha^{(k)} = \sum_{k=1}^{3} \int_{h_k} \sigma_\alpha^{(k)} \, dz, M_\alpha \equiv \sum_{k=1}^{3} M_\alpha^{(k)} = \sum_{k=1}^{3} \int_{h_k} \sigma_\alpha^{(k)} z \, dz,$$

$$H_\alpha = M_\alpha^{(3)} + c\left(T_\alpha^{(1)} - T_\alpha^{(2)}\right), Q = \int_{-c}^{c} \sigma_{rz}^{(3)} \, dz, \tag{6.50}$$

where $\sigma_\alpha^{(k)}$ are the components of the stress tensor ($\alpha = r, \varphi$).

Let us assume that the arbitrarily distributed loads are applied on the outer surfaces of the bearing layers, while the forces and moments—on the plate contour. The work of the external surface load will vary as follows ($dS = r dr d\varphi$):

$$\delta A_1 = \iint_S (q\delta w + p\delta u) r \, dr \, d\varphi \cdot \tag{6.51}$$

Variation of the work of the contour forces will be $T_r^0$, $H_r^0$, $M_r^0$, $Q^0$

$$\delta A_2 = \int_0^{2\pi} (T_r^0 \delta u + H_r^0 \delta\psi + M_r^0 \delta w,_r + Q^0 \delta w) \, d\varphi. \tag{6.52}$$

Variation of the work of the elastic forces is

$$\delta W = \iint_S \left[ \sum_{k=1}^{3} \int_{h_k} (\sigma_r^{(k)} \delta\varepsilon_r^{(k)} + \sigma_\varphi^{(k)} \delta\varepsilon_\varphi^{(k)}) \, dz + \int_{-c}^{c} \sigma_{rz}^{(3)} \delta\psi \, dz \right] r \, dr \, d\varphi. \tag{6.53}$$

The double integral is distributed here all over the filler middle surface $S$. Since the displacement variations in the layers are

$$\begin{aligned}
\delta u^{(1)} &= \delta u + c\delta\psi - z\delta w,_r, & c \le z \le c + h_1, \\
\delta u^{(3)} &= \delta u + z\delta\psi - z\delta w,_r, & -c \le z \le c, \\
\delta u^{(2)} &= \delta u - c\delta\psi - z\delta w,_r, & -c - h_2 \le z \le -c,
\end{aligned}$$

the deformation variations will be:

$$\delta\varepsilon_r^{(1)} = \delta u,_r + c\delta\psi,_r - z\delta w,_{rr}, \delta\varepsilon_\phi^{(1)} = \frac{1}{r}(\delta u + c\delta\psi - z\delta w,_r),$$

$$\delta\varepsilon_r^{(3)} = \delta u,_r + z\delta\psi,_r - z\delta w,_{rr}, \delta\varepsilon_\phi^{(3)} = \frac{1}{r}(\delta u + z\psi - z\delta w,_r),$$

$$\delta\varepsilon_r^{(2)} = \delta u,_r - c\delta\psi,_r - z\delta w,_{rr}, \delta\varepsilon_\phi^{(2)} = \frac{1}{r}(\delta u - c\delta\psi - z\delta w,_r),$$

$$\delta\varepsilon_{rz}^{(3)} = \tfrac{1}{2}\delta\psi. \tag{6.54}$$

Let us consider the total integral across the layers being an integral part of the virtual work of the elastic forces (6.53). Using (6.54) for the radial components in the first layer, we obtain

$$\int_{h_1} \sigma_r^{(1)}\delta\varepsilon_r^{(1)}\,\mathrm{d}z = \int_{h_1} \sigma_r^{(1)}(\delta u,_r + c\delta\psi,_r - z\delta w,_{rr})\,\mathrm{d}z = T_r^{(1)}\delta u,_r + cT_r^{(1)}\delta\psi,_r - M_r^{(1)}\delta w,_{rr}.$$

Similarly,

$$\int_{h_2} \sigma_r^{(2)}\delta\varepsilon_r^{(2)}dz = T_r^{(2)}\delta u,_r \text{-} cT_r^{(2)}\delta\psi,_r \text{-} M_r^{(2)}\delta w,_{rr},$$

$$\int_{h_3} \sigma_r^{(3)}\delta\varepsilon_r^{(3)}\,\mathrm{d}z = T_r^{(3)}\delta u,_r + M_r^{(3)}\delta\psi,_r - M_r^{(3)}\delta w,_{rr},$$

$$\int_{h_1} \sigma_\phi^{(1)}\delta\varepsilon_\phi^{(1)}\,\mathrm{d}z = \int_{h_1} \sigma_\phi^{(1)}\frac{1}{r}(\delta u + c\delta\psi - z\delta w,_r)\,\mathrm{d}z = \frac{1}{r}\left(T_\phi^{(1)}\delta u + cT_\phi^{(1)}\delta\psi - M_\phi^{(1)}\delta w,_r\right),$$

$$\int_{h_2} \sigma_\phi^{(2)}\delta\varepsilon_\phi^{(2)}\,\mathrm{d}z = \frac{1}{r}\left(T_\phi^{(2)}\delta u - cT_\phi^{(2)}\delta\psi - M_\phi^{(2)}\delta w,_r\right),$$

$$\int_{h_3} \sigma_\phi^{(3)}\delta\varepsilon_\phi^{(3)}\,\mathrm{d}z = \frac{1}{r}\left(T_\phi^{(3)}\delta u + M_\phi^{(3)}\delta\psi - M_\phi^{(3)}\delta w,_r\right).$$

If we sum up:

$$\sum_{k=1}^{3} \int_{h_k} (\sigma_r^{(k)}\delta\varepsilon_r^{(k)} + \sigma_\phi^{(k)}\delta\varepsilon_\phi^{(k)})\,\mathrm{d}z + \int_{-c}^{c} \sigma_{rz}^{(3)}\delta\varepsilon\,\mathrm{d}z$$

$$= T_r\,\delta u,_r + H_r\delta\psi,_r - M_r\delta w,_{rr} + Q\delta\psi + \frac{1}{r}\left(T_\phi\,\delta u + H_\phi\delta\psi - M_\phi\delta w,_r\right),$$

where the internal forces $T_\alpha$, $M_\alpha$, $H_\alpha$ and $Q$ are introduced by relations (6.50).

The variation of the potential energy of deformation is obtained by substitution of the previous expression into (6.53) and integration in polar coordinates. Then,

$$\delta W = \int_r \int_\phi \big[r(T_r\,\delta u,_r + H_r\delta\psi,_r - M_r\delta w,_{rr} + Q\delta\psi)$$

$$+T_\varphi \delta u + H_\varphi \delta\psi - M_\varphi \delta w,_r \big] \mathrm{d}\varphi \, \mathrm{d}r .$$

Let us transform the integrand expressing its summands as follows:

$$rT_r \delta u,_r = (rT_r \delta u),_r - (rT_r),_r \delta u, \; rH_r \delta\psi,_r = (rH_r \delta\psi),_r - (rH_r),_r \delta\psi ,$$

$$rM_r \delta w,_{rr} = (rM_r \delta w,_r),_r - [(rM_r),_r \delta w],_r + (rM_r),_{rr} \delta w ,$$

$$M_\varphi \delta w,_r = (M_\varphi \delta w),_r - M_\varphi,_r \delta w .$$

So, the expression for the stress work variation $\delta W$ may be divided into two integrals by factoring the differentiation in the former, and grouping the terms at identical virtual displacements in the latter:

$$\delta W = \int_r \int_\varphi \left\{ rT_r \delta u + rH_r \delta\psi - rM_r \delta w,_r + \left[ (rM_r),_r - M_\varphi \right] \delta w \right\},_r \mathrm{d}\varphi \, \mathrm{d}r$$

$$- \int_r \int_\varphi \left\{ \left[ (rT_r),_r - T_\varphi \right] \delta u + \left[ (rH_r),_r - H_\varphi - rQ \right] \delta\psi - \left[ (rM_r),_{rr} - M_\varphi,_r \right] \delta w \right\} \mathrm{d}\varphi dr .$$

It follows that

$$\delta W = \int_0^{2\pi} \left\{ rT_r \delta u + rH_r \delta\psi - rM_r \delta w,_r + [(rM_r),_r - M_\varphi] \delta w \right\} \mathrm{d}\varphi$$

$$- \int_r \int_\varphi \left\{ \left[ (rT_r),_r - T_\varphi \right] \delta u + \left[ (rH_r),_r - H_\varphi - rQ \right] \delta\psi - \left[ (rM_r),_{rr} - M_\varphi,_r \right] \delta w \right\} \mathrm{d}\varphi dr .$$

Now we equate the expression for the work of external and contour forces (6.51), (6.52) and require its abeyance at any values of the varying displacements. This is possible if the coefficients are zero when the variations of the sought functions are independent. Wherefrom, we obtain a system of differential equations of equilibrium in the forces describing deformation of a circular sandwich plate

$$T_r,_r + \frac{1}{r}(T_r - T_\varphi) = -p , \quad H_r,_r + \frac{1}{r}(H_r - H_\varphi) - Q = 0 ,$$

$$M_r,_r + \frac{1}{r}(2M_r,_r - M_\varphi,_r) = -q . \qquad (6.55)$$

The conditions for the forces on the boundary $r = 1$are to be met

$$T_r = T_r^0 , \; H_r = H_r^0 , \; M_r = M_r^0 , \; M_r,_r + \frac{1}{r}(M_r - M_\varphi) = Q^0 . \qquad (6.56)$$

It is supposed that the relationship between stresses and strains in the layers is described by the relations of the linear elasticity theory (without summation over $k$)

$$s_\alpha^k = 2G_k \, \text{э}_\alpha^k, \sigma^k = K_k \theta^k .$$

After substitution of deformations (6.49) thereinto and using (6.50), we obtain an expression for generalized forces $T_\alpha$, $M_\alpha$ and $H_\alpha$ in terms of three unknown functions: $u(r)$, $\psi(r)$, $w(r)$. Let us write a detailed conclusion for $T_r$:

$$T_r \equiv \sum_{k=1}^{3} T_r^{(k)} = \sum_{k=1}^{3} \int_{h_k} \sigma_r^{(k)} dz = \sum_{k=1}^{3} \int_{h_k} (2G_k \, \text{э}_r^{(k)} + 3K_k \varepsilon^{(k)}) \, \mathrm{d} z$$

$$= \sum_{k=1}^{3} \int_{h_k} [2G_k (\varepsilon_r^{(k)} - \varepsilon^{(k)}) + 3K_k \varepsilon^{(k)}] \, \mathrm{d} z$$

$$= \sum_{k=1}^{3} \int_{h_k} [2G_k \varepsilon_r^{(k)} + (K_k - \tfrac{2}{3} G_k)(\varepsilon_r^{(k)} + \varepsilon_\varphi^{(k)})] \, \mathrm{d} z$$

$$= \sum_{k=1}^{3} \int_{h_k} [(K_k + \tfrac{4}{3} G_k)\varepsilon_r^{(k)} + (K_k - \tfrac{2}{3} G_k)\varepsilon_\varphi^{(k)}] \, \mathrm{d} z .$$

For the sake of convenience, let us denote that

$$K_k + \tfrac{4}{3} G_k \equiv K_k^+ , \quad K_k - \tfrac{2}{3} G_k \equiv K_k^- .$$

Then, by writing the latter sum layer by layer, we have

$$T_r = \int_{c}^{c+h_1} \left[ K_1^+ (u,_r + c\psi,_r - zw,_{rr}) + \frac{1}{r} K_1^- (u + c\psi - zw,_r) \right] \mathrm{d} z$$

$$+ \int_{-c-h_2}^{-c} \left[ K_2^+ (u,_r - c\psi,_r - zw,_{rr}) + \frac{1}{r} K_2^- (u - c\psi - zw,_r) \right] \mathrm{d} z$$

$$+ \int_{-c}^{c} \left[ K_3^+ (u,_r + z\psi,_r - zw,_{rr}) + \frac{1}{r} K_3^- (u + z\psi - zw,_r) \right] \mathrm{d} z . \qquad (6.57)$$

Let us write out the simplest integrals, needed hereafter:

$$\int_{h_k} \mathrm{d} z = h_k , \quad \int_{c}^{c+h_1} z \, \mathrm{d} z = h_1 \left( c + \frac{h_1}{2} \right), \quad \int_{-c}^{c} z \, \mathrm{d} z = 0, \quad \int_{-c-h_2}^{-c} z \, \mathrm{d} z = -h_2 \left( c + \frac{h_2}{2} \right),$$

$$\int_{c}^{c+h_1} z^2 \, \mathrm{d} z = h_1 \left( c^2 + c h_1 + \frac{h_1^2}{3} \right),$$

$$\int_{-c}^{c} z^2\, \mathrm{d}z = \frac{2}{3}c^3, \int_{-c-h_2}^{-c} z^2\, \mathrm{d}z = h_2\left(c^2 + ch_2 + \frac{h_2^2}{3}\right). \tag{6.58}$$

With the help of relations (6.58), after integration of expression (6.57) and reduction of similar terms, we obtain

$$T_r = \sum_{k=1}^{3} h_k (K_k^+ u,_r + \frac{u}{r} K_k^-) + c(K_1^+ h_1 - K_2^+ h_2)\psi,_r + c(K_1^- h_1 - K_2^- h_2)\frac{\psi}{r}$$

$$-\left[K_1^- h_1\left(c + \frac{h_1}{2}\right) - K_2^- h_2\left(c + \frac{h_2}{2}\right)\right]\frac{w,_r}{r}. \tag{6.59}$$

The expression for $T_\varphi$ may be obtained from $T_r$, after interchanging positions of $K_k^+$ and $K_k^-$,

$$T_\varphi = \sum_{k=1}^{3} \int_{h_k} (K_k^- \varepsilon_r^{(k)} + K_k^+ \varepsilon_\varphi^{(k)})\,\mathrm{d}z = \sum_{k=1}^{3} h_k (K_k^- u,_r + \frac{u}{r} K_k^+)$$

$$+ c(K_1^- h_1 - K_2^- h_2)\psi,_r + c(K_1^+ h_1 - K_2^+ h_2)\frac{\psi}{r} \tag{6.60}$$

$$-\left[K_1^- h_1\left(c + \frac{h_1}{2}\right) - K_2^- h_2\left(c + \frac{h_2}{2}\right)\right] w,_{rr} - \left[K_1^+ h_1\left(c + \frac{h_1}{2}\right) - K_2^+ h_2\left(c + \frac{h_2}{2}\right)\right]\frac{w,_r}{r}.$$

In a similar way, the generalized moments $M_\alpha$, $H_\alpha$ can be obtained

$$M_r = \left[K_1^+ h_1\left(c + \frac{h_1}{2}\right) - K_2^+ h_2\left(c + \frac{h_2}{2}\right)\right] u,_r + \left[K_1^- h_1\left(c + \frac{h_1}{2}\right) - K_2^- h_2\left(c + \frac{h_2}{2}\right)\right]\frac{u}{r}$$

$$+\left[cK_1^+ h_1\left(c + \frac{h_1}{2}\right) + cK_2^+ h_2\left(c + \frac{h_2}{2}\right) + \frac{2}{3}c^3 K_3^+\right]\psi,_r$$

$$+\left[cK_1^- h_1\left(c + \frac{h_1}{2}\right) + cK_2^- h_2\left(c + \frac{h_2}{2}\right) + \frac{2}{3}c^3 K_3^-\right]\frac{\psi}{r}$$

$$-\left[K_1^+ h_1\left(c^2 + ch_1 + \frac{h_1^2}{3}\right) + K_2^+ h_2\left(c^2 + ch_2 + \frac{h_2^2}{3}\right) + \frac{2}{3}c^3 K_3^+\right] w,_{rr}$$

$$-\left[K_1^- h_1\left(c^2 + ch_1 + \frac{h_1^2}{3}\right) + K_2^- h_2\left(c^2 + ch_2 + \frac{h_2^2}{3}\right) + \frac{2}{3}c^3 K_3^-\right]\frac{w,_r}{r}.$$

$$H_r = c(K_1^+ h_1 - K_2^+ h_2) u,_r + c(K_1^- h_1 - K_2^- h_2)\frac{u}{r}$$

$$+\left[c^2(K_1^+ h_1 + K_2^+ h_2) + \frac{2}{3}c^3 K_3^+\right]\psi,_r + \left[c^2(K_1^- h_1 + K_2^- h_2) + \frac{2}{3}c^3 K_3^-\right]\frac{\psi}{r}$$

$$-\left[c\left(K_1^+h_1(c+\frac{h_1}{2})+K_2^+h_2(c+\frac{h_2}{2})\right)+\frac{2}{3}c^3K_3^+\right]w,_{rr}$$

$$\left[c\left(K_1^-h_1(c+\frac{h_1}{2})+K_2^-h_2(c+\frac{h_2}{2})\right)+\frac{2}{3}c^3K_3^-\right]\frac{w,_r}{r}.\ Q=2cG_3\psi. \tag{6.61}$$

The relations for $M_\varphi$, $H_\varphi$ follow from $M_r$, $H_r$, if to interchange positions of $K_k^+$ and $K_k^-$, so we may neglect them.

After substitution of obtained expressions for the generalized internal forces and moments in (6.55), we obtain the following system of differential displacement equations of equilibrium for a circular sandwich plate:

$$\mathrm{L}_2(a_1u+a_2\psi-a_3w,_r)=-p,\ \mathrm{L}_2(a_2u+a_4\psi-a_5w,_r)-2cG_3\psi=0\ ,$$

$$\mathrm{L}_3(a_3u+a_5\psi-a_6w,_r)=-q\ . \tag{6.62}$$

Here, coefficients $a_i$ and differential operators $L_2$ (*Bessel*[3] operator), $L_3$ can be presented as:

$$a_1=\sum_{k=1}^{3}h_kK_k^+,\ a_2=c(h_1K_1^+-h_2K_2^+)\ ,$$

$$a_3=h_1\left(c+\frac{h_1}{2}\right)K_1^+-h_2\left(c+\frac{h_2}{2}\right)K_2^+,\ a_4=c^2\left(h_1K_1^++h_2K_2^++\frac{2}{3}cK_3^+\right),$$

$$a_5=c\left[h_1\left(c+\frac{h_1}{2}\right)K_1^++h_2\left(c+\frac{h_2}{2}\right)K_2^++\frac{2}{3}c^2K_3^+\right],$$

$$a_6=h_1\left(c^2+ch_1+\frac{h_1^2}{3}\right)K_1^++h_2\left(c^2+ch_2+\frac{h_2^2}{3}\right)K_2^++\frac{2}{3}c^3K_3^+\ ,$$

$$\mathrm{L}_2(g)\equiv\left(\frac{1}{r}(rg),_r\right),_r\equiv g,_{rr}+\frac{g,_r}{r}-\frac{g}{r^2},$$

$$\mathrm{L}_3(g)\equiv\frac{1}{r}(rL_2(g)),_r\equiv g,_{rrr}+\frac{2g,_{rr}}{r}-\frac{g,_r}{r^2}+\frac{g}{r^3}. \tag{6.63}$$

The problem of defining functions $u(r)$, $\psi(r)$, $w(r)$ is solved by adding boundary conditions, for example, (6.56)–(6.62).

After a single integration, it follows from the third equation (6.62) that

$$\mathrm{L}_2(a_3u+a_5\psi-a_6w,_r)=-\frac{1}{r}\int qr\,\mathrm{d}r+\frac{C_1}{r}\ .$$

The integration constant $C_1$ will be found later.

Using this equation and the first one from (6.62), the radial displacement $u(r)$ and deflection $w(r)$ may be excluded from the second equation of the same system. As a result, we have a heterogeneous modified *Bessel's equation* for functions $\psi(r)$

$$\mathrm{L}_2(\psi)-\beta^2\psi=f\ . \tag{6.64}$$

Here,

$$\beta^2 = \frac{2cb_3G_3}{b_1b_3 - b_2^2}, b_1 = \frac{a_1a_4 - a_2^2}{a_1}, b_2 = \frac{a_1a_5 - a_2a_3}{a_1}, b_3 = \frac{a_1a_6 - a_3^2}{a_1},$$

$$f(r) = \frac{b_3}{b_1b_3 - b_2^2}\left[\frac{p(a_2b_3 - a_3b_2)}{a_1b_3} + \frac{b_2}{b_3r}\left(\int qr\,\mathrm{d}r - C_1\right)\right]. \tag{6.65}$$

The corresponding numerical factors are included into the integration constant $C_1$.

The obtained equation may be presented as a sum of a general solution of the corresponding homogeneous equation $\psi_0$ and a partial solution $\psi_r$ of the heterogeneous equation (6.64)

$$\psi = \psi_0 + \psi_r, \tag{6.66}$$

in which connection,

$$\mathrm{L}_2(\psi_0) - \beta^2\psi_0 = 0.$$

The solution of this equation is known [29]:

$$\psi_0 = C_2 I_1(\beta r) + C_3 K_1(\beta r). \tag{6.67}$$

Where $I_1(\beta r)$ is a *modified* first-order *Besselian function*, $K_1(\beta r)$ is the first-order *Macdonald's function*, which in a general case at $z > 0$ may be presented by a series [81]:

$$I_n(z) = \sum_{k=0}^{\infty} \frac{(z/2)^{n+2k}}{k!\Gamma(k+n+1)},$$

$$K_n(z) = (-1)^{n+1} I_n(z) \ln\frac{z}{2} + \frac{1}{2}\sum_{k=0}^{n-1} \frac{(-1)^k (n-k-1)!}{k!}\left(\frac{z}{2}\right)^{2k-n}$$

$$+\frac{1}{2}(-1)^n \sum_{k=0}^{\infty} \frac{(z/2)^{n+2k}}{k!(k+n)!}[\varphi(n+k+1) + \varphi(k+1)],$$

$$\Gamma(z) = \int_0^{\infty} e^{-t} t^{z-1}\,\mathrm{d}t, \Gamma(n+1) = n!, \varphi(z) = \Gamma'(z)/\Gamma(z). \tag{6.68}$$

Here, $\Gamma(z)$ is a gamma function. When $n = 0$, the first of the sums in $K_n(z)$ should be assumed equal to zero, the prime means a derivative of $z$.

As it is known from the theory of the linear differential second-order equations, the partial solution of equation (6.64) may be expressed in terms of two linearly independent solutions of a corresponding homogeneous equation

$$\psi_r = K_1(\beta r)\int \frac{I_1(\beta r) f(r)}{W}\mathrm{d}r - I_1(\beta r)\int \frac{K_1(\beta r) f(r)}{W}\mathrm{d}r,$$

where $W$—*Wronskian*[4] *determinant*. In our case? with account of (6.68),

$$W\{I_1, K_1\} = K_{1,r} I_1 - I_{1,r} K_1 = -\frac{1}{r}.$$

Thus,

$$\psi_r = -K_1(\beta r)\int I_1(\beta r) f(r) r \, dr + I_1(\beta r)\int K_1(\beta r) f(r) r \, dr. \quad (6.69)$$

By summing up (6.67) and (6.69), we obtain the desired solution for the displacement (6.66). Afterwards, the deflection and the radial displacement appear from the remaining equations of system (6.62)

$$w = \frac{b_2}{b_3}\int \psi \, dr - \frac{a_3}{b_3 a_1}\int L_2^{-1}(p)\, dr + \frac{1}{b_3}\int L_3^{-1}(q)\, dr - \frac{1}{4b_3} C_1 r^2(\ln r - 1)$$

$$+\frac{C_5 r^2}{4b_3} + C_6 \ln r + C_4,$$

$$u = \frac{a_3}{a_1} w,_r - \frac{a_2}{a_1}\psi - \frac{1}{a_1} L_2^{-1}(p) + \frac{C_7 r}{2} + \frac{C_8}{r}. \quad (6.70)$$

$L_2^{-1}$, $L_3^{-1}$ are the linear integral operators inverse to the differentiation operators (6.63)

$$L_2^{-1}(f) \equiv \frac{1}{r}\int r \int f \, dr \, dr, \; L_3^{-1}(f) \equiv \frac{1}{r}\int r \int \frac{1}{r}\int rf \, dr \, dr \, dr.$$

Based on the condition of smoothness in the center of the plate ($r = 0$), for continuous plates it is necessary to set for (6.65), (6.69) and (6.70), that

$$C_1 = C_3 = C_6 = C_8 = 0.$$

The rest four integration constants are defined by the boundary conditions.

As a result, an exact solution of the elasticity problem for deformation of a continuous circular sandwich plate is

$$\psi = C_2 I_1(\beta r) + \psi_r,$$

$$u = \frac{a_3}{a_1 a_6 - a_3^2}\left[L_3^{-1}(q) - \frac{a_6}{a_3} L_2^{-1}(p) + \left(a_5 - \frac{a_2 a_6}{a_3}\right)\psi + C_7 r\right], \quad (6.71)$$

$$w = \frac{1}{b_3}\left[b_2\left(\frac{C_2}{\beta} I_0(\beta r) + \int \psi_r \, dr\right) - \int\left(\frac{a_3}{a_1} L_2^{-1}(p) - L_3^{-1}(q)\right) dr + \frac{C_5 r^2}{4} + C_4\right].$$

If the plate contour is fixed ($u = \psi = w = w,_r = 0$ at $r = 1$), the integration constants will be:

$$C_2 = -\left.\frac{\psi_r}{I_1(\beta)}\right|_{r=1},$$

$$C_4 = -b_2\left(\frac{C_2}{\beta}I_0(\beta) + \int \psi_r\, dr\Big|_{r=1}\right) + \int\left(\frac{a_3}{a_1}L_2^{-1}(p) - L_3^{-1}(q)\right)dr\Big|_{r=1} - \frac{C_5}{4},$$

$$C_5 = 2\left(\frac{a_3}{a_1}L_2^{-1}(p) - L_3^{-1}(q)\right)\Big|_{r=1}, C_7 = \left(\frac{a_6}{a_3}L_2^{-1}(p) - L_3^{-1}(q)\right)\Big|_{r=1}. \qquad (6.72)$$

In the case the plate is bent by a uniformly distributed transverse load ($p=0, q=\text{const}$), it follows from (6.65) and (6.69) that

$$f(r) = \frac{b_2 q}{2(b_1 b_3 - b_2^2)} r, \psi_r = -\frac{b_2 q}{4cb_3 G_3} r.$$

After substitution of these expressions into formulas (6.71), (6.72), we obtain a solution of the elasticity problem for bending of a circular sandwich plate with the pinched edges,

$$\psi = \frac{b_2 q}{4cb_3 G_3}\left(\frac{I_1(\beta r)}{I_1(\beta)} - r\right),$$

$$u = \frac{a_3}{a_1 a_6 - a_3^2}\left[\left(a_5 - \frac{a_2 a_6}{a_3}\right)\psi + \frac{qr}{16}(r^2 - 1)\right],$$

$$w = \frac{b_2^2 q}{4cb_3^2 G_3}\left(\frac{I_0(\beta r) - I_0(\beta)}{\beta I_1(\beta)} - \frac{1}{2}(r^2 - 1)\right) + \frac{q}{64 b_3}(r^2 - 1)^2.$$

If a hinged plate ($u = \psi = w = M_r = 0$ at $r = 1$) is taken, the following integration constants will participate:

$$C_2 = \frac{b_2 q}{4cb_3 G_3 I_1(\beta)}, C_4 = -\frac{b_2^2 q}{4cb_3 G_3}\left(\frac{I_0(\beta)}{\beta I_1(\beta)} - \frac{1}{2}\right) - \frac{q}{64} - \frac{C_5}{4}$$

$$C_5 = \frac{2b_3}{a_6 + a_7}\left[a_3 u,_r(1) + \psi,_r(1)\left(a_5 - \frac{a_6 b_2}{b_3}\right) - \frac{q}{16 b_3}(3a_6 - a_7)\right],$$

$$C_7 = -\frac{q}{16}, \psi,_r(1) = \frac{b_2 q}{4cb_3 G_3}\left(\frac{\beta I_0(\beta)}{I_1(\beta)} - 2\right),$$

$$u,_r(1) = \frac{a_3}{a_1 a_6 - a_3^2}\left[\left(a_5 - \frac{a_2 a_6}{a_3}\right)\psi,_r(1) + \frac{q}{8}\right],$$

$$a_7 = h_1(c^2 + ch_1 + \frac{h_1^2}{3})K_1^- + h_2(c^2 + ch_2 + \frac{h_2^2}{3})K_2^- + \frac{2}{3}c^3 K_3^-. \qquad (6.73)$$

For the case when the plate edges are free, the integration constants are defined by conditions of the type (6.13), (6.73).

Let us take some examples for *local axially axisymmetric* of loads of various kinds exposure of a circular 3-layer plate. For convenience of the analytical notation we shall use a zero-order Heaviside function from the family (13.17):

$$H_0(x)=\begin{cases}1, & x\geq 0,\\ 0, & x<0.\end{cases} \tag{6.74}$$

The numerical calculation is performed for a restrained plate whose layers are composed of such materials as D-16T–fluorocarbon polymer–D-16T. The corresponding mechanical characteristics of the materials are given in Ch. 11. The value of the surface load intensity is assumed $q_0 = 30$ MPa. The geometric parameters of the plate are related to its radius $r_0$; relative thickness of the layers is: $h_1 = h_2 = 0.04$, $h_3 = 0.4$.

The load uniformly distributed over a circle $[0, b]$. A local vertical rectangular in shape surface load uniformly distributed over a circle with a relative radius $r \leq b$ affects the plate under consideration. Using function (6.74), we may write:

$$q(r)=q_0H_0(b-r),\quad p=0. \tag{6.75}$$

To transform the solution of (6.71) for the case of the load (6.75), we should make some integral operations by using preliminary Heaviside function. Thus, the integral operator of the third degree, being a part of solution (6.71), will be

$$\mathrm{L}_3^{-1}\left(H_0(b-r)\right)=\left(\frac{r^3}{16}-\frac{b^4}{16r}-\frac{b^2r}{4}\ln\left(\frac{r}{b}\right)\right)H_0(b-r)\cdot \tag{6.76}$$

The function (6.65), required for determining relative shear in the filler, will become under the load (6.75):

$$f(r)=\frac{\gamma_1}{r}\left(\int q_0H_0(b-r)r\,\mathrm{d}r-C_1\right)=\frac{\gamma_1}{r}\left(\frac{q_0\left(r^2-b^2\right)}{2}H_0(b-r)-C_1\right),$$

$$\gamma_1=\frac{b_2}{b_1b_3-b_2^2}\,.$$

If to introduce it into the integrals, which are the members of the partial solution (4.18), we obtain

$$\int I_1(\beta r)g(r)r\,\mathrm{d}r=\frac{\gamma_1q_0}{2\beta}\left(r^2I_2(\beta r)-b^2I_2(\beta b)\right)H_0(b-r)$$

$$-\frac{\gamma_1q_0b^2}{2\beta}\left(I_0(\beta r)-I_0(\beta b)\right)H_0(b-r)-\frac{C_1\gamma_1}{\beta}I_0(\beta r)$$

$$=\frac{\gamma_1q_0}{2\beta}H_0(b-r)\left[\left(r^2-b^2\right)I_0(\beta r)+\frac{2}{\beta}\left(bI_1(\beta b)-rI_1(\beta r)\right)\right]-\frac{C_1\gamma_1}{\beta}I_0(\beta r),$$

$$\int K_1(\beta r)g(r)r\,\mathrm{d}r=\frac{\gamma_1q_0}{2\beta}\left(r^2K_2(\beta r)-b^2K_2(\beta b)\right)H_0(b-r)$$

$$-\frac{\gamma_1 q_0 b^2}{2\beta}(K_0(\beta b)-K_0(\beta r))H_0(b-r)+\frac{C_1\gamma_1}{\beta}K_0(\beta r)=\frac{\gamma_1 q_0}{2\beta}H_0(b-r)$$

$$\times\left[\left(b^2-r^2\right)K_0(\beta r)+\frac{2}{\beta}(bK_1(\beta b)-rK_1(\beta r))\right]+\frac{C_1\gamma_1}{\beta}K_0(\beta r).$$

As a result, the partial solution (6.69) takes the form:

$$\psi_r=\frac{\gamma_1 q_0}{2\beta^2}H_0(b-r)\left[\frac{b^2}{r}-r+2b(K_1(\beta b)I_1(\beta r)-I_1(\beta b)K_1(\beta r))\right]+\frac{C_1\gamma_1}{\beta^2 r}. \quad (6.77)$$

If we substitute (6.76) into expressions (6.71) with due regard of (6.77), we obtain the sought displacements in the sandwich - circular plate:

$$\psi=C_2 I_1(\beta r)+C_3 K_1(\beta r)+\psi_r,$$

$$w=\frac{b_2}{b_3}\int\psi\,\mathrm{d}r+\frac{1}{b_3}\int \mathrm{L}_3^{-1}(q)\,\mathrm{d}r-\frac{C_1}{4b_3}r^2(\ln r-1)+\frac{C_5 r^2}{4b_3}+C_6\ln r+C_4,$$

$$u=\frac{a_3}{a_1}w_{,r}-\frac{a_2}{a_1}\psi+\frac{C_7 r}{2}+\frac{C_8}{r}, \quad (6.78)$$

where

$$\int \mathrm{L}_3^{-1}(q)\,\mathrm{d}r=q_0\int\left(\frac{r^3}{16}-\frac{b^4}{16r}-\frac{b^2 r}{4}\ln\left(\frac{r}{b}\right)\right)H_0(b-r)\,\mathrm{d}r=q_0\left(\frac{r^4-5b^4}{64}\right.$$

$$\left.-\frac{b^4}{16}\ln\left(\frac{r}{b}\right)-\frac{b^2r^2}{8}\ln\left(\frac{r}{b}\right)+\frac{b^2r^2}{16}\right)H_0(b-r),$$

$$\int\psi\,\mathrm{d}r=\frac{C_2 I_0(\beta r)}{\beta}-\frac{C_3 K_0(\beta r)}{\beta}+\frac{C_1\gamma_1}{\beta^2}\ln(r)+\frac{\gamma_1 q_0}{2\beta^2}H_0(b-r)$$

$$\times\left[\frac{b^2-r^2}{2}+b^2\ln\left(\frac{r}{b}\right)+\frac{2b}{\beta}(K_1(\beta b)I_0(\beta r)+I_1(\beta b)K_0(\beta r))-\frac{2}{\beta^2}\right].$$

Let us define the integration constants in the assumption that the plate is fixed and the smoothness condition of the solution in its center is obeyed. The smoothness condition implies that

$$C_1=-\frac{q_0 b^2}{2},\ C_3=\frac{q_0\gamma_1 b I_1(\beta b)}{\beta^2},\ C_6=\frac{q_0 b^4}{16b_3},\ C_8=0.$$

In the case the outer edge of the plate is fixed at $r = 1$, we should meet the boundary conditions $u=\psi=w=w_{,r}=0$. This means that

$$C_2=\frac{\gamma_1 q_0}{\beta^2 I_1(\beta)}\left(\frac{b^2}{2}-bK_1(\beta)I_1(\beta b)\right),\ C_5=\frac{q_0 b^2}{8}\left(2-b^2\right),$$

$$C_4 = -\frac{b_2 \gamma_1 q_0}{b_3 \beta^4 I_1(\beta)} \left( \frac{b^2 I_0(\beta)\beta}{2} - b I_1(\beta b) \right) + \frac{q_0 b^4}{32 b_3} + \frac{q_0 b^2}{16 b_3}, C_7 = 0 \cdot$$

Figure 6.12a, b shows the changes in the filler shear and deflection of the elastic circular sandwich plate over its radius. The curves for various spot radii of the local uniformly distributed surface load are presented: *1* – $b = 0.25$, *2* – $b = 0.5$, *3* – $b = 0.75$, *4* – $b = 1$. The maximal displacements (*4*) are observed under the load effect on the entire surface of the plate.

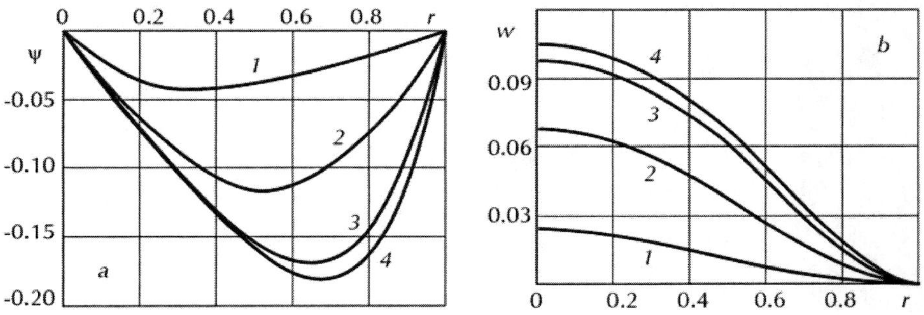


**Figure 6.12.**

Figure 6.13 shows the dependence of the maximum deflection of the circular plate with a spot radius $b$ of the local uniformly distributed surface load. As the radius increases, the deflection grows non-linearly reaching the maximum value at a load acting on the entire surface of the plate.

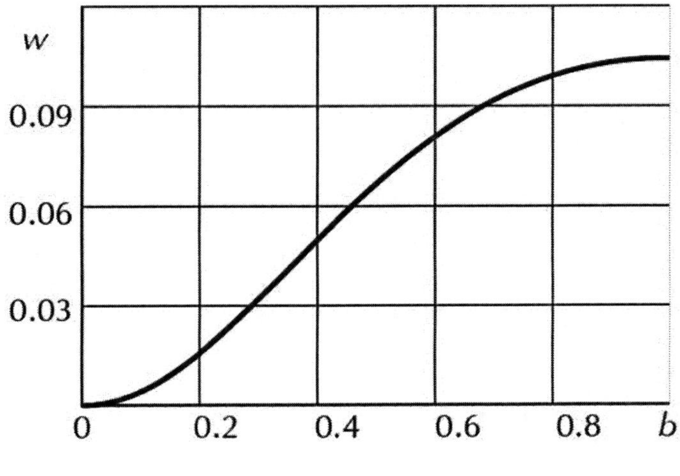


**Figure 6.13.**

The load uniformly distributed over the circle $[a, b]$. The local load uniformly distributed over the circle with a relative radius $a \le r \le b$ is acting on the plate. So, the difference between two loads (6.75) is:

$$q = q_0(H_0(b-r) - H_0(a-r)\ . \tag{6.79}$$

In view of linearity of the problem the desired solution will be obtained as a difference of two solutions (6.78) corresponding to the components of the load (6.79). As a result, the shear in the filler is

$$\begin{aligned}\psi &= C_2 I_1(\beta r) + C_3 K_1(\beta r)\\ &+ \frac{\gamma_1 q_0}{2\beta^2} H_0(b-r)\left[\frac{b^2}{r} - r + 2b\big(K_1(\beta b) I_1(\beta r) - I_1(\beta b) K_1(\beta r)\big)\right]\\ &- \frac{\gamma_1 q_0}{2\beta^2} H_0(a-r)\left[\frac{a^2}{r} - r + 2a\big(K_1(\beta a) I_1(\beta r) - I_1(\beta a) K_1(\beta r)\big)\right] + \frac{C_1\gamma_1}{\beta^2 r}.\end{aligned} \tag{6.80}$$

The deflection $w(r)$ and radial displacement $u(r)$ of the sandwich circular plate are kept formally according to (6.78). However, the integrals introduced take the following form:

$$\begin{aligned}\int \mathrm{L}_3^{-1}(q)\,\mathrm{d}r &= q_0\left[\frac{r^4 - 5b^4}{64} - \frac{b^4}{16}\ln\left(\frac{r}{b}\right) - \frac{b^2 r^2}{8}\ln\left(\frac{r}{b}\right) + \frac{b^2 r^2}{16}\right] H(b-r)\\ &- q_0\left[\frac{r^4 - 5a^4}{64} - \frac{a^4}{16}\ln\left(\frac{r}{a}\right) - \frac{a^2 r^2}{8}\ln\left(\frac{r}{a}\right) + \frac{a^2 r^2}{16}\right] H(a-r),\end{aligned}$$

$$\begin{aligned}\int \psi\,\mathrm{d}r &= \frac{C_2 I_0(\beta r)}{\beta} - \frac{C_3 K_0(\beta r)}{\beta} + \frac{C_1\gamma_1}{\beta^2}\ln(r) + \frac{\gamma_1 q_0}{2\beta^2} H_0(b-r)\\ &\times\left[\frac{b^2 - r^2}{2} + b^2\ln\left(\frac{r}{b}\right) + \frac{2b}{\beta}\big(K_1(\beta b) I_0(\beta r) + I_1(\beta b) K_0(\beta r)\big) - \frac{2}{\beta^2}\right] - \frac{\gamma_1 q_0}{2\beta^2}\\ &\times H_0(a-r)\left[\frac{a^2 - r^2}{2} + a^2\ln\left(\frac{r}{a}\right) + \frac{2a}{\beta}\big(K_1(\beta a) I_0(\beta r) + I_1(\beta a) K_0(\beta r)\big) - \frac{2}{\beta^2}\right].\end{aligned}$$

The integration constants for the restrained plate are presented as:

$$C_1 = -\frac{q_0(b^2 - a^2)}{2},\ C_3 = \frac{q_0\gamma_1}{\beta^2}\big(bI_1(\beta b) - aI_1(\beta a)\big),$$

$$C_2 = \frac{\gamma_1 q_0}{\beta^2 I_1(\beta)}\left(\frac{b^2 - a^2}{2} - bK_1(\beta) I_1(\beta b) + aK_1(\beta) I_1(\beta a)\right),$$

$$C_4 = -\frac{b_2\gamma_1 q_0}{b_3\beta^4 I_1(\beta)}\left(\frac{I_0(\beta)\beta\left(b^2 - a^2\right)}{2} - bI_1(\beta b) + aI_1(\beta a)\right)$$

$$+\frac{q_0\left(b^4-a^4\right)}{32b_3}+\frac{q_0\left(b^2-a^2\right)}{16b_3}, C_5=\frac{q_0}{8}\left(2b^2-b^4-2a^2+a^4\right),$$

$$C_6=\frac{q_0}{16b_3}\left(b^4-a^4\right), C_7=\frac{a_3q_0}{4b_3a_1}\left(b^2-a^2\right), C_8=0.$$

Figure 6.14 shows the changes in the plate deflection as dependent on the loading spot movement towards the edge. Thickness the circle is assumed to be $d = b - a = 0.25$. At $a = 0$ the load is distributed over the circle with radius $b$, while at $a = 0.75$ the loading circle approaches the plate edge. The maximum deflection is observed round the circle domain with $a = 0.25$.

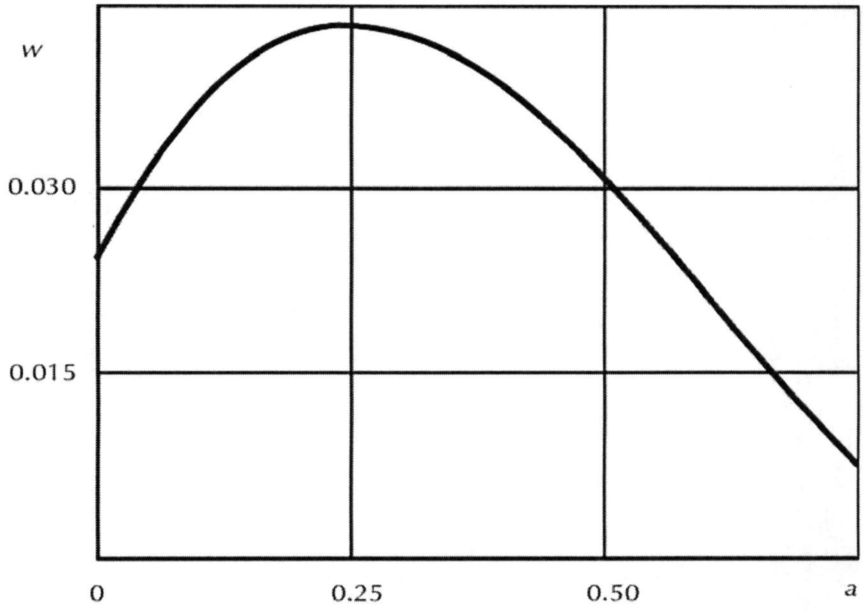


**Figure 6.14.**

Shear force. The linear shear load $Q(r)$ of a constant intensity $Q_0$ applied over the circle $r = a$ is acting on a circular sandwich plate. Its closed form may be assumed as:

$$Q(r)=Q_0H_0(r-a)H_0(a-r). \tag{6.81}$$

We shall find the solution using the results obtained for a uniformly distributed surface load $q_0$, distributed over a circle $a - \xi \leq r \leq a + \xi$. Let us assume that the circle is thin ($\xi < 1$), make a substitution of $q_0 = Q_0/(2\xi)$ into function (6.80) and let parameter $\xi$ tend towards zero, keeping value $Q_0$ constant. For a relative shear, we obtain the following expression:

$$\psi=\lim_{\xi\to 0}\left[C_2 I_1(\beta r)+C_3 K_1(\beta r)+\frac{\gamma_1 Q_0}{4\xi\beta^2}H_0(a+\xi-r)\right.$$

$$\times\left[\frac{(a+\xi)^2}{r}-r+2(a+\xi)\big(K_1(\beta(a+\xi))I_1(\beta r)-I_1(\beta(a+\xi))K_1(\beta r)\big)\right]$$

$$-\frac{\gamma_1 Q_0}{4\xi\beta^2}H_0(a-\xi-r)\left[\frac{(a-\xi)^2}{r}-r+2(a-\xi)\big(K_1(\beta(a-\xi))I_1(\beta r)\right.$$

$$\left.\left.-I_1(\beta(a-\xi))K_1(\beta r)\big)\right]+\frac{C_1\gamma_1}{\beta^2 r}\right].$$

After exhausting the limit, the solution of the circular sandwich-type plate deflection under the influence of a linear load will be:

$$\psi=C_2 I_1(\beta r)+C_3 K_1(\beta r)$$

$$+\frac{\gamma_1 Q_0 a}{\beta}H_0(a-r)\left(\frac{1}{\beta r}-K_1(\beta r)I_0(\beta a)-I_1(\beta r)K_0(\beta a)\right)+\frac{C_1\gamma_1}{\beta^2 r},$$

$$w=\frac{b_2}{b_3}\int\psi\,\mathrm{d}r+\frac{1}{b_3}\int \mathrm{L}_3^{-1}(Q_0)\,\mathrm{d}r-\frac{C_1}{4b_3}r^2(\ln r-1)+\frac{C_5 r^2}{4b_3}+C_6\ln r+C_4,$$

$$u=\frac{a_3}{a_1}w_{,r}-\frac{a_2}{a_1}\psi+\frac{C_7 r}{2}+\frac{C_8}{r},\qquad (6.82)$$

where

$$\int \mathrm{L}_3^{-1}(Q_0)\,\mathrm{d}r=\lim_{\xi\to 0}\left[\frac{Q_0}{2\xi}\left(\frac{r^4-5(a+\xi)^4}{64}-\frac{(a+\xi)^4}{16}\ln\left(\frac{r}{a+\xi}\right)\right.\right.$$

$$\left.-\frac{(a+\xi)^2 r^2}{8}\ln\left(\frac{r}{a+\xi}\right)+\frac{(a+\xi)^2 r^2}{16}\right)H(a+\xi-r)$$

$$-\left(\frac{r^4-5(a-\xi)^4}{64}-\frac{(a-\xi)^4}{16}\ln\left(\frac{r}{a-\xi}\right)-\frac{(a-\xi)^2 r^2}{8}\ln\left(\frac{r}{a-\xi}\right)+\frac{(a-\xi)^2 r^2}{16}\right)$$

$$\left.\times H_0(a-\xi-r)\big)\right]=\frac{Q_0 a}{4}\left(r^2-a^2-a^2\ln\left(\frac{r}{a}\right)-r^2\ln\left(\frac{r}{a}\right)\right)H_0(a-r),$$

$$\int\psi\,\mathrm{d}r=\lim_{\xi\to 0}\left(\frac{C_2 I_0(\beta r)}{\beta}-\frac{C_3 K_0(\beta r)}{\beta}+\frac{C_1\gamma_1}{\beta^2}\ln(r)+\frac{\gamma_1 Q_0}{4\xi\beta^2}H_0(a+\xi-r)\right.$$

$$\times\left(\frac{(a+\xi)^2-r^2}{2}+(a+\xi)^2\ln\left(\frac{r}{a+\xi}\right)+\frac{2(a+\xi)}{\beta}\big(K_1(\beta(a+\xi))I_0(\beta r)\right.$$

$$\left.+I_1(\beta(a+\xi))K_0(\beta r)\big)-\frac{2}{\beta^2}\right)-\frac{\gamma_1 Q_0}{4\xi\beta^2}H_0(a-\xi-r)\left(\frac{(a-\xi)^2-r^2}{2}+(a-\xi)^2\right.$$

$$\left.\left.\times\ln\left(\frac{r}{a-\xi}\right)+\frac{2(a-\xi)}{\beta}\big(K_1(\beta(a-\xi))I_0(\beta r)+I_1(\beta(a-\xi))K_0(\beta r)\big)-\frac{2}{\beta^2}\right)\right)$$

$$=\frac{C_2 I_0(\beta r)}{\beta}-\frac{C_3 K_0(\beta r)}{\beta}+\frac{C_1\gamma_1}{\beta^2}\ln(r)+\frac{\gamma_1 Q_0 a}{\beta^2}H_0(a-r)$$

$$\times\left(\ln\left(\frac{r}{a}\right)+I_0(\beta a)K_0(\beta r)-K_0(\beta a)I_0(\beta r)\right).$$

The integration constants in case of restrained plate are:

$$C_1=-Q_0 a,\, C_2=\frac{\gamma_1 Q_0 a}{\beta^2 I_1(\beta)}(1-\beta K_1(\beta)I_0(\beta a)),\, C_3=\frac{\gamma_1 Q_0 a I_0(\beta a)}{\beta},$$

$$C_4=-\frac{b_2\gamma_1 Q_0 a}{b_3\beta^3 I_1(\beta)}(I_0(\beta)-I_0(\beta a))+\frac{Q_0 a(a^2+1)}{8b_3},\, C_5=\frac{Q_0 a}{2}\left(1-a^2\right),$$

$$C_6=\frac{Q_0 a^3}{4b_3},\, C_7=0, C_8=0\,.$$

Figure 6.15a, b shows the changes in the shear of the filler and the circular sandwich-type plate deflection over its radius. The curves are plotted in accordance with formulas (6.82) at various radii of the loaded circle (6.81): *1* – $a$ = 0.25, *2* – $a$ = 0.5, *3* – $a$ = 0.75. The intensity of the linear force has been assumed $Q_0 = 3\cdot 10^6$ N/m. The displacements reach the maximal values when the loaded circle radius equals to a half-radius of the plate.

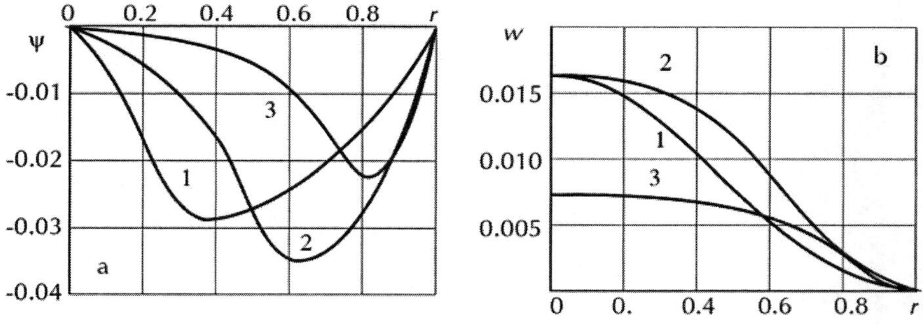


**Figure 6.15.**

Variations of the deflection in the plate center in response to the loaded-circle radius are shown in Fig. 6.16. The deflection shown in Fig. 6.16a corresponds to solution (6.82). The maximum value is observed in the domain $b$ = 0.38. Further motion of the load towards the edge increases its resultant, while the deflection decreases down to zero.

Formulas (6.82) do not reflect deformations of the plate by a concentrated load acting in its center, since the circle degenerates at $b$ = 0 and the resultant of the linear force becomes zero. To avoid this, let us assume that a specified resultant $Q = 2\pi a Q_0$ remains constant at a changing radius of the circle onto which it has been applied. This may be true if intensity $Q_0$ varies and compensates changes in $a$.

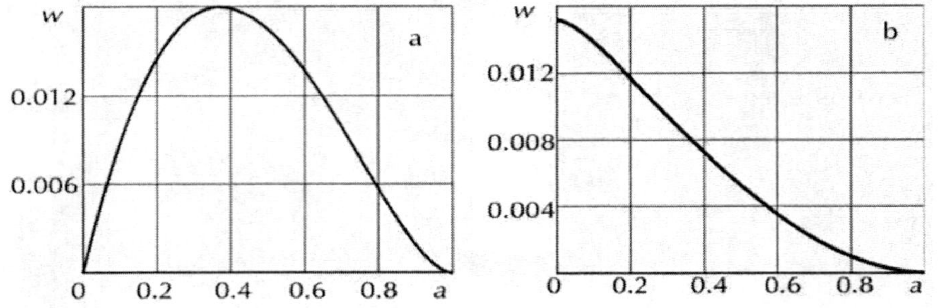


**Figure 6.16.**

To obtain a corresponding solution let us make a substitution of $Q_0 = 2\pi a/Q$ in (6.82). Then we obtain the following displacements:

$$\psi = C_2 I_1(\beta r) + C_3 K_1(\beta r) +$$

$$+\frac{\gamma_1 Q}{2\pi\beta} H_0(a-r)\left(\frac{1}{\beta r} - K_1(\beta r) I_0(\beta a) - I_1(\beta r) K_0(\beta a)\right) + \frac{C_1\gamma_1}{\beta^2 r},$$

$$w = \frac{b_2}{b_3}\int \psi\,dr + \frac{1}{b_3}\int \mathrm{L}_3^{-1}(Q)\,dr - \frac{C_1}{4b_3} r^2(\ln r - 1) + \frac{C_5 r^2}{4b_3} + C_6 \ln r + C_4,$$

$$u = \frac{a_3}{a_1} w_{,r} - \frac{a_2}{a_1}\psi + \frac{C_7 r}{2} + \frac{C_8}{r}, \tag{6.83}$$

where

$$\int \mathrm{L}_3^{-1}(Q)\,dr = \frac{Q}{8\pi}\left(r^2 - a^2 - (r^2 + a^2)\ln\left(\frac{r}{a}\right)\right) H_0(a-r),$$

$$\int \psi\,dr = \frac{C_2 I_0(\beta r)}{\beta} - \frac{C_3 K_0(\beta r)}{\beta} + \frac{C_1\gamma_1}{\beta^2}\ln(r) + \frac{\gamma_1 Q}{2\pi\beta^2} H_0(a-r)$$

$$\times\left(\ln\left(\frac{r}{a}\right) + I_0(\beta a) K_0(\beta r) - K_0(\beta a) I_0(\beta r)\right).$$

The integration constants in the case of a restrained plate are as follows:

$$C_1 = -\frac{Q}{2\pi},\; C_2 = \frac{\gamma_1 Q}{2\pi\beta^2 I_1(\beta)}\left(1 - \beta K_1(\beta) I_0(\beta a)\right),\; C_3 = \frac{\gamma_1 Q I_0(\beta a)}{2\pi\beta},$$

$$C_4 = -\frac{b_2\gamma_1 Q}{2\pi b_3 \beta^3 I_1(\beta)}\left(I_0(\beta) - I_0(\beta a)\right) + \frac{Q(a^2+1)}{16\pi b_3},\; C_5 = \frac{Q}{4}\left(1 - a^2\right),$$

$$C_6 = \frac{Qa^2}{8\pi b_3},\; C_7 = 0,\; C_8 = 0.$$

The changes in the maximum deflection (6.83) of the plate as a function of the radius of the circle, round which the shearing linear force is applied with a constant resultant, are shown in Fig. 5.16b. The maximum is observed here at $b = 0$, which corresponds to the concentrated force applied in the center of the plate. Under a load moving towards the edge, the deflection decreases down to zero.

**The moment.** A linear moment of intensity $M_0$ = const distributed over a circle with radius $r = a$ is acting on the plate under study:

$$m(r)=M_0H_0(r-a)H_0(a-r)\cdot \tag{6.84}$$

Let us find a solution of this problem using the difference of solutions (6.82) for two identical linear shear forces acting on closely approximating circles with radii $r = a - \xi$ and $r = a + \xi$. In this difference, we substitute $Q_0 = M_0/(2\xi)$ and assume $c$ as tending to zero. Then,

$$\psi=\lim_{\xi\to 0}\left[C_2I_1(\beta r)+C_3K_1(\beta r)+\frac{\gamma_1M_0a}{2\xi\beta}H_0(a+\xi-r)\left[\frac{1}{\beta r}-(K_0(\beta(a+\xi))\right.\right.$$

$$\left.\times I_1(\beta r)+I_0(\beta(a+\xi))K_1(\beta r))\right]-\frac{\gamma_1M_0a}{2\xi\beta}H_0(a-\xi-r)$$

$$\left.\times\left[\frac{1}{\beta r}-\Big(K_0(\beta(a-\xi))I_1(\beta r)+I_0(\beta(a-\xi))K_1(\beta r)\Big)\right]+\frac{C_1\gamma_1}{\beta^2r}\right].$$

When calculating the limit, we come to

$$\psi=C_2I_1(\beta r)+C_3K_1(\beta r)+\frac{\gamma_1M_0}{\beta}H_0(a-r)\left(\frac{1}{\beta r}-K_1(\beta r)(I_0(\beta a)\right.$$

$$\left.+I_1(\beta a)\beta a)-I_1(\beta r)(K_0(\beta a)-K_1(\beta a)\beta a)\right)+\frac{C_1\gamma_1}{\beta^2r},$$

$$w=\frac{b_2}{b_3}\int\psi\,\mathrm{d}r+\frac{1}{b_3}\int \mathrm{L}_3^{-1}(M_0)\,\mathrm{d}r-\frac{C_1}{4b_3}r^2(\ln r-1)+\frac{C_5r^2}{4b_3}+C_6\ln r+C_4,$$

$$u=\frac{a_3}{a_1}w_{,r}-\frac{a_2}{a_1}\psi+\frac{C_7r}{2}+\frac{C_8}{r}, \tag{6.85}$$

where

$$\int \mathrm{L}_3^{-1}(M_0)\,\mathrm{d}r=\lim_{\xi\to 0}\left[\frac{M_0}{8\xi}\left(r^2(a+\xi)-(a+\xi)^3-(a+\xi)^3\ln\left(\frac{r}{a+\xi}\right)\right.\right.$$

$$\left.\left.-r^2(a+\xi)\ln\left(\frac{r}{a+\xi}\right)\right)H(a+\xi-r)-\frac{M_0}{8\xi}\right[r^2(a-\xi)-(a-\xi)^3$$

$$\left.\left.-(a-\xi)^3\ln\left(\frac{r}{a-\xi}\right)-r^2(a-\xi)\ln\left(\frac{r}{a-\xi}\right)\right]H_0(a-\xi-r)\right]$$

$$=\frac{M_0}{4}\left(2r^2-2a^2-3a^2\ln\left(\frac{r}{a}\right)-r^2\ln\left(\frac{r}{a}\right)\right)H_0(a-r),$$

$$\int\psi\,\mathrm{d}r=\lim_{\xi\to 0}\left(\frac{C_2I_0(\beta r)}{\beta}-\frac{C_3K_0(\beta r)}{\beta}+\frac{C_1\gamma_1}{\beta^2}\ln(r)+\frac{\gamma_1M_0(a+\xi)}{2\xi\beta^2}H_0(a+\xi-r)\right.$$

$$\times\left(\ln\left(\frac{r}{a+\xi}\right)+\left(I_0(\beta(a+\xi))K_0(\beta r)-K_0(\beta(a+\xi))I_0(\beta r)\right)\right)-\frac{\gamma_1M_0(a-\xi)}{2\xi\beta^2}$$

$$\left.\times H_0(a-\xi-r)\left(\ln\left(\frac{r}{a+\xi}\right)+\left(I_0(\beta(a+\xi))K_0(\beta r)-K_0(\beta(a+\xi))I_0(\beta r)\right)\right)\right)$$

$$=\frac{C_2I_0(\beta r)}{\beta}-\frac{C_3K_0(\beta r)}{\beta}+\frac{C_1\gamma_1}{\beta^2}\ln(r)+\frac{\gamma_1M_0}{\beta^2}H_0(a-r)$$

$$\times\left(\ln\left(\frac{r}{a}\right)-1+K_0(\beta r)\left(I_0(\beta a)+I_1(\beta a)\beta a\right)-I_0(\beta r)\left(K_0(\beta a)-K_1(\beta a)\beta a\right)\right).$$

The integration constants in the case of a fixed plate are:

$$C_1=-M_0,\ C_2=\frac{\gamma_1M_0}{\beta^2I_1(\beta)}\left(1-\beta K_1(\beta)\left(I_0(\beta a)+I_1(\beta a)\beta a\right)\right),$$

$$C_3=\frac{\gamma_1M_0}{\beta}\left(I_0(\beta a)+I_1(\beta a)\beta a\right),$$

$$C_4=-\frac{b_2\gamma_1M_0}{b_3\beta^3I_1(\beta)}\left(I_0(\beta)-\left(I_0(\beta a)-I_1(\beta a)\beta a\right)\right)+\frac{3M_0a^2}{8b_3}+\frac{M_0}{8b_3},$$

$$C_5=\frac{M_0}{2}\left(1-3a^2\right),\ C_6=\frac{3M_0a^2}{4b_3},\ C_7=0,\ C_8=0.$$

Functions (6.85) describe formally a case of the effect of a concentrated moment applied in the plate center, since the solution does not degenerate at $a = 0$.

Figure 6.17a, b shows a shear variation in the filler and in the elastic circular sandwich plate deflection over its radius. The curves are plotted proceeding from formulas (6.85) at various radii of a torque circle (6.84): *1* – $a = 0.25$, *2* – $a = 0.5$, *3* – $a = 0.75$, *4* – $a = 0.99$.

The intensity of the linear moments was assumed $M_0 = 3\cdot 10^5$ N. As the torque circle moves towards the plate edge the deflection decreases in magnitude and changes its sign. The shear maxima presented by the peaks coincide with the radii of applied linear moments.

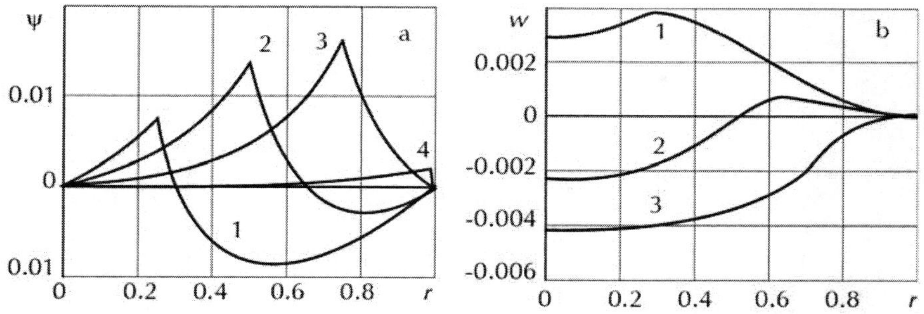


**Figure 6.17.**

The changes in the central deflection of the plate as dependent on the torque circle radius are shown in Fig. 6.18. The deflection shown in Fig. 6.18a agrees with solution (6.85). The maximum here is observed at $a = 0$. Further movement of the load towards the edge results in the deflection approach towards zero.

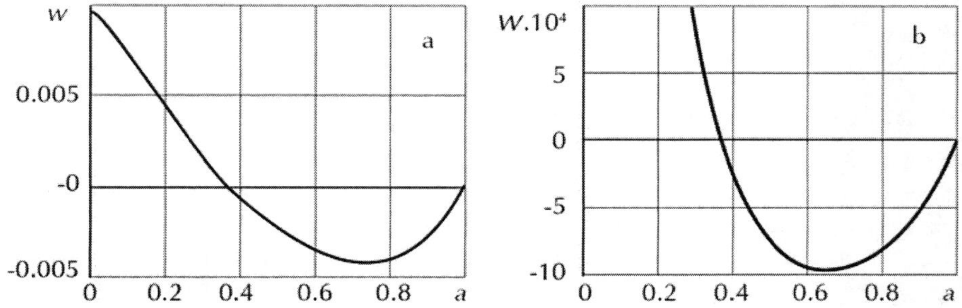


**Figure 6.18.**

We shall accept the resultants of the linear moments $M = 2\pi a M_0$ to be constant at any radii of the circle they are acting on. The intensity $M_0$ is variable and compensates variations in $a$. So, based on (6.85), and making substitution of $M_0 = M/(2\pi a)$, we obtain an expression for the sought functions:

$$\psi = C_2 I_1(\beta r) + C_3 K_1(\beta r) + \frac{C_1 \gamma_1}{\beta^2 r} + \frac{\gamma_1 M}{2\pi a \beta} H_0(a-r)$$

$$\times\left(\frac{1}{\beta r} - K_1(\beta r)(I_0(\beta a) + I_1(\beta a)\beta a) - I_1(\beta r)(K_0(\beta a) - K_1(\beta a)\beta a)\right),$$

$$w = \frac{b_2}{b_3}\int \psi \, \mathrm{d}r + \frac{1}{b_3}\int \mathrm{L}_3^{-1}(M)\,\mathrm{d}r - \frac{C_1}{4b_3} r^2(\ln r - 1) + \frac{C_5 r^2}{4b_3} + C_6 \ln r + C_4,$$

$$u = \frac{a_3}{a_1} w,_r - \frac{a_2}{a_1}\psi + \frac{C_7 r}{2} + \frac{C_8}{r}, \qquad (6.86)$$

where

$$\int \mathrm{L}_3^{-1}(M_0)\,\mathrm{d}r = \frac{M}{8\pi a}\left(2r^2 - 2a^2 - 3a^2\ln\left(\frac{r}{a}\right) - r^2\ln\left(\frac{r}{a}\right)\right)H_0(a-r),$$

$$\int \psi\,\mathrm{d}r = \frac{C_2 I_0(\beta r)}{\beta} - \frac{C_3 K_0(\beta r)}{\beta} + \frac{C_1\gamma_1}{\beta^2}\ln(r) + \frac{\gamma_1 M}{2\pi a\beta^2}H_0(a-r)$$

$$\times\left(\ln\left(\frac{r}{a}\right) - 1 + K_0(\beta r)\left(I_0(\beta a) + I_1(\beta a)\beta a\right) - I_0(\beta r)\left(K_0(\beta a) - K_1(\beta a)\beta a\right)\right).$$

The integration constants for the case of the restrained plate are as follows:

$$C_1 = -Q_0 a,\ C_2 = \frac{\gamma_1 Q_0 a}{\beta^2 I_1(\beta)}(1 - \beta K_1(\beta) I_0(\beta a)),\ C_3 = \frac{\gamma_1 Q_0 a I_0(\beta a)}{\beta},$$

$$C_4 = -\frac{b_2\gamma_1 Q_0 a}{b_3\beta^3 I_1(\beta)}\left(I_0(\beta) - I_0(\beta a)\right) + \frac{Q_0 a(a^2+1)}{8b_3},\ C_5 = \frac{Q_0 a}{2}\left(1 - a^2\right),$$

$$C_6 = \frac{Q_0 a^3}{4b_3},\ C_7 = 0,\ C_8 = 0.$$

Variations in the deflection (6.86) in the plate center as a function of to the circle radius onto which a linear moment load is applied with a constant resultant is shown in Fig. 6.18b. The extremum is observed here at $b = 0.65$. Motion of the load to the plate center causes a deflection unbounded in magnitude.

## AN ANNULAR THREE-LAYER PLATE ON AN ELASTIC FOUNDATION

The statement and solution of the problem are made in cylindrical coordinates $r$, $\varphi$, and $z$. Kirchhoff's hypotheses are assumed for the isotropic load-carrying layers of $h_1$ and $h_2$ thicknesses. The filler incompressible across thickness ($h_3 = 2c$) is light, that is, we can neglect the work of stresses $\sigma_{rz}$ in the tangential direction. A local vertical load $q_0(r)$ operates on the external surface of the plate (Fig. 6.19). The displacements at the layer interfaces are continuous. On the external and internal contours of the plate, having radii $r_0$ and $r_1$, there are rigid diaphragms to prevent relative shifts of the layers.

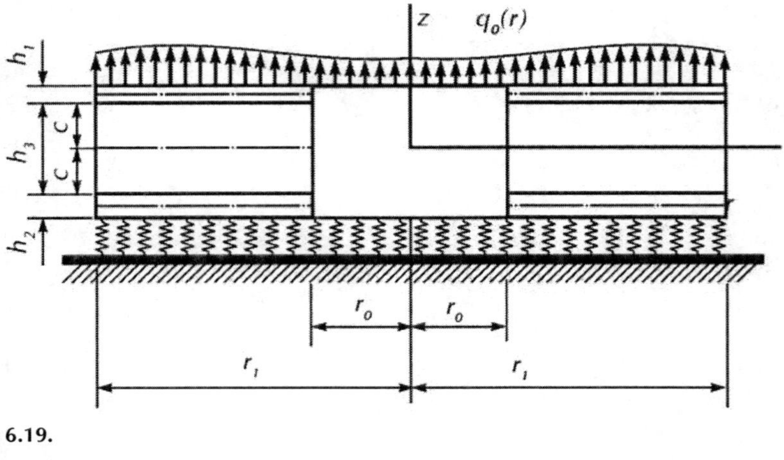


**Figure 6.19.**

Due to symmetry of the load, the tangential displacements in the layers are absent, $u_{\varphi}^{(k)} = 0$ ($k$ is the layer number), and deflection $w(r)$ of the plate, relative shift $\psi(r)$ in the filler, and radial displacement $u(r)$ of the midsurface are independent of coordinate $\varphi$. In what follows, these functions are regarded as the sought-for ones hence, the displacements and strains in the layers follow from equations (6.48), (6.49), so a system of differential equations in forces, which describes equilibrium of a circular sandwich plate on an elastic foundation, is

$$T_{r},_{r}+\frac{1}{r}(T_{r}-T_{\varphi})=-p\,,\quad H_{r},_{r}+\frac{1}{r}(H_{r}-H_{\varphi})=0\,,$$

$$M_{r},_{r}+\frac{1}{r}(2M_{r},_{r}-M_{\varphi},_{r})=-q_{0}+q_{R}\,. \qquad (6.87)$$

The conditions of forces on the plate contours ($r = r_0$ and $r = r_1$)

$$T_{r}=T_{r}^{n},\ H_{r}=H_{r}^{n},\ M_{r}=M_{r}^{n},\ M_{r},_{r}+\frac{1}{r}(M_{r}-M_{\varphi})=Q^{n}$$

must be fulfilled.

The relation between the reaction of the foundation and the deflection of the plate is assumed based on the Winkler model, according to which

$$q_{R}=\kappa_{0}w\,, \qquad (6.88)$$

where $\kappa_0$ is the coefficient of rigidity of the elastic foundation (a modulus of subgrade reaction).

After substitution of equations (6.59)–(6.61), and (6.87) into equation (6.88), we come to a system of differential equations in displacements, which describes bending of an annular sandwich plate with light filler on an elastic foundation:

$$\mathrm{L}_{2}(a_{1}u+a_{2}\psi-a_{3}w,_{r})=0\,,\quad \mathrm{L}_{2}(a_{2}u+a_{4}\psi-a_{5}w,_{r})=0\,,$$

$$\mathrm{L}_{3}(a_{3}u+a_{5}\psi-a_{6}w,_{r})-\kappa_{0}w=-q_{0}\,, \qquad (6.89)$$

where $q_0$ is the intensity of the external distributed load, coefficients and operations are defined by expressions (6.63).

The problem of searching for the functions $u(r)$, $\psi(r)$, and $w(r)$ is closed by supplementing equation (6.89) with boundary conditions. For rigidly fixed contours of the plate ($r = r_0, r_1$) we have,

$$u=\psi=w=w,_{r}=0\,; \qquad (6.90)$$

for the hinge-supported contours ($r = r_0, r_1$),

$$u=\psi=w=M_{r}=0\,. \qquad (6.91)$$

System (6.89) can be solved in the form

$$u=b_{1}w_{r}+C_{1}r+C_{2}/r,\ \psi=b_{2}w_{r}+C_{3}r+C_{4}/r\,,$$

$$w=C_{5}\,\mathrm{ber}(\kappa r)+C_{6}\,\mathrm{bei}(\kappa r)+C_{7}\,\mathrm{ker}(\kappa r)+C_{8}\,\mathrm{kei}(\kappa r)+w_{0}\,, \qquad (6.92)$$

where $C_1, C_2, \ldots, C_8$ are the constants of integration; $\kappa^4 = \kappa_0 D$, $q = q_0 D$,

$$D = \frac{a_1(a_1a_4 - a_2^2)}{(a_1a_6 - a_3^2)(a_1a_4 - a_2^2) - (a_1a_5 - a_2a_3)^2},$$

$$b_1 = \frac{a_3a_4 - a_2a_5}{a_1a_4 - a_2^2}, \quad b_2 = \frac{a_1a_5 - a_2a_3}{a_1a_4 - a_2^2}.$$

In relations (6.92), Kelvin's functions (13.28)–(13.30)

$$\psi_n(\kappa r) = \mathrm{ber}(\kappa r), \mathrm{bei}(\kappa r), \mathrm{ker}(\kappa r), \mathrm{kei}(\kappa r)$$

form a fundamental system of solutions; a particular solution $w_0$ can be obtained by using Cauchy kernel $K(r, s)$:

$$w_0(r) = \int_{r_0}^{r} K(r,s)q(s)\mathrm{d}s, \tag{6.93}$$

$$K(r,s) = C_1(s)\psi_1(r) + C_2(s)\psi_2(r) + C_3(s)\psi_3(r) + C_4(s)\psi_4(r).$$

Here, the functions $C_n(s)$ are determined by the ratios

$$C_1(s) = \frac{W_1(s)}{W(s)}, \quad C_2(s) = \frac{W_2(s)}{W(s)}, \quad C_3(s) = \frac{W_3(s)}{W(s)}, \quad C_4(s) = \frac{W_4(s)}{W(s)},$$

where

$$W(r) = \begin{vmatrix} \psi_1(r) & \psi_2(r) & \psi_3(r) & \psi_4(r) \\ \psi_1'(r) & \psi_2'(r) & \psi_3'(r) & \psi_4'(r) \\ \psi_1''(r) & \psi_2''(r) & \psi_3''(r) & \psi_4''(r) \\ \psi_1'''(r) & \psi_2'''(r) & \psi_3'''(r) & \psi_4'''(r) \end{vmatrix}, \quad W_1(r) = \begin{vmatrix} 0 & \psi_2(r) & \psi_3(r) & \psi_4(r) \\ 0 & \psi_2'(r) & \psi_3'(r) & \psi_4'(r) \\ 0 & \psi_2''(r) & \psi_3''(r) & \psi_4''(r) \\ 1 & \psi_2'''(r) & \psi_3'''(r) & \psi_4'''(r) \end{vmatrix},$$

$$W_2(r) = \begin{vmatrix} \psi_1(r) & 0 & \psi_3(r) & \psi_4(r) \\ \psi_1'(r) & 0 & \psi_3'(r) & \psi_4'(r) \\ \psi_1''(r) & 0 & \psi_3''(r) & \psi_4''(r) \\ \psi_1'''(r) & 1 & \psi_3'''(r) & \psi_4'''(r) \end{vmatrix}, \quad W_3(r) = \begin{vmatrix} \psi_1(r) & \psi_2(r) & 0 & \psi_4(r) \\ \psi_1'(r) & \psi_2'(r) & 0 & \psi_4'(r) \\ \psi_1''(r) & \psi_2''(r) & 0 & \psi_4''(r) \\ \psi_1'''(r) & \psi_2'''(r) & 1 & \psi_4'''(r) \end{vmatrix},$$

$$W_4(r) = \begin{vmatrix} \psi_1(r) & \psi_2(r) & \psi_3(r) & 0 \\ \psi_1'(r) & \psi_2'(r) & \psi_3'(r) & 0 \\ \psi_1''(r) & \psi_2''(r) & \psi_3''(r) & 0 \\ \psi_1'''(r) & \psi_2'''(r) & \psi_3'''(r) & 1 \end{vmatrix}$$

the primes mean differentiation with respect to $r$.

At a *rigid fixation of the boundary contours* of the plate, solution (6.92) must be substituted into equation (6.90). As a result, with account of the fact that the derivatives of deflections on the boundary contours of the plate are equal to zero and $w_0(r_0)=w_0'(r_0)=0$ according to equation (6.93), we come to a linear system of eight algebraic equations ($q_0$ = const):

$$C_1 r_1 + C_2 / r_1 = 0,$$

$$C_1 r_0 + C_2 / r_0 = 0,$$

$$C_3 r_1 + C_4 / r_1 = 0,$$

$$C_3 r_0 + C_4 / r_0 = 0,$$

$$C_5 ber(\kappa r_1) + C_6 bei(\kappa r_1) + C_7 \ker(\kappa r_1) + C_8 kei(\kappa r_1) = -q_0 / \kappa_0,$$

$$C_5 ber(\kappa r_0) + C_6 bei(\kappa r_0) + C_7 \ker(\kappa r_0) + C_8 kei(\kappa r_0) = -q_0 / \kappa_0,$$

$$b_3 C_5 + b_4 C_6 + b_{30} C_7 + b_{40} C_8 = 0,$$

$$b_{31} C_5 + b_{41} C_6 + b_{32} C_7 + b_{42} C_8 = 0, \qquad (6.94)$$

where

$$b_3 = \frac{\kappa\sqrt{2}}{2}[\mathrm{ber}_1 \kappa + \mathrm{bei}_1 \kappa], b_4 = \frac{\kappa\sqrt{2}}{2}[-\mathrm{ber}_1 \kappa + \mathrm{bei}_1 \kappa],$$

$$b_{30} = \frac{\kappa\sqrt{2}}{2}[\mathrm{ker}_1 \kappa + \mathrm{kei}_1 \kappa],$$

$$b_{40} = \frac{\kappa\sqrt{2}}{2}[-\mathrm{ker}_1 \kappa + \mathrm{kei}_1 \kappa], b_{31} = \frac{\kappa\sqrt{2}}{2}[\mathrm{ber}_1(\kappa r_0) + \mathrm{bei}_1(\kappa r_0)],$$

$$b_{41} = \frac{\kappa\sqrt{2}}{2}[-\mathrm{ber}_1(\kappa r_0) + \mathrm{bei}_1(\kappa r_0)],$$

$$b_{32} = \frac{\kappa\sqrt{2}}{2}[\mathrm{ker}_1(\kappa r_0) + \mathrm{kei}_1(\kappa r_0)], b_{42} = \frac{\kappa\sqrt{2}}{2}[-\mathrm{ker}_1(\kappa r_0) + \mathrm{kei}_1(\kappa r_0)].$$

System (6.94) is solved like

$$C_1 = C_2 = C_3 = C_4 = 0, \quad C_5 = \frac{\Delta_5}{\Delta},$$

$$C_6 = \frac{\Delta_6}{\Delta}, \quad C_7 = \frac{\Delta_7}{\Delta}, \quad C_8 = \frac{\Delta_8}{\Delta}, \qquad (6.95)$$

where

$$\Delta = \begin{vmatrix} \mathrm{ber}(\kappa r_1) & \mathrm{bei}(\kappa r_1) & \mathrm{ker}(\kappa r_1) & \mathrm{kei}(\kappa r_1) \\ \mathrm{ber}(\kappa r_0) & \mathrm{bei}(\kappa r_0) & \mathrm{ker}(\kappa r_0) & \mathrm{kei}(\kappa r_0) \\ b_3 & b_4 & b_{30} & b_{40} \\ b_{31} & b_{41} & b_{32} & b_{42} \end{vmatrix},$$

is a determinant of the system, and the remaining determinants are found by replacing the corresponding columns with the column of free terms of the last four equations of the system.

If *both plate contours are hinge-supported*, solution (6.92) must be substituted into equation (6.91). In this case, $w_0''(r_0)=0$. As a result, we obtain a system of eight linear algebraic equations for determining the integration constants.

Numerical investigations of solution (6.92)–(6.95) were carried out for a plate made of D16T-fluoroplastic-D16T. The corresponding mechanical characteristics of the materials used are presented in (Starovoitov, 2004). The relative geometrical parameters of the plate referred to its radius $r_1$, are as follows: $r_0 = 0.2$; $r_1 = 1$; $h_1 = h_2 = 0.04$ and $h_3 = 0.4$. The local distributed load is specified by using the Heaviside $H_0(x)$ and Dirac delta $\delta(x)$ functions.

*A uniformly distributed load* $q_0$ = const. In this case, the particular solution has the form $w_0 = q_0/\kappa_0$. The data in Fig. 6.20 illustrate the maximum deflection of the annular plate in relation to the coefficient of rigidity $\kappa_0$ of the foundation. For the foundations with a low-rigidity ($\kappa_0 \leq 10$ MPa/m), the influence of rigidity on the deflection is insignificant. For the medium-rigidity foundations, $\kappa_0 = 10$–1000 MPa/m), the deflection decreases sharply. For the foundations of high rigidity, $\kappa_0 > 1000$ MPa/m), the deflection is small and stable. The curves are calculated at different intensities of the distributed load $q_0$ (MPa): *1* – $q_0 = -10$, *2* – $q_0 = -7$, *3* – $q_0 = -4$.

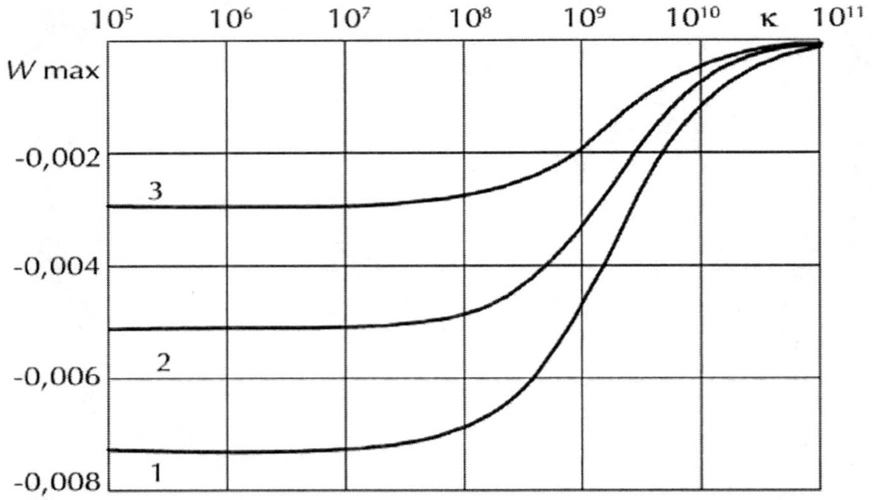


**Figure 6.20.**

*A local load distributed uniformly* over a force ring $a \leq r \leq b$ $(r_0 \leq a,\ b \leq 1)$:

$$q = q_0\left[H_0(b-r) - H_0(a-r)\right]. \tag{6.96}$$

In this case, according to equation (6.93), a particular solution and its values on the boundary contours of the plate under load (6.96) have the form

$$w_0(r) = Dq_0 \int_{r_0}^{r} K(r,s)\left[H_0(b-r) - H_0(a-r)\right] \mathrm{d}s,$$

$$w_0(r_1) = Dq_0 \int_{a}^{b} K(r_1,s)\,\mathrm{d}s,\; w_0(r_0) = w_0'(r_0) = 0$$

$$w_0'(r_1) = Dq_0 \int_{r_0}^{r} \frac{\partial K(r,s)}{\partial r}\left[H_0(b-s) - H_0(a-s)\right] \mathrm{d}s \Bigg|_{r=r_1} = Dq_0 \int_{a}^{b} \frac{\partial K(r,s)}{\partial r}\Bigg|_{r=r_1} \mathrm{d}s,$$

Figure 6.21 shows variations of the plate deflections and shear strains of the filler along the radius at different radius values of the force ring $b$. A change of the force ring area leads to a disproportionate growth of displacements. With increasing radius b, the loading area grows faster, while the deflection growth slows down. In this case, the form of the curves does not change much, but the extremes are displaced to the mid contour of the plate.

*Transverse force* $Q(r)$ operating along a circle of radius $r = a$ ($r_0 \le a \le 1$):

$$Q(r) = Q_0 \delta(r-a),\; Q_0 = const. \tag{6.97}$$

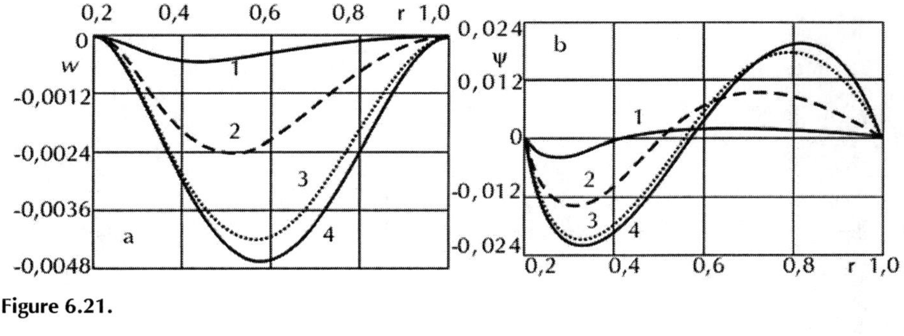


**Figure 6.21.**

A particular solution (6.93) and its values on the boundary contours of the plate under load (6.97) are

$$w_0(r) = DQ_0 \int_{0}^{r} K(r,s)\delta(s-a)\,\mathrm{d}s = DQ_0 K(r,a),$$

$$w_0'(r_1) = DQ_0 \frac{\partial K(r,a)}{\partial r}\Bigg|_{r=r_1},\; w_0(r_1) = DQ_0 K(r_1,a).$$

Variations of $w$ and $\psi$ along radius of the annular plate are shown in Fig. 6.22. The curves are constructed for different values of radius $a$ of the force circle: *1* – 0.4; *2* – 0.6; *3* – 0.8 ($Q_0$ = 1000 kN/м, $\kappa_0$ = 1000 MPa/м). The maximum values of displacements are reached at $a$ = 0.6.

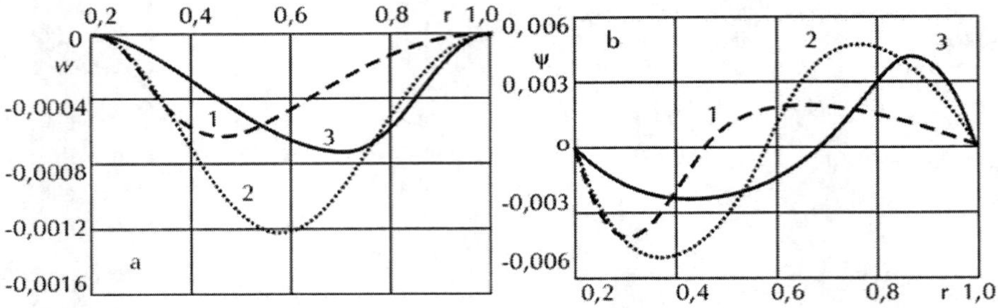


**Figure 6.22.**

*Linear bending moments* of intensity $M_0$ = const distributed along a circle of radius $r = a$ ($r_0 \le a \le 1$):

$$M(r) = M_0 \delta'(r-a). \tag{6.98}$$

A corresponding partial solution (6.93) on the contour under the load (6.98) has the form

$$w_0(r) = DM_0 \int_{r_0}^{r} K(r,s)\delta'(s-a)\,\mathrm{d}s = -DM_0 \left.\frac{\partial K(r,s)}{\partial s}\right|_{s=a},$$

$$w_0'(r_1) = -DM_0 \left.\frac{\partial^2 K(r,s)}{\partial r\,\partial s}\right|_{r=r_1,\,s=a}, \quad w_0(r_1) = -DM_0 \left.\frac{\partial K(r_1,s)}{\partial s}\right|_{s=a}.$$

Figure 6.23 shows variations in deflections $w$ of the plate and in shear strains of the filler along the radius of the annular sandwich plate. The curves are constructed for different values of radius $a$ of the circle: *1* – $a = 0.4$, *2* – $a = 0.6$, *3* – $a = 0.8$. As the torque circle (the intensity of linear moments $M_0 = 1000$ kN) approaches the plate contour, the deflection changes its sign. For the shear strains, the maxima in the form of peaks occur in the points of application in the linear moments.

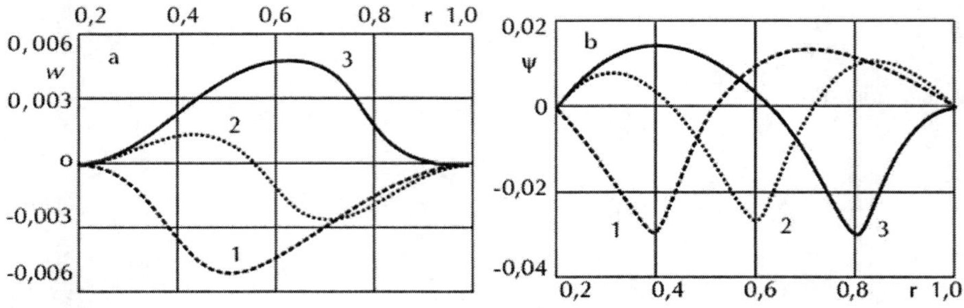


**Figure 6.23.**

**Conclusions.** The general solution (6.92) presented in this study can be used for investigation of any case of symmetric bending loads of annular sandwich composite plates with a light filler on an elastic foundation under the local of different ways of restricting its external and internal contours.

In more detail the problems of the applied elasticity theory are considered in works (Aleksandrov and Potapov, 1990, Demidov, 1979, Gorshkov and Tarlakovskii, 1995, Green and Zerna, 1968, Koltunov et al., 1983, Kupradze et al., 1968, Novozhilov, 1958, Southwell, 1948, Starovoitov, 1987, 1988, 2001, 2002, 2004, 2006, Starovoitov et al., 2002, Starovoitov et al., 2003, 2006, Starovoitov and Zeyad, 2007).

## NOTES

1. *Winkler, E.* (1835–1888), Austrian and German scientist and civil engineer.
2. *Bryan, G.H.* (1864–1928), English mathematical, astronomer and mechanical engineer.
3. *Bessel, F.W.* (1784–1846), German astronomer.
4. *Wroński, Y.* (1776–1853), Polish mathematician.

## KEYWORDS

- **Antisymmetrical**
- **Boundary conditions**
- **Kirhhoff's hypothesis**
- **Median plane**
- **Trigonometric**

# Chapter 7

## Deformation of a Half-Space and Contact Problems

### AN ELASTIC HALF-SPACE AFFECTED BY SURFACE FORCES

Let us take a sizeable isotropic body restricted by a plane. This semi-infinite body is called a *half-space*. We presume for definiteness the half-space to be orientated so that the plane restricting it is horizontal and the half-space body is located below the boundary plane. Orientation of the half-space is conditioned by Cartesian coordinates $Ox_1x_2x_3$, where $Ox_3$ axis is directed vertically downwards.

Let us find a stress–strain state (SSS) of a linearly elastic homogeneous isotropic half-space $x_3 \geq 0$ in the absence of the body forces and preset loading on the boundary plane (plane $Ox_1x_2$). Hence, for boundary conditions (3.5), we should accept that $l_1 = l_2 = 0$, $l_3 = -1$:

$$\sigma_{k3}\big|_{x_3=0} = -R_k(x_1, x_2)\,. \tag{7.1}$$

A closed system of equations describing equilibrium of a medium includes Lame's equations (3.6), Cauchy's relations (1.27), and Hooke's law (2.6):

$$(\lambda + \mu)\theta,_k + \mu\Delta u_k = 0\,, \tag{7.2}$$

$$\varepsilon_{kl} = \tfrac{1}{2}(u_k,_l + u_l,_k)\,, \tag{7.3}$$

$$\sigma_{kl} = \lambda\theta\delta_{kl} + 2\mu\varepsilon_{kl}\,. \tag{7.4}$$

The boundary problem (7.1)–(7.4) is closed by the boundedness condition of SSS components at infinity.

A particular case of a surface load (7.1) is a concentrated force

$$\sigma_{k3}\big|_{x_3=0} = -P_k\delta(x_1)\delta(x_2)\,, \tag{7.5}$$

where $\delta(x)$ is *Dirac's*[1] *delta function*

The problem with the boundary conditions of this kind is called in a general case *Mindlin*[2] *problem*, or *Boussinesq problem* when $P_2 = P_3 = 0$, and *Cerrutti*[3] *problem* when $P_1 = P_3 = 0$ or $P_1 = P_2 = 0$.

In addition to practical applications, the solutions of these problems on the influence of concentrated forces on a half-space infer the following meaning. Instead of forces (7.5) we may consider a problem with the following boundary conditions:

$$\sigma_{k3}\big|_{x_3=0} = \delta_{jk}\delta(x_1)\delta(x_2)\,, \tag{7.6}$$

where $\delta_{jk}$—Kronecker's delta.

In fact three problems obey conditions (7.6): at a fixed number $j = 1, 2, 3$ only one stress on the half-space surface is other than zero. Let us introduce a complementary index in designations of SSS for the components corresponding to the problem number

$$u_k = u_{jk},\ \varepsilon_{kl} = \varepsilon_{jkl},\ \sigma_{kl} = \sigma_{jkl}. \tag{7.7}$$

Then, in view of linearity of the displacement problem, strains and stresses corresponding to an arbitrary load $R_j$ (7.1) are defined in terms of functions (7.7) (summation over repeated indices from 1 till 3):

$$u_k = -R_j ** u_{jk},\ \varepsilon_{kl} = -R_j ** \varepsilon_{jkl},\ \sigma_{kl} = -R_j ** \sigma_{jkl}. \tag{7.8}$$

Double asterisk denotes here convolution of the functions with respect to coordinates $x_1$ and $x_2$ (double integration is performed throughout the coordinate plane):

$$f(x_1,x_2,x_3) = g**h = \iint g(\xi_1,\xi_2)h(x_1-\xi_1,x_2-\xi_2,x_3)\,d\xi_1\,d\xi_2$$
$$= \iint g(x_1-\xi_1,x_2-\xi_2)h(\xi_1,\xi_2,x_3)\,d\xi_1\,d\xi_2. \tag{7.9}$$

The functions in (7.8) are known as *fundamental solutions* (*dominant functions, Green surface functions*), and a combination of these values—as *fundamental tensors* or *Green tensors*.

It follows from (7.9) that to solve a problem with any boundary conditions of the type (7.1), it is suffice to find the dominant functions. To find them, let us use the integral Fourier[4] transform over derivatives $x_1$ and $x_2$ (sign « $F$ » indicates a transform, $q_1$ and $q_2$ —transformation parameters)

$$g^F(q_1,q_2) = \iint g(x_1,x_2)e^{iq_lx_l}\,dx_1\,dx_2,$$
$$g(x_1,x_2) = \frac{1}{4\pi^2}\iint g^F(q_1,q_2)e^{-iq_lx_l}\,dq_1\,dq_2\ (l=1,2). \tag{7.10}$$

In the transformation space (7.10), equations (7.2)–(7.4) become: ($k,l=1,2$)

$$-iq_k(\lambda+\mu)\theta^F + \mu\left(-q^2u_k^F + u_{k,33}^F\right) = 0,$$
$$(\lambda+\mu)\theta_{,3}^F + \mu\left(-q^2u_3^F + u_{3,33}^F\right) = 0,\ q = \sqrt{q_1^2+q_1^2}, \tag{7.11}$$

$$\varepsilon_{kl}^F = -\frac{i}{2}(q_lu_k^F + q_ku_l^F),\quad \varepsilon_{k3}^F = \frac{1}{2}(-iq_ku_3^F + u_{k,3}^F),\ \varepsilon_{33}^F = u_{3,3}^F,$$
$$\theta^F = -iq_ku_k^F + u_{3,3}^F, \tag{7.12}$$
$$\sigma_{kl}^F = \lambda\theta^F\delta_{kl} + 2\mu\varepsilon_{kl}^F. \tag{7.13}$$

Here and further in intermediate calculations, the problem number $j$ in SSS designations of the components is omitted to shorten the record.

The boundary conditions (7.3) transfer into the next equalities

$$\sigma_{k3}^F\Big|_{x_3=0} = \delta_k. \tag{7.14}$$

It is possible now to construct directly a general solution for a series of common differential equations (7.11), although it is more convenient to use their result (3.11)

$$\Delta\Delta u_k = 0\cdot \tag{7.15}$$

After using Fourier's transform, we come to three independent equations

$$\frac{\partial^4 u_k^F}{\partial x_3^4} - 2q^2 \frac{\partial^2 u_k^F}{\partial x_3^2} + q^4 u_k^F = 0\,. \tag{7.16}$$

Solutions of these equations meeting the boundedness conditions at infinity will be as follows:

$$u_k^F = (A_k + x_3 B_k)e^{-qx_3}, \tag{7.17}$$

where $A_k$, $B_k$ are arbitrary constants.

After substitution of (7.17) into (7.12) and (7.13), we find the corresponding representations for strains and stresses ($\alpha = 1, 2$; summation over the repeated Greek indices is omitted)

$$\begin{aligned}
&\varepsilon_{\alpha\alpha}^F = -iq_\alpha\left(A_\alpha + x_3 B_\alpha\right)e^{-qx_3},\\
&\varepsilon_{\alpha 3}^F = \frac{1}{2}\left[-qA_\alpha - iq_\alpha A_3 + B_\alpha - x_3\left(qB_\alpha + iq_\alpha B_3\right)\right]e^{-qx_3},\\
&\varepsilon_{12}^F = -\frac{i}{2}\left[q_2 A_1 + q_1 A_2 + x_3\left(q_2 B_1 + q_1 B_2\right)\right]e^{-qx_3},\\
&\varepsilon_{33}^F = \left[-qA_3 + \left(1 - qx_3\right)B_3\right]e^{-qx_3},\ \theta^F = -2(1-2\nu)B_3 e^{-qx_3},\\
&\sigma_{\alpha\alpha}^F = -2\mu\left[iq_\alpha A_\alpha + 2\nu B_3 + ix_3 q_\alpha B_\alpha\right]e^{-qx_3},\\
&\sigma_{\alpha 3}^F = \mu\left[-qA_\alpha - iq_\alpha A_3 + B_\alpha - x_3\left(qB_\alpha + iq_\alpha B_3\right)\right]e^{-qx_3},\\
&\sigma_{12}^F = -i\mu\left[q_2 A_1 + q_1 A_2 + x_3\left(q_2 B_1 + q_1 B_2\right)\right]e^{-qx_3},\\
&\sigma_{33}^F = 2\mu\left[-qA_3 + \left(1 - 2\nu - qx_3\right)B_3\right]e^{-qx_3}.
\end{aligned} \tag{7.18}$$

We have taken into account here the relationship between Lame's constants and other elasticity constants of the material (see 2.2).

Solutions of system (7.15) present a subset of a series of solutions of equations (7.11). We find the relation between integration constants by substituting solutions (7.17) in system (7.11). After canceling an exponent and equating coefficients at $x_3^0$ and $x_3^1$ ($x_3$—arbitrary number), we obtain a homogeneous system of linear algebraic equations with respect to $A_k$ and $B_k$ ($m,n=1,2$), consisting of two groups.

$$\begin{aligned}
&q_m(-q_n A_n + iqA_3) - 2(1-2\nu)qB_m - iq_m B_3 = 0,\\
&q(iq_n A_n + qA_3) - iq_n B_n - 4(1-\nu)qB_3 = 0,\\
&q_m(-q_n B_n + iqB_3) = 0,
\end{aligned} \tag{7.19}$$

$$q(iq_nB_n + qB_3) = 0. \tag{7.20}$$

The analysis of equations (7.19) and (7.20) show that they are equivalent to three equations ( $m,n=1,2$ )

$$q_nA_n - iqA_3 + 2iq(1-\nu)B_3 = 0, \quad qB_m = iq_mB_3. \tag{7.21}$$

It is closed by three relations obtained from boundary conditions (7.14) with regard to equalities (7.18)

$$-qA_m - iq_mA_3 + B_m = \frac{1}{\mu}\delta_{jm},$$

$$-qA_3 + (1-2\nu)B_3 = \frac{1}{2\mu}\delta_{j3}. \tag{7.22}$$

Solution of the system (7.21)–(7.22) takes the following form ( $m,n=1,2$ )

$$A_m = \frac{1}{2\mu q}\left[2\nu\frac{iq_mq_n}{q^2}\delta_{jn} - 2\delta_{jm} + (1-2\nu)\frac{iq_m}{q}\delta_{j3}\right],$$

$$A_3 = -\frac{1}{2\mu q}\left[(1-2\nu)\frac{iq_n}{q}\delta_{jn} + 2(1-\nu)\delta_{j3}\right],$$

$$B_m = \frac{1}{2\mu q}\left(\frac{q_mq_n}{q}\delta_{jn} - iq_m\delta_{j3}\right), \quad B_3 = -\frac{1}{2\mu}\left(\frac{q_n}{q}\delta_{jn} + \delta_{j3}\right). \tag{7.23}$$

Substitution of expressions (7.23) into (7.17) and (7.18) results in the following formulas for representing displacements and stresses ( $\alpha,n=1,2$ ):

$$u_\alpha^F = \frac{1}{2\mu q}\left[\left(\frac{2\nu}{q} + x_3\right)\frac{q_\alpha q_n}{q}\delta_{jn} - 2\delta_{j\alpha} + iq_\alpha\left(\frac{1-2\nu}{q} - x_3\right)\delta_{j3}\right]e^{-qx_3},$$

$$u_3^F = -\frac{1}{2\mu}\left[\left(\frac{1-2\nu}{q} + x_3\right)\frac{iq_n}{q}\delta_{jn} + \left(2\frac{1-\nu}{q} + x_3\right)\delta_{j3}\right]e^{-qx_3},$$

$$\sigma_{\alpha\alpha}^F = \left\{\frac{2iq_\alpha}{q}\delta_{j\alpha} + \left[2\nu\left(1 - \frac{q_\alpha^2}{q^2}\right) - x_3\frac{q_\alpha^2}{q}\right]\frac{iq_n}{q}\delta_{jn}\right.$$

$$\left. + \left[2\nu + (1-2\nu)\frac{q_\alpha^2}{q^2} - x_3\frac{q_\alpha^2}{q}\right]\delta_{j3}\right\}e^{-qx_3}, \tag{7.24}$$

$$\sigma_{12}^F = \frac{1}{q}\left[i\left(q_2\delta_{j1} + q_1\delta_{j2}\right) - \left(\frac{2\nu}{q} + x_3\right)\frac{iq_1q_2q_n}{q}\delta_{jn} + \left(\frac{1-2\nu}{q} - x_3\right)q_1q_2\delta_{j3}\right]e^{-qx_3},$$

$$\sigma_{\alpha 3}^F = \left[\delta_{j\alpha} - x_3q_\alpha\left(\frac{q_n}{q}\delta_{jn} - i\delta_{j3}\right)\right]e^{-qx_3}, \quad \sigma_{33}^F = \left[iq_n\delta_{jn}x_3 + (1+qx_3)\delta_{j3}\right]e^{-qx_3}.$$

Wherefrom we find the corresponding precursors (in designations of displacements and stresses we return index $j$ of the problem number)

$$u_{j\alpha} = -\frac{1}{4\mu\delta}\left\{\left(\frac{1}{r} + \frac{1-2\nu}{r+x_3}\right)\delta_{\alpha j} + \left[\frac{1}{r^2} - \frac{1-2\nu}{(r+x_3)^2}\right]\frac{x_\alpha x_n\delta_{jn}}{r} + \left(\frac{x_3}{r^2} - \frac{1-2\nu}{r+x_3}\right)\frac{x_\alpha\delta_{j3}}{r}\right\},$$

$$u_{j3} = -\frac{1}{4\mu\partial r}\left\{\left(\frac{1-2\nu}{r+x_3}+\frac{x_3}{r^2}\right)x_n\delta_{jn}+\left[2(1-\nu)+\frac{x_3^2}{r^2}\right]\delta_{j3}\right\}, \tag{7.25}$$

$$\sigma_{j\alpha\alpha} = \frac{1}{2\partial r}\left\langle 2(1-2\nu)\frac{x_\alpha\delta_{j\alpha}}{(r+x_3)^2}+\left\{\frac{3x_\alpha^2}{r^4}-(1-2\nu)\left[\frac{1}{r^2}-\frac{1}{(r+x_3)^2}\right]\right\}x_n\delta_{jn}\right.$$

$$\left.+\left\{\frac{1-2\nu}{r+x_3}\left[1-\frac{x_\alpha^2}{r}\left(\frac{1}{r}+\frac{1}{r+x_3}\right)\right]-\frac{x_3}{r^2}\left(1-2\nu-\frac{3x_\alpha^2}{r^2}\right)\right\}\delta_{j3}\right\rangle,$$

$$\sigma_{j12} = \frac{1}{2\partial r}\left\langle\left\{\frac{1}{r^2}+\frac{x_3}{r^2(r+x_3)^2}\left[2r+x_3-\frac{x_1^2(8r^2+8rx_3+3x_3^2)}{r^2(r+x_3)}\right]\right.\right.$$

$$\left.+\frac{2\nu}{(r+x_3)^2}\left[\frac{x_1^2(3r+x_3)}{r^2(r+x_3)}-1\right]\right\}x_2\delta_{j1}+x_1\delta_{j2}\left\{\frac{1}{r^2}+\frac{x_3}{r^2(r+x_3)^2}\right[2r+x_3$$

$$\left.-\frac{x_2^2(8r^2+8rx_3+3x_3^2)}{r^2(r+x_3)}\right]+\frac{2\nu}{(r+x_3)^2}\left[\frac{x_2^2(3r+x_3)}{r^2(r+x_3)}-1\right]\right\}x_1\delta_{j2}$$

$$\left.+\frac{x_1x_2}{r^2}\left[\frac{3x_3}{r^2}-(1-2\nu)\frac{2r+x_3}{(r+x_3)^2}\right]\delta_{j3}\right\rangle,$$

$$\sigma_{j\alpha3} = \frac{3x_\alpha x_3}{2\pi r^5}\left(x_n\delta_{jn}+x_3\delta_{j3}\right),\quad \sigma_{j33} = \frac{3x_3^2}{2\pi r^5}\left(x_n\delta_{jn}+x_3\delta_{j3}\right). \tag{7.26}$$

Notice that stresses $\sigma_{jk3}$ are independent of elastic parameters of the medium. From (7.25) and (7.26) we obtain solutions of Boussinesq and Cerrutti problems.

- Displacements and stresses in *Boussinesq problem* ($j = 3$).

$$u_{3\alpha} = \frac{x_\alpha}{4\mu\partial r}\left(\frac{1-2\nu}{r+x_3}-\frac{x_3}{r^2}\right),\quad u_{33} = -\frac{1}{4\mu\partial r}\left[2(1-\nu)+\frac{x_3^2}{r^2}\right],$$

$$\sigma_{3\alpha\alpha} = \frac{1}{2\partial r}\left\{\frac{3x_\alpha^2x_3}{r^4}+\frac{1-2\nu}{r^2(r+x_3)}\left[r^2-rx_3-x_3^2-\frac{x_\alpha^2(2r+x_3)}{r+x_3}\right]\right\},$$

$$\sigma_{312} = \frac{x_1x_2}{2\partial r^3}\left[\frac{3x_3}{r^2}-(1-2\nu)\frac{2r+z}{(r+z)^2}\right],\quad \sigma_{3\alpha3} = \frac{3x_\alpha x_3^2}{2\pi r^5},\quad \sigma_{j33} = \frac{3x_3^3}{2\pi r^5}.$$

- Displacements and stresses in *Cerrutti problem* ($j = 1$)

$$u_{11} = -\frac{1}{4\mu\partial}\left\{\frac{1}{r}+\frac{x_1^2}{r^3}+\frac{1-2\nu}{r+x_3}\left[1-\frac{x_1^2}{r(r+x_3)}\right]\right\},$$

$$u_{12} = -\frac{x_1x_2}{4\mu\partial r}\left[\frac{1}{r^2}-\frac{1-2\nu}{(r+x_3)^2}\right],\quad u_{13} = -\frac{x_1}{4\mu\partial r}\left(\frac{1-2\nu}{r+x_3}+\frac{x_3}{r^2}\right),$$

$$\sigma_{111} = \frac{x_1}{2\partial r^3}\left[\frac{3x_1^2}{r^2}+(1-2\nu)\frac{2r^2-2rx_3-x_3^2}{(r+x_3)^2}\right],\quad \sigma_{122} = \frac{x_1}{2\partial r^3}\left[\frac{3x_2^2}{r^2}-(1-2\nu)\frac{x_3(2r+x_3)}{(r+x_3)^2}\right],$$

$$\sigma_{112} = \frac{x_2}{2\eth r}\left\{\frac{1}{r^2} + \frac{x_3}{r^2\left(r+x_3\right)^2}\left[2r + x_3 - \frac{x_1^2\left(8r^2+8rx_3+3x_3^2\right)}{r^2\left(r+x_3\right)}\right]\right.$$

$$\left.+\frac{2\nu}{\left(r+x_3\right)^2}\left[\frac{x_1^2\left(3r+x_3\right)}{r^2\left(r+x_3\right)}-1\right]\right\},\ \sigma_{1\alpha 3} = \frac{3x_1x_\alpha x_3}{2\pi r^5},\ \sigma_{j33} = \frac{3x_1x_3^2}{2\pi r^5}.$$

## FUNDAMENTAL SOLUTIONS FOR AN ELASTIC HALF-PLANE

Let us consider a problem on the influence of load ($k = 1, 3$) on a half-plane boundary $x_3 \geq 0$ in terms of above stated discourse.

$$\sigma_{k3}\big|_{x_3=0} = -R_k(x_1)\ . \tag{7.27}$$

In the assumption of (7.27) the components of SSS are independent of coordinate $x_2$, and the displacement is $u_2 \equiv 0$. Thus, we have a plane strain state (see 5.1). The closed system of equations includes Lame's equations, Cauchy's relations, and Hooke's law. The boundary problem is also closed by the boundedness condition of the components of SSS at infinity.

In much the same way to a 3D case, the solution of this problem may be presented as follows (all indices will take 1 and 3 values):

$$u_k = -R_j * u_{jk},\ \varepsilon_{kl} = -R_j * \varepsilon_{jkl},\ \sigma_{kl} = -R_j * \sigma_{jkl}\,. \tag{7.28}$$

The asterisk in (7.28) indicates convolution of the functions about coordinate $x_1$

$$f(x_1, x_3) = g * h = \int_{-\infty}^{+\infty} g(\xi)h(x_1-\xi, x_3)\,d\xi = \int_{-\infty}^{+\infty} g(x_1-\xi)h(\xi, x_3)\,d\xi\,.$$

Functions $u_{jk}$, $\varepsilon_{jkl}$ and $\sigma_{jkl}$ on the half-plane boundary meet the boundary conditions

$$\sigma_{k3}\big|_{x_3=0} = \delta_{jk}\delta(x_1)\,, \tag{7.29}$$

and are fundamental solutions (components of Green tensors) for an elastic half-plane.

The corresponding problem for $j = 3$ is *Flamant problem*, and if $j = 1$, it is called *Cerrutti problem* like previously.

By analogy with a 3D case, we shall use a unidimensional Fourier transform with respect to coordinate $x_1$ (sign « $F$ » indicates the transform, $q$—transformation parameter)

$$g^F(q) = \int_{-\infty}^{+\infty} g(x_1)e^{iqx_1}\,dx_1,\ g(x_1) = \frac{1}{2\pi}\int_{-\infty}^{+\infty} g(x_1)e^{-iqx_1}\,dq\,. \tag{7.30}$$

The transforms of the sought functions follow from the results of the preceding section, if to assume in (7.24) and (7.25) that $q_2 = 0$, and substitute $q_1$ and $q$ by $q$ and $|q|$, respectively.

Let us show another way of solving the plane problem bringing to a simpler series of algebraic equations as compared to systems (7.21), (7.22). For this purpose, we shall use statement of the problem in stresses and omit problem number $j$ in temporary calculations.

We shall use Fourier transform (7.30) in the first stage for equilibrium equations

$$\sigma_{mn,n}=0 \quad (m,n=1,3).$$

As a result,

$$-iq\sigma_{11}^F+\frac{\partial\sigma_{13}^F}{\partial x_3}=0,\quad -iq\sigma_{13}^F+\frac{\partial\sigma_{33}^F}{\partial x_3}=0\,. \tag{7.31}$$

This series of equations is not closed. Therefore, we should additionally account for the biharmonic property of the stress tensor components (3.14). In particular, for stress $\sigma_{33}$ we have the equation

$$\Delta\Delta\sigma_{33}=0,\quad \Delta=\frac{\partial^2}{\partial x_1^2}+\frac{\partial^2}{\partial x_3^2},$$

which in the transformation space is equivalent to the equality (see also (7.16))

$$\frac{\partial^4\sigma_{33}^F}{\partial x_3^4}-2q^2\frac{\partial^2\sigma_{33}^F}{\partial x_3^2}+q^4\sigma_{33}^F=0\,.$$

Its general solution meeting the boundedness condition at infinity is similar to (7.17) and is given by

$$\sigma_{33}^F=(A+x_3B)e^{-|q|x_3}, \tag{7.32}$$

where $A$ and $B$ are arbitrary constants.

After substitution of equation (7.32) in (7.31), we obtain representations of other stresses in a series

$$\sigma_{13}^F=-\frac{i}{q}\left(B-|q|A-x_3|q|B\right)e^{-|q|x_3},\ \sigma_{11}^F=\left(\frac{2}{|q|}B-A-x_3B\right)e^{-|q|x_3}.$$

These relations along with representation of boundary conditions (7.29)

$$\left.\sigma_{k3}^F\right|_{x_3=0}=\delta_{jk},$$

serve to calculate constants $A$ and $B$:

$$A=\delta_{j3},\quad B=iq\delta_{j1}+|q|\delta_{j3}.$$

Finally, representations of stresses and strains become of the form:

$$\sigma_{11}^F=\left[i\left(2\operatorname{sign}q-qx_3\right)\delta_{j1}+\left(1-|q|x_3\right)\delta_{j3}\right]e^{-|q|x_3},$$

$$\sigma_{13}^F = \left[\left(1-|q|x_3\right)\delta_{j1} + iqx_3\delta_{j3}\right]e^{-|q|x_3},\quad \sigma_{33}^F = \left[iqx_3\delta_{j1} + \left(1+|q|x_3\right)\delta_{j3}\right]e^{-|q|x_3},$$

$$\sigma_{11}^F + \sigma_{33}^F = 2\left(i\,\mathrm{sign}q\,\delta_{j1} + \delta_{j3}\right)e^{-|q|x_3},$$

$$\varepsilon_{11}^F = \frac{1}{2\mu}\left[\sigma_{11}^F - \nu\left(\sigma_{11}^F + \sigma_{33}^F\right)\right]$$

$$= \frac{1}{2\mu}\left\{i\left[2(1-\nu)\mathrm{sign}q - qx_3\right]\delta_{j1} + \left(1-2\nu-|q|x_3\right)\delta_{j3}\right\}e^{-|q|x_3},$$

$$\varepsilon_{13}^F = \frac{\sigma_{13}^F}{2\mu} = \frac{1}{2\mu}\left[\left(1-|q|x_3\right)\delta_{j1} + iqx_3\delta_{j3}\right]e^{-|q|x_3},$$

$$\varepsilon_{33}^F = \frac{1}{2\mu}\left[\sigma_{33}^F - \nu\left(\sigma_{11}^F + \sigma_{33}^F\right)\right]$$

$$= \frac{1}{2\mu}\left[i\left(qx_3 - 2\nu\,\mathrm{sign}q\right) + \left(1-2\nu+|q|x_3\right)\delta_{j3}\right]e^{-|q|x_3}. \tag{7.33}$$

In view of Cauchy's relations, representations of displacements obey the equations (see (7.12))

$$-iqu_1^F = \varepsilon_{11}^F,\quad -iqu_3^F + \frac{\partial u_1^F}{\partial x_3} = 2\varepsilon_{13}^F,\quad \frac{\partial u_3^F}{\partial x_3} = \varepsilon_{33}^F. \tag{7.34}$$

Proceeding from the first two relations (7.34) and taking into account equations (7.33), we find

$$u_1^F = -\frac{1}{2\mu}\left\{\left[\frac{2(1-\nu)}{|q|} - x_3\right]\delta_{j1} - i\left(\frac{1-2\nu}{q} - x_3\,\mathrm{sign}q\right)\delta_{j3}\right\}e^{-|q|x_3},$$

$$u_3^F = -\frac{1}{2\mu}\left\{i\left(\frac{1-2\nu}{q} + x_3\,\mathrm{sign}q\right)\delta_{j1} - \left[\frac{2(1-\nu)}{|q|} + x_3\right]\delta_{j3}\right\}e^{-|q|x_3}. \tag{7.35}$$

The check of the latter equality in (7.34) shows that it is fulfilled similarly.

Based on relations (7.33), (7.35), we can find the original displacements and stresses (return index $j$ to designations of displacements and stresses)

$$u_{j1} = \frac{1}{2\pi\mu}\left\{\left[2(1-\nu)(\ln r + \mathrm{C}) + \frac{x_3^2}{r^2}\right]\delta_{j1} + \left[(1-2\nu)\,\mathrm{arctg}\frac{x_1}{x_3} - \frac{x_1x_3}{r^2}\right]\delta_{j3}\right\}, \tag{7.36}$$

$$u_{j3} = -\frac{1}{2\pi\mu}\left\{\left[(1-2\nu)\,\mathrm{arctg}\frac{x_1}{x_3} + \frac{x_1x_3}{r^2}\right]\delta_{j1} + \left[2(1-\nu)(\ln r + \mathrm{C}) - \frac{x_3^2}{r^2}\right]\delta_{j3}\right\}, \tag{7.37}$$

$$r = \sqrt{x_1^2 + x_3^2},\quad \sigma_{jkl} = \frac{2x_jx_kx_l}{\pi r^4}. \tag{7.38}$$

It should be noted, that all stresses in the plane problem are independent of the elastic characteristics of the medium. Besides, the following symmetry property is to be obeyed: $\sigma_{113} = \sigma_{311}$, $\sigma_{133} = \sigma_{313}$.

Based on (7.36)–(7.38), we obtain solutions for the Flamant and Cerrutti problems:

$$u_{31} = \frac{1}{2\pi\mu}\left[(1-2\nu)\operatorname{arctg}\frac{x_1}{x_3} - \frac{x_1 x_3}{r^2}\right],$$

$$u_{33} = -\frac{1}{2\pi\mu}\left[2(1-\nu)(\ln r + \mathbf{C}) - \frac{x_3^2}{r^2}\right],$$

$$u_{11} = \frac{1}{2\pi\mu}\left[2(1-\nu)(\ln r + \mathbf{C}) + \frac{x_3^2}{r^2}\right],$$

$$u_{13} = -\frac{1}{2\pi\mu}\left[(1-2\nu)\operatorname{arctg}\frac{x_1}{x_3} + \frac{x_1 x_3}{r^2}\right], \tag{7.39}$$

$$\sigma_{113} = \sigma_{311} = \frac{2x_1^2 x_3}{\pi r^4}, \quad \sigma_{133} = \sigma_{313} = \frac{2x_1 x_3^2}{\pi r^4},$$

$$\sigma_{333} = \frac{2x_3^3}{\pi r^4}, \quad \sigma_{111} = \frac{2x_1^3}{\pi r^4}. \tag{7.40}$$

In the polar system of coordinates

$$x_3 = r\cos\vartheta, \quad x_1 = r\sin\vartheta \quad \left(r \ge 0, -\frac{\pi}{2} \le \vartheta \le \frac{\pi}{2}\right),$$

the formulas for stresses are

$$\sigma_{113} = \sigma_{311} = \frac{2}{\pi r}\sin^2\vartheta\cos\vartheta, \quad \sigma_{133} = \sigma_{313} = \frac{2}{\pi r}\sin\vartheta\cos^2\vartheta,$$

$$\sigma_{333} = \frac{2}{\pi r}\cos^3\vartheta, \quad \sigma_{111} = \frac{2}{\pi r}\sin^3\vartheta.$$

Figures 7.1a–d show the lines of level $\sigma_{jkl} = C_n = \text{const}$, being (following from 7.40) the quartic algebraic curves.

We can finally find the principal axes and stresses corresponding to fundamental solutions (7.37) for the elastic half-plane. The principal stresses with regard to the 2D problem are the red by (1.15)

$$\begin{vmatrix} \sigma_{j11} - \sigma & \sigma_{j13} \\ \sigma_{j13} & \sigma_{j33} - \sigma \end{vmatrix} = \sigma^2 - J_1\sigma = 0, \; J_1 = \sigma_{j11} + \sigma_{j33} = \frac{2x_j}{\pi r^2}.$$

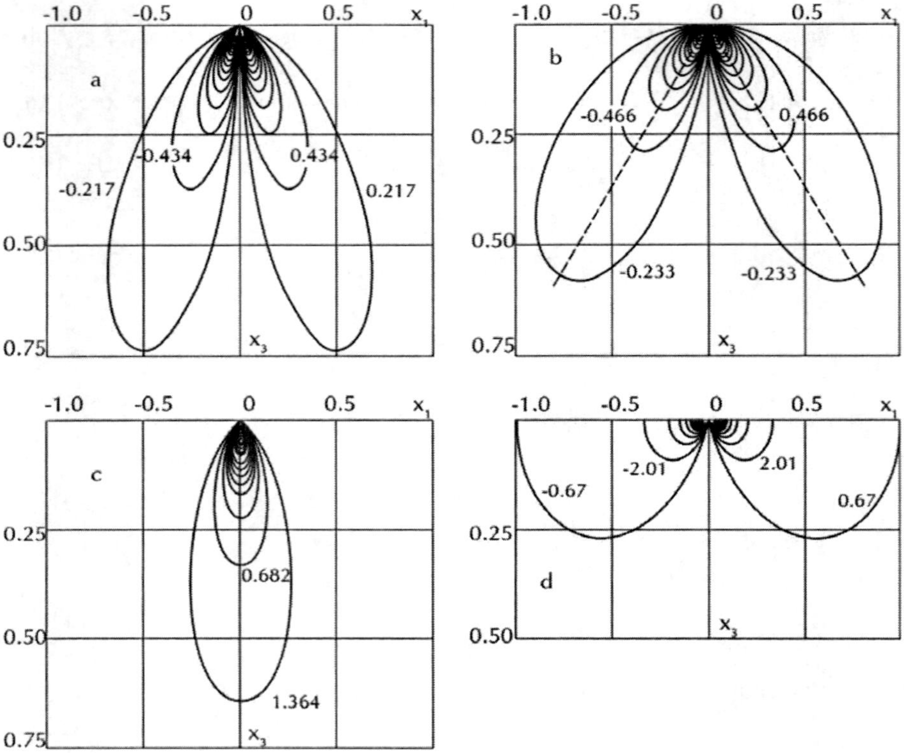


**Figure 7.1.**

This implies that

$$\sigma_1 = 0, \quad \sigma_2 = \frac{2x_j}{\pi r^2}.$$

Their order depends on the sign of $x_j$. The corresponding principal axes are defined by the equation (see to (1.14))

$$\sigma_{j11} l_1 + \sigma_{j13} l_3 = 0,$$

hence,

$$\nu_1 = e_J = \frac{1}{r}(x_1 e_1 - x_3 e_3), \quad \nu_2 = e_r = \frac{r}{r} = \frac{1}{r}(x_1 e_1 + x_3 e_3).$$

Therefore, the stresses on the beam areas ( $\vartheta = \text{const}$ ) of the polar system are zero, while on the circle areas ( $r = \text{const}$ ) only normal stresses $\sigma_2$ exist. Their distribution for the Flamant and Cerrutti problems is shown in Figs. 7.2 a and b, respectively.

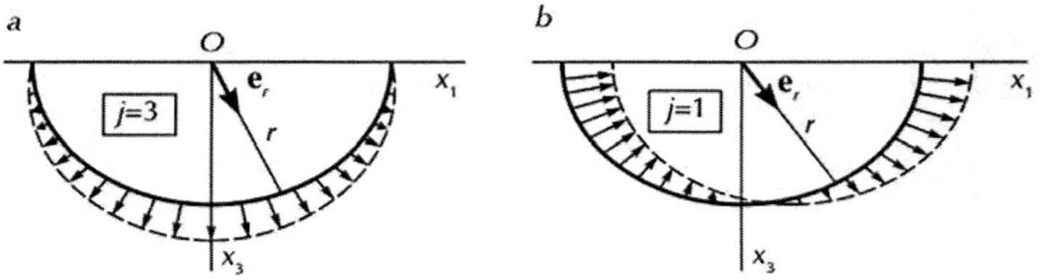


**Figure 7.2.**

## THE PROBLEM OF A PUNCH ON AN ELASTIC HALF-PLANE

Above stated fundamental solutions are used to solve the problem with mixed boundary conditions (the third boundary problem, see 3.1). These are so-called *contact problems*, in which a part of boundaries of two deformed bodies (one of which may be absolutely rigid) is common. The interacting areas are called *contact surfaces*.

Let us consider a simplest plane problem. We assume that a flat symmetric undeforming punch is pressed vertically without friction into a homogeneous isotropic linearly elastic half-plane $x_3 \geq 0$ (Fig. 7.3).

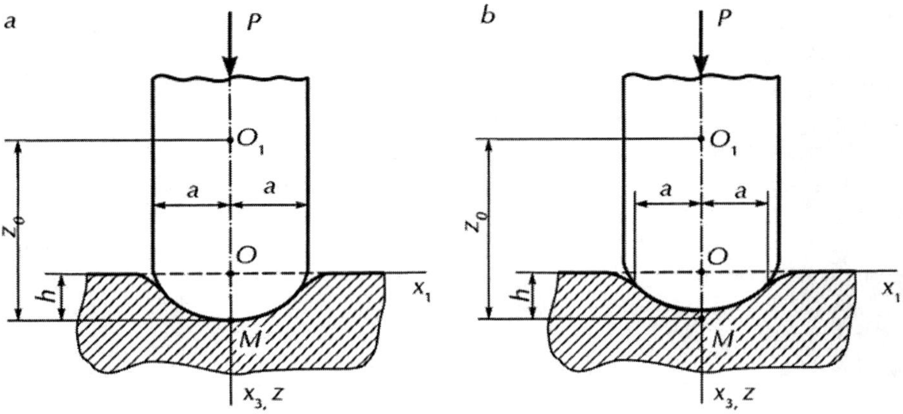


**Figure 7.3.**

Axis $Ox_3$ is superposed with the symmetry axis of the punch and the origin of coordinates $Ox_1x_3$ is located on the half-plane boundary. The lower end of the punch is restricted by a base being a smooth nonconcave surface. Its equation in coordinates $O_1x_1z$ ($O_1z$ coincides with $Ox_3$, where $z_0$--distance between $O_1$ and head point $M$—base apex) related with the punch, is as follows:

$$z = f(x_1)\,. \tag{7.41}$$

here, $f(x_1)$—even function, $f(0) = z_0$ and $f''(x_1) \leq 0$.

By denoting depth of indenter penetration (motion of the indenter head point along $Ox_3$ axis) by $h$, taking into account (7.41) and the relation of coordinates $z = x_3 + z_0 - h$ (Fig. 7.3), we obtain the equation for the base in the system $Ox_Ix_3$

$$x_3 = f(x_1) + h - z_0 . \tag{7.42}$$

The right-hand side of equation (7.42) presets normal displacements of the half-plane boundary in the contact area.

We shall assume that force $P$, directed along $Ox_3$ axis is acting on the punch. The tangential stresses on the contact surface are equal to zero (the contact takes place under conditions of *free slippage*), and the half-plane boundary is free outside this surface. The corresponding boundary problem includes the equations in some form that describe a flat strain state, boundedness condition of the solution at infinity, as well as the conditions on the half-plane boundary.

$$u_3\big|_{x_3=0} = f(x_1) + h - z_0 \ \left(|x_1| \le a\right), \ \sigma_{33}\big|_{x_3=0} = 0 \ \left(|x_1| > a\right),$$

$$\sigma_{13}\big|_{x_3=0} = 0 \ \left(-\infty < x_1 < +\infty\right). \tag{7.43}$$

here $h$—unknown depth of punch penetration.

In view of linearity of the problem, conditions (7.43) are brought onto the unstrained boundary of the half-plane. The contact surface is substituted by the *contact area*—segment $|x_1| \le a$. Besides, one more approximation corresponding to the linear theory is made in (7.43), namely, the tangential components of displacements of material points have not been taken into account. It means that the punch and the half-space boundary points found on the same vertical straight line at the initial moment of interaction, remain on this straight line when in contact as well. Consequently, normal displacements in the contact area agree with the right-hand side of equation (7.42). Depending on the problem geometry, the radius in the contact area $a$ is either set (Fig. 7.3a) or is unknown (Fig.7.3b).

The problem is closed by the equilibrium condition of the punch

$$\int_{-a}^{a} p(\xi)\,d\xi = P, \ \ p(x_1) = -\sigma_3\ (x_1, 0), \tag{7.44}$$

where $p$ is contact pressure.

Based on properties of (7.28) of the fundamental problems and conditions (7.43), we can bring the boundary problem to the integral equation for the contact pressure

$$\int_{-a}^{a} p(\xi) K(x_1 - \xi)\,d\xi = f(x_1) + h - z_0 , \tag{7.45}$$

where kernel $K(x_1)$ is defined according to (7.39) in the following way

$$K(x_1) = -u_{33}(x_1, 0) = \frac{1-\nu}{\pi\mu}\left(\ln|x_1| + \mathbf{C}\right).$$

By differentiating (7.45) with respect to $x_1$, we come to the equation

$$\int_{-a}^{a} \tilde{p}(\xi)K_1(x_1-\xi)\,d\xi = f'(x_1),\quad K_1(x_1)=\frac{1}{x_1},\quad \tilde{p}=\frac{1-\nu}{\pi\mu}p. \tag{7.46}$$

here, the integral has a singular feature (nonintegrable integrand), and is regarded as the principal value according to Cauchy. An equation of such kind is present in the theory of a finite-span wing, which is called the *Prandtl*[5] *equation.*

By direct verification [46] we can prove that the function

$$\tilde{p}(x_1)=\frac{1}{\sqrt{a^2-x_1^2}}\left[\frac{1}{\pi^2}\int_{-a}^{a}\frac{\sqrt{a^2-\xi^2}}{x_1-\xi}f'(\xi)\,d\xi+A\right],$$

is a solution of equation (7.46) at any value of constant $A$. This constant is found from condition (7.42)

$$\pi A_1 = P,\quad A_1=\frac{\pi\mu}{1-\nu}A,\quad \int_{-a}^{a}\frac{dx_1}{\sqrt{a^2-x_1^2}}\int_{-a}^{a}\frac{\sqrt{a^2-\xi^2}}{x_1-\xi}f'(\xi)\,d\xi=0$$

So, the contact pressure is calculated by the following equation

$$p(x_1)=\frac{1}{\pi\sqrt{a^2-x_1^2}}\left(P+\frac{\mu}{1-\nu}\int_{-a}^{a}\frac{\sqrt{a^2-\xi^2}}{x_1-\xi}f'(\xi)\,d\xi\right). \tag{7.47}$$

Remembering that derivative $f'(x_1)$ is odd, we obtain that $p(x_1)$ is an even function. It is not surprising in view of symmetry of the problem.

If the radius of the contact area $a$ is known (Fig. 7.3a), it will be the pressure (7.47), which is the solution of the problem. We can prove that in this case, the contact pressure at the ends $x_1 = \pm a$ is unbounded at any function $f(x_1)$.

If the radius of the contact area $a$ is unknown (Fig. 7.3b), the condition of boundedness of the contact pressure should be used (normal pressure on the half-plane boundary).

Since $f(0) = z_0$, from the integral equation (7.45) with regard to evenness of the contact pressure, and equation (7.44), we obtain the following formula for the punch penetration depth:

$$h=\frac{1-\nu}{\pi\mu}\left(CP+2\int_{0}^{a}p(\xi)\ln\xi\,d\xi\right). \tag{7.48}$$

Displacements and stresses in any point of the half-plane in accordance with (7.28), (7.43), and (7.44) are calculated by a straightforward integration ($k, l = 1, 3$)

$$u_k(x_1,x_3)=-\int_{-a}^{a}p(\xi)u_{3k}(x_1-\xi,x_3)\,d\xi,$$

$$\sigma_{kl} = -\int_{-a}^{a} p(\xi)\sigma_{3kl}(x_1 - \xi, x_3)\,\mathrm{d}\xi$$

Let us consider specific examples of giving the base shape of the punch.

1. *A parabolic punch* whose radius of the contact area is known. The base shape in this case shall be set as follows ($R$—radius of the curvature of the punch base in point $x_1 = 0$)

$$f(x_1) = -\frac{x_1^2}{2R}, \; R > 0, \; z_0 = 0. \tag{7.49}$$

After calculation of the integral in (7.47), we come to the following expression for the contact pressure:

$$p(x_1) = \frac{1}{\sqrt{a^2 - x_1^2}}\left[\frac{P}{\pi} + \frac{\mu\left(a^2 - 2x_1^2\right)}{2R(1-\nu)}\right]. \tag{7.50}$$

It is obvious that there is an integrable exponential singularity of the order 1/2 at the boundaries of the contact area.

The penetration depth is found by substituting (7.50) in (7.48)

$$h = \frac{1-\nu}{\pi\mu}\left(C + \ln\frac{a}{2}\right)P - \frac{a^2}{4R}. \tag{7.51}$$

2. *A parabolic punch* whose radius of the contact area is unknown. The contact pressure is bounded at $x_1 = \pm a$, if the bracket in (7.50) becomes zero. Wherefrom, we define the contact area radius

$$a = \sqrt{2PR\frac{1-\nu}{\pi\mu}}, \tag{7.52}$$

and find the pressure

$$p(x_1) = \frac{P}{\pi}\sqrt{a^2 - x_1^2}. \tag{7.53}$$

This implies that the pressure is restricted and falls down to zero on the boundary of the contact area, reaching the maximum $p_{\max} = Pa/\pi$ in the central point.

The punch penetration depth in this case is defined in the following way:

$$h = \frac{1-\nu}{\pi\mu}\left[\mathrm{C} + \frac{a^2}{2}\left(\ln\frac{a}{2} - \frac{1}{2}\right)\right]P. \tag{7.54}$$

3. *A punch with rectangular base* (Fig. 7.4). In this case, the simplest way of solution is to pass over to the limit when $R \to \infty$ in (7.50) and (7.51)

$$p(x_1) = \frac{P}{\pi\sqrt{a^2 - x_1^2}}, \; h = \frac{1-\nu}{\pi\mu}\left(C + \ln\frac{a}{2}\right)P. \tag{7.55}$$

As a result, pressure on the boundaries of the contact area acquires an integrable exponential singularity of the order 1/2 (Fig. 7.5).

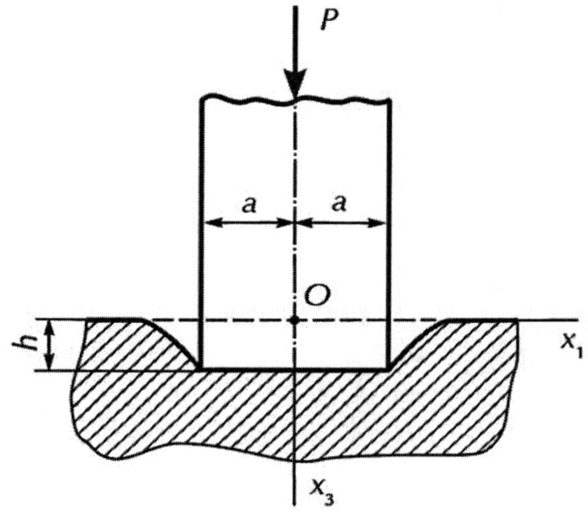


Figure 7.4.

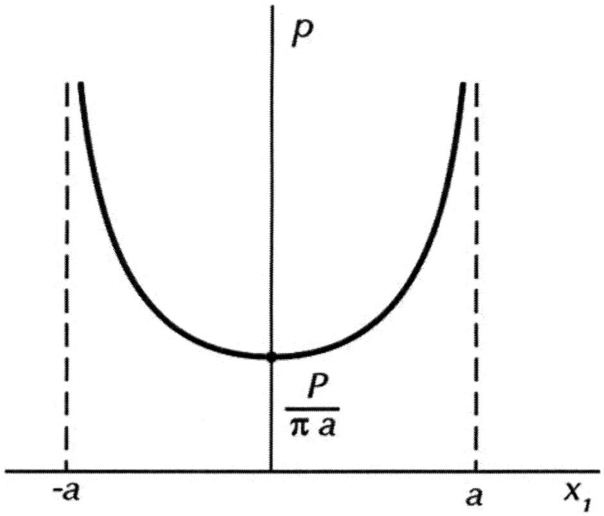


Figure 7.5.

## THE PLANE CONTACT PROBLEM FOR TWO ELASTIC BODIES

The methodology of the contact problem solution for a half-plane stated above is used for the approximated solution of the plane contact problem for two elastic bodies.

Two infinitely long symmetrical cylindrical bodies are assumed to be bounded by smooth convex surfaces. The equations for these surfaces in coordinates $O_jx_1x_2z_j$ and ($j$ = 1, 2) related to the bodies are of the type

$$\Pi_j : z_j = f_j(x_1),$$

$$f_j(x_1) = f_j(-x_1)\quad f_j(0) = -z_{0j},\ f_j'(0) = 0,\ f_j''(x_1) \geq 0. \tag{7.56}$$

Axes $O_jz_j$ coincide and lie in the planes of the body symmetry. Axes $O_jx_2$ are parallel to the generatrix, $z_{0j}$--distances between $O_j$ and head points of the director boundary surfaces (Fig. 7.6).

Let us introduce in addition a general system of coordinates $Ox_1x_2x_3$, where axis $Ox_2$ coincides with a common generatrix of cylindrical surfaces at the initial moment of interaction, $Ox_3$ coincides with $O_jz_j$ and is directed inside the second body ($j$ = 2), while $Ox_1$ is, respectively, perpendicular to the symmetry plane (Fig. 7.6). In this system of coordinates the unstrained boundary surfaces are set by equations (Fig. 7.7, (7.42))

$$\Pi_1 :\ x_3 = -f_1(x_1) - z_{01} + h_1,$$

$$\Pi_2 :\ x_3 = f_2(x_1) + z_{02} + h_2.$$

where $h_j$ are displacements along $\nu_j$ axis of undeforming bodies.

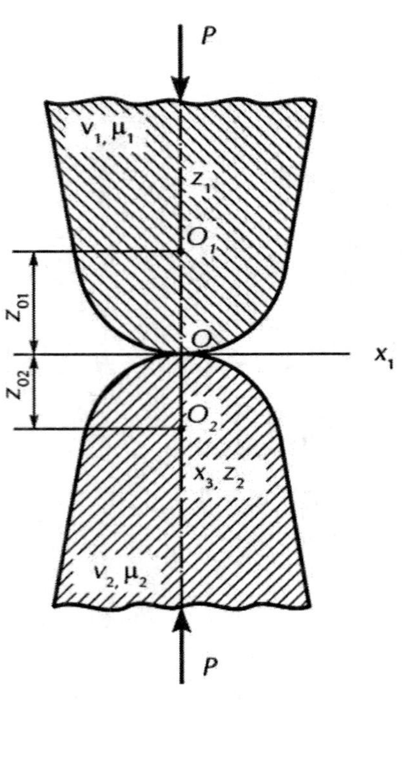


Figure 7.6.

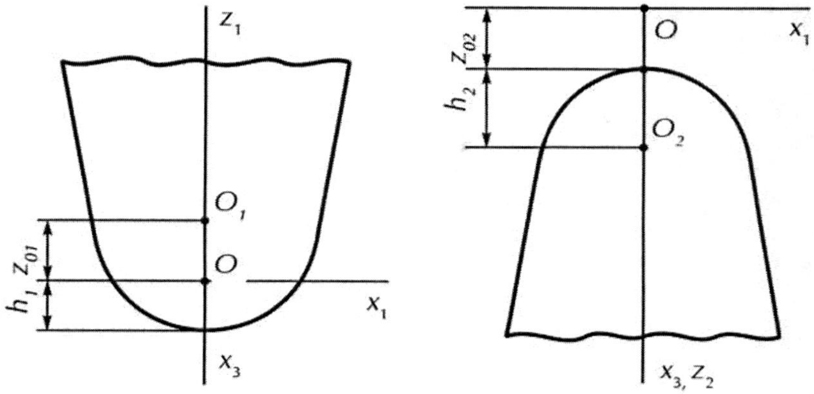


**Figure 7.7.**

The bodies are filled with homogeneous isotropic linearly elastic media characterized by constants $\nu_j$ and $\mu_j$. The contact occurs under conditions of free slippage, and the boundaries of the bodies are free from stresses outside the contact surface. The corresponding boundary problem includes the equations in one form or another that describe the plane strain state of both bodies (see 5.1), boundedness conditions of the solution at infinity, as well as boundary conditions in the contact area $|x_1| \leq a$ transferred onto $Ox_1$ axis

$$
\begin{aligned}
& w_1\big|_{x_3=0} + w_2\big|_{x_3=0} = -f(x_1) - h, \quad \sigma_{33}^{(1)}\big|_{x_3=0} = \sigma_{33}^{(2)}\big|_{x_3=0} \quad \left(|x_1| \leq a\right), \\
& f(x_1) = f_1(x_1) + z_{01} + f_2(x_1) + z_{02}, \quad h = h_2 - h_1, \\
& \sigma_{33}^{(j)}\big|_{x_3=0} = 0 \left(|x_1| > a\right), \quad \sigma_{13}^{(j)}\big|_{x_3=0} = 0 \left(-\infty < x_1 < +\infty\right). \qquad (7.57)
\end{aligned}
$$

Where $h$—unknown relative displacement of nondeformable bodies, $w_j$—elastic displacement of the $j$th body along $O_j z_j$ axis, $\sigma_{mn}^{(j)}$—corresponding components of stress tensors.

The first equality in (7.57) is obtained from the condition of matched deformed boundary surfaces

$$
\begin{aligned}
& \Pi_{1*} : x_3 = -f_1(x_1) - z_{01} + h_1 - w_1\big|_{x_3=0}, \\
& \Pi_{2*} : x_3 = f_2(x_1) + z_{02} + h_2 + w_2\big|_{x_3=0}.
\end{aligned}
$$

on the contact surface. The second equality is the condition for continuity of normal stresses.

It is supposed that the bodies are under the influence of forces $P$ uniformly distributed along axis $Ox_2$ and directed oppositely along axis $Ox_3$. So, the problem is closed by the equilibrium condition of the system.

$$\int_{-a}^{a} p(\xi)d\xi = P, \quad p(x_1) = -\sigma_{33}^{(j)}\Big|_{x_3=0},$$

where $p$ is contact pressure.

The exact solution of the stated problem is connected with consideration of corresponding boundary conditions for each body (development of fundamental solutions similar to stated in 6.2). The corresponding mathematical problems depend essentially on the shape of the boundary. Hence, we accept two simplifying assumptions proceeding from smallness of the contact area and Saint-Venant's principle:

- Fundamental solutions for a half-plane are used for each interacting body, that is, it is assumed that the surface curvature effects but negligibly SSS;
- The boundary surfaces are substituted for parabolic cylinders, being equivalent to retaining the second-order terms in expansions of functions (7.56) into power series in terms of $x_1$ (remembering that $f_j'(0) = 0$):

$$\Pi_j: \; z_j = f_j(x_1) \approx -z_{0j} + \frac{x_1^2}{2R_j}, \quad \frac{1}{R_j} = f_j''(0),$$

where $R_j$—curvature radius of $\Pi_j$ surface in point $x_1 = 0$ (to be compared to (7.49)).

According to the second assumption, we should assume in the kinematic part of the boundary conditions (7.57) that

$$f(x_1) = \frac{x_1^2}{2R_{12}}, \quad \frac{1}{R_{12}} = \frac{1}{R_1} + \frac{1}{R_2}. \tag{7.58}$$

Based on the first assumption, and using properties of (7.28) for the fundamental solutions along with conditions (7.57) by analogy with (7.45), we may bring the problem to the integral equation in respect to the contact pressure.

$$\frac{\gamma}{\pi}\int_{-a}^{a} p(\xi)\left(\ln|x_1 - \xi| + C\right)d\xi = f(x_1) + h, \quad \gamma = \frac{1-\nu_1}{\mu_1} + \frac{1-\nu_2}{\mu_2}. \tag{7.59}$$

Hence, the solution is given by formulas (7.52)–(7.54), in which $(1-\nu)/\mu$ should be substituted by $\gamma$, and $R$ shall be understood as a linear dimension $R_{12}$ specified in (7.58). Then, the radius of the contact area is

$$a = \sqrt{2PR_{12}\frac{\gamma}{\pi}},$$

the contact pressure:

$$p(x_1) = \frac{P}{\pi}\sqrt{a^2 - x_1^2},$$

and the relative displacement of the undeforming bodies:

$$h = \frac{\gamma}{\pi}\left[C + \frac{a^2}{2}\left(\ln\frac{a}{2} - \frac{1}{2}\right)\right]P.$$

Despite the fact that the contact pressure is not explicitly dependent on the elastic properties of the body, this dependence is indirectly included into the formula for the contact area radius.

There exist several particular cases of the problem considered. If two bodies with equal curvature radii $R_1 = R_2 = R$ come into contact, then $R_{12} = R/2$. At equal elastic properties of the bodies ($\nu_1 = \nu_2 = \nu$ and $\mu_1 = \mu_2 = \mu$), we have $\gamma = 2(1 - \nu)/\mu$. If both conditions are met simultaneously, the radius of the contact area will be like that of a nondeformable punch contact with a half-plane, and the relative displacement will be twice as large as the penetration depth of the punch.

When $R_1 = R$ and $R_2 = \infty$, there is a problem on penetration of an elastic smooth punch $R_{12} = R$ into the elastic half-plane. If to consider additionally the first body as undeforming, which is equivalent to equality $(1 - \nu_1)/\mu_1 = 0$, we come to the problem solution (7.52)–(7.54).

## INTERACTION OF A PUNCH ON AN ELASTIC HALF-SPACE

Let us consider a 3D contact problem on a perfectly rigid punch in the form of elliptic paraboloid whose axis is perpendicular to plane $x_3 = 0$ restricting a half-space $x_3 \geq 0$. We shall place the origin of coordinates $O_1x_1x_2x_3$ on the boundary half-plane. Its axes $Ox_1$ and $Ox_2$ are selected so that the parabolic equation in coordinates $O_1x_1x_2z$ (point $O_1$ coincides with the paraboloid vertex, axis $O_1z$ is co-directed with $Ox_3$, see Fig. 7.7, where $z_0 = 0$) is brought to a canonical form.

$$z = -Ax_1^2 - Bx_2^2,\ A, B \geq 0.$$

Hence, taking into account the relationship of coordinates $x_3$, $z$, and the indenter penetration depth $h$

$$x_3 = z + h,$$

we obtain the equation for the punch boundary in coordinates $Ox_1x_2x_3$

$$x_3 = -Ax_1^2 - Bx_2^2 + h. \tag{7.60}$$

The corresponding boundary problem includes the elasticity theory equations, boundedness condition of solution at infinity and conditions on the half-space boundary.

$$u_3\big|_{x_3=0} = -Ax_1^2 - Bx_2^2 + h\ \left((x_1, x_2) \in \Omega\right),\ \sigma_{33}\big|_{x_3=0} = 0\ \left((x_1, x_2) \notin \Omega\right),$$

$$\sigma_{13}\big|_{x_3=0} = \sigma_{23}\big|_{x_3=0} = 0\ \left((x_1, x_2) \in R^2\right). \tag{7.61}$$

Similar to Ch. 4.10, the approximations were drawn in accordance with the linear approximation theory. In particular, conditions (7.61) are carried onto the unstrained boundary of the half-space, and the contact area is substituted for a plane area $\Omega$.

If the contact area $\Omega$ is unknown (Fig. 7.3b), a supplementary condition is required for its determination. The latter should specify this area as a set on which normal stresses are nonpositive, and the boundary presents an intersection line of the strained half-space surface and the paraboloid. This, however, leads to a complex mathematical problem. In this connection, we shall confine to a complementary simplifying assumption that the contact area is simply connected and its boundary is an ellipse given by

$$\partial\Omega: \frac{x_1^2}{a^2}+\frac{x_2^2}{b^2}=1. \tag{7.62}$$

Based on this approach, the unknowns in (7.61) are the penetration depth $h$ and semi-axes $a$ and $b$ of the ellipse. The problem is closed by the equilibrium condition of the punch

$$\iint_{\Omega} p(x_1,x_2)\,\mathrm{d}x_1\,\mathrm{d}x_2 = P, \quad p(x_1,x_2)=\sigma_{33}^{(j)}\Big|_{x_3=0}, \tag{7.63}$$

where $p$ is contact pressure.

By using the properties of (7.8) of the fundamental solutions and conditions (7.61), the problem is reduced to the integral equation in respect to the contact pressure

$$\iint_{\Omega} p(\xi_1,\xi_2)K(x_1-\xi_1,x_2-\xi_2)\,\mathrm{d}\xi_1\,\mathrm{d}\xi_2 = -Ax_1^2 - Bx_2^2 + h, \tag{7.64}$$

where kernel $K(x_1,x_2)$, according to (7.25), is found from:

$$K(x_1,x_2) = -u_{33}(x_1,x_2,0) = \frac{1-\nu}{\pi\mu r_2}, \quad r_2=\sqrt{x_1^2+x_2^2}.$$

The integral in the left-hand side of equation (7.64) is the potential value of a simple surface layer $x_3=0$, $(x_1,x_2)\in\Omega$ at $x_3=0$

$$U(x_1,x_2)=\iint_{\Omega}\frac{\rho(\xi_1,\xi_2)\,\mathrm{d}\xi_1\,\mathrm{d}\xi_2}{\sqrt{(x_1-\xi_1)^2+(x_2-\xi_2)^2}}, \tag{7.65}$$

with density

$$\rho(x_1,x_2)=\frac{1-\nu}{\pi\mu}p(x_1,x_2). \tag{7.66}$$

We can select such function $p(x_1,x_2)$ that the result of integration in (7.64) would be a multinomial of the same type as the right-hand side of this equation. For this aim, according to Shtaerman (1994), the density should be of the form

$$\rho(x_1,x_2)=\frac{a_0}{\sqrt{1-\frac{x_1^2}{a^2}-\frac{x_2^2}{b^2}}}+a_1\sqrt{1-\frac{x_1^2}{a^2}-\frac{x_2^2}{b^2}}. \tag{7.67}$$

After substitution of this expression into (7.65), and enough cumbersome transformations, we have:

$$U(x_1,x_2)=\pi ab\left\{(2a_0+a_1)I_0(a,b)-a_1\left[I_1(a,b)x_1^2+I_1(b,a)x_2^2\right]\right\},$$

$$I_0(a,b)=\frac{2}{a+b}K(k),\ k=\frac{|a-b|}{a+b},$$

$$I_1(a,b)=\frac{1}{a^2(a^2-b^2)}\left[2aK(k)-(a+b)E(k)\right], \tag{7.68}$$

where $K(k)$ and $E(k)$ are complete elliptic beta and gamma integrals of modulus $k$.

The comparison of the right-hand members of equations (7.64) and (7.68) gives the following relations:

$$h=\pi ab(2a_0+a_1)I_0(a,b), \tag{7.69}$$

$$\pi aba_1I_1(a,b)=A,\ \pi aba_1I_1(b,a)=B\cdot \tag{7.70}$$

As it follows from (7.66) and (7.67), the contact pressure will be:

$$p(x_1,x_2)=\frac{\pi\mu}{1-\nu}\left[\frac{a_0}{\sqrt{1-\frac{x_1^2}{a^2}-\frac{x_2^2}{b^2}}}+a_1\sqrt{1-\frac{x_1^2}{a^2}-\frac{x_2^2}{b^2}}\right]. \tag{7.71}$$

By substitution of this equality in (7.63), we obtain a representation of the resultant force (when calculating double integrals, the generalized polar coordinates $x_1=ar\cos\varphi,\ x_2=br\sin\varphi$ are used)

$$P=\frac{\pi\mu ab}{1-\nu}\int_{-\pi}^{\pi}d\phi\int_0^1 r\left(\frac{a_0}{\sqrt{1-r^2}}+a_1\sqrt{1-r^2}\right)dr=\frac{2\pi^2\mu ab}{1-\nu}\left(a_0+\frac{a_1}{3}\right). \tag{7.72}$$

The relations derived allow us to obtain the solutions for two basic problems.

1. *Area $\Omega$ is unknown.* In view of boundedness of the contact pressure the coefficient $a_0=0$. Then, proceeding from (7.72), we have

$$a_1=\frac{3P(1-\nu)}{2\pi^2\mu ab}.$$

Substitution of obtained coefficients into (7.71) gives the contact pressure

$$p(x_1,x_2)=\frac{3P}{2\pi ab}\sqrt{1-\frac{x_1^2}{a^2}-\frac{x_2^2}{b^2}}\cdot \tag{7.73}$$

This implies that similarly to the corresponding plane problem (see (7.53)), the pressure is bounded, reduces to zero on the boundary of the contact area and reaches the maximum $p_{\max} = 3P/(2\pi ab)$ in the central point.

Based on (7.69) and taking into account (7.68), we obtain a formula for the penetration depth

$$h = \frac{3P(1-\nu)}{2\pi\mu(a+b)} K(k). \tag{7.74}$$

Semi-axes $a$ and $b$ present a solution of the transcendental series of equations following (7.70) with regard to the expression for coefficient $a_1$

$$I_1(a,b) = \frac{2\pi\mu A}{3P(1-\nu)}, \quad I_1(b,a) = \frac{2\pi\mu B}{3P(1-\nu)}. \tag{7.75}$$

If the punch is a paraboloid of revolution ($A = B = 1/(2R)$, $R$—curvature radius of meridian in the central point), then $I_1(a,b) = I_1(b,a)$. Taking into account (7.68), after some transformations, we obtain the equation for modulus $k$

$$2abK(k) - (a^2 + b^2)E(k) = 0 .$$

It is easy to show that there is a unique solution, $k = 0$. Therefore, the contact area is a circle of radius $r_0 = a = b$. It is found by any of two relations in (7.75). For this purpose, let us find a limiting value of the left-hand side, for example, $I_1(a,b)$ at $a \to b$. Assuming that $a = b + \varepsilon$ and regarding that

$$k = O(\varepsilon),\ K(k) = \frac{\pi}{2}[1 + o(\varepsilon)],\ \ E(k) = \frac{\pi}{2}[1 + o(\varepsilon)],\ \varepsilon \to 0 ,$$

we obtain

$$\lim_{\varepsilon \to 0} I_1(a,b) = \frac{\pi}{b^3} = \frac{\pi}{r_0^3} .$$

Consequently, the contact area radius is

$$r_0^3 = \frac{3PR(1-\nu)}{4\mu}.$$

Correspondingly, based on (7.73) and (7.74), we have for contact pressure and penetration depth ($K(0) = \pi/2$)

$$p(x_1, x_2) = \frac{3P}{2\pi r_0^3}\sqrt{r_0^2 - r_2^2}, \quad r_2 = \sqrt{x_1^2 + x_2^2}\ , h = \frac{r_0^2}{2R} = \frac{1}{4}\sqrt[3]{\frac{9P^2(1-\nu)^2}{2R\mu^2}}. \tag{7.76}$$

2. *A flat-based punch* ($A = B = 0$). In this case, the right-hand side of integral equation (7.64) is constant. So, according to (7.68), $a_1 = 0$. Coefficient $a_0$ shall be found from (7.72)

$$a_0 = \frac{P(1-\nu)}{2\pi^2 \mu ab}.$$

The contact pressure and penetration depth are:

$$p(x_1, x_2) = \frac{P}{2\pi ab\sqrt{1 - \frac{x_1^2}{a^2} - \frac{x_2^2}{b^2}}},\ h = \frac{2P(1-\nu)}{\pi\mu(a+b)} K(k). \tag{7.77}$$

Similarly to the plane problem (see (7.55)), the pressure on the boundary of the contact area is not confined but has a power singularity of the order 1/2.

In the case of a circular punch with radius $r_0 = a = b$, formulas (7.77) transfer into the following ones (value $r_2$ was found in (7.76)):

$$p(x_1, x_2) = \frac{P}{2\pi\sqrt{r_0^2 - r_2^2}},\ h = \frac{P(1-\nu)}{2\mu r_0}.$$

They serve to calculate the pressure under a circular punch and its penetration depth.

## HERTZIAN PROBLEM

Let us consider a 3D problem of two contacting elastic bodies analogous to the one stated in 4.11. These bodies possessing symmetry axes and restricted by convex surfaces are assumed to enter into the contact at the initial moment of interaction in point $O$. Their symmetry axes pass through point $O$ and are perpendicular to a common tangential plane (Fig. 7.6). Let us introduce a stationary coordinate system $Ox_1x_2x_3$, where axes $Ox_1$ and $Ox_2$ lie in the tangential plane (supplementary requirements to their arrangement will be stated below), and axis $Ox_3$ is directed inside the second body ($j$ = 2). The equations for the surfaces restricting the bodies in coordinate systems $O_jx_1x_2z_j$, and related to the bodies, are of the type

$$\Pi_j : z_j = f_j(x_1, x_2),\ f_j(0,0) = -z_{0j},\ df_j(0,0) = 0,\ d^2 f_j(0,0) \geq 0. \tag{7.78}$$

Axes $O_jz_j$ are the symmetry axes, that is, they coincide with a straight line $Ox_3$, and are directed inside the corresponding bodies, $z_{0j}$—distances between $O_j$ and head points of the boundary surfaces (Fig. 7.6).

The unstrained boundary surfaces are set in coordinates $Ox_1x_2x_3$ by the equations (Fig. 7.7, (7.60))

$$\Pi_1 : x_3 = -f_1(x_1, x_2) - z_{01} + h_1,$$

$$\Pi_2 : x_3 = f_2(x_1, x_2) + z_{02} + h_2,$$

where $h_j$ are displacements of the undeforming bodies along $Ox_3$ axis.

The bodies are filled with homogeneous isotropic linearly elastic media characterized by constants $\nu_j$ and $\mu_j$. The contact occurs under conditions of a free slippage. Outside the contact surface the boundaries of the bodies are free from stresses. A corresponding boundary problem includes the equations taken in one form or another, and describing the plane strain state of both bodies and boundedness condition of the solution at infinity. Analogous to (7.5) we also add boundary conditions in the contact area $\Omega$ carried onto $Ox_1x_2$ axis

$$w_1\big|_{x_3=0}+w_2\big|_{x_3=0}=-f(x_1,x_2)-h,\ \sigma_{33}^{(1)}\big|_{x_3=0}=\sigma_{33}^{(2)}\big|_{x_3=0}\quad\left((x_1,x_2)\in\Omega\right),$$

$$f(x_1,x_2)=f_1(x_1,x_2)+z_{01}+f_2(x_1,x_2)+z_{02},\ h=h_2-h_1,$$

$$\sigma_{33}^{(j)}\big|_{x_3=0}=0\ \left((x_1,x_2)\notin\Omega\right),\ \sigma_{13}^{(j)}\big|_{x_3=0}=\sigma_{23}^{(j)}\big|_{x_3=0}=0\ \left((x_1,x_2)\in R^2\right). \tag{7.79}$$

The same designations as in (7.57) have used. It should be noted, that the contact area is unknown at all.

Forces $P$ directed oppositely along $Ox_3$ axis, are assumed to effect the bodies. So, the problem is closed by the equilibrium condition of system (7.63), where $p=-\sigma_{33}^{(j)}\big|_{x_3=0}$ —contact pressure.

On the grounds specified in items 4.11 and 4.12, we accept three simplifying assumptions, namely:

- the fundamental solutions for a half-space are used for each interacting body, wherefore the surface curvature is assumed to effect SSS insignificantly;
- the boundary surfaces are substituted for elliptic paraboloids, which is equivalent to retaining second-order terms in the power series expansions of functions (7.78) in two variables about point (0,0) (where $df_j(0,0)=0$)

$$\Pi_j:\ z_j=f_j(x_1,x_2)\approx -z_{0j}+A_jx_1^2+B_jx_2^2+2C_jx_1,x_2,$$

$$A_j=f_{j,11}(0,0),\ B_j=f_{j,22}(0,0),\ C_j=f_{j,12}(0,0); \tag{7.80}$$

- the contact area is simply connected and its boundary is an ellipse (7.62).

According to the second assumption, it should be assumed in the kinematic part of boundary conditions (7.79) that

$$f(x_1,x_2)=Ax_1^2+Bx_2^2+2Cx_1,x_2,$$

$$A=A_1+A_2,\ B=B_1+B_2,\ C=C_1+C_2.$$

The arrangement of axes $Ox_1$ and $Ox_2$ by revolving coordinate system $Ox_1x_2x_3$ about axis $Ox_3$ may be selected so that coefficient $C$ of the homogeneous quadratic polynomial in (7.80) will be equal to zero (the corresponding quadratic form is canonical)

$$f(x_1,x_2)=Ax_1^2+Bx_2^2.$$

The former designations are left for new coordinates.

Further we shall confine to considering such boundary surfaces, for which $A,B\geq 0$ and $AB\neq 0$ (the corresponding quadratic form is defined positively or semi-defined).

In accordance with the first two assumptions and using the properties of (7.8) for the fundamental solutions, and conditions (7.79) similarly to (7.64), we may reduce the problem to the integral equation on the contact pressure (value $\gamma$ is defined in (7.59))

$$\frac{\gamma}{\pi}\iint_{\Omega}\frac{p(\xi_1,\xi_2)}{\sqrt{(x_1-\xi_1)^2+(x_2-\xi_2)^2}}\,d\xi_1\,d\xi_2=-Ax_1^2-Bx_2^2+h\,.$$

Consequently, the problem solution is given by formulas (7.73)–(7.75), in which $(1-\nu)/\mu$ should be substituted for value $\gamma$, while the contact pressure is set by the same expression as in (7.73), and the relative displacement of the undeforming bodies by

$$h=\frac{3P\gamma}{2\pi(a+b)}K(k),$$

and the equation for semi-axes of the contact area is:

$$I_1(a,b)=\frac{2\pi A}{3P\gamma},\quad I_1(b,a)=\frac{2\pi B}{3P\gamma}.$$

Similarly to the plane problem, the dependence of the contact pressure on the elastic properties of the bodies is displayed in a corresponding dependence of semi-axes in the contact area.

If both bodies are bounded by the surfaces of revolution, then $A = B = 1/(2R)$. The pressure is defined by the first equality in (7.76). For the contact area radius and the depth of penetration, we have

$$r_0^3=\frac{3}{4}PR\gamma,\; h=\frac{1}{4}\sqrt[3]{\frac{9P^2\gamma^2}{2R}}\,.$$

If the properties of the elastic bodies are identical ($\nu_1=\nu_2=\nu$ and $\mu_1=\mu_2=\mu$) we have $\gamma=2(1-\nu)/\mu$.

If the second body ($j = 2$) is a half-space ($A_2 = B_2 = 0$), then $A = A_1$, $B = B_1$. If there is an additional requirement for dimensional stability of the first body, which is equivalent to the equality $(1-\nu_1)/\mu_1 = 0$, we obtain the problem solution (7.73)–(7.75).

Summarizing the chapters devoted to the elasticity theory, the following publications may be recommended for a more comprehensive study of this problem: (Bezuhov, 1968, Galin, 1953, Gorshkov et al., 2002, Gorshkov and Tarlakovskii, 1995, Filonenko-Borodich, 1947, Green and Zerna, 1968, Ilyushin, 1948, Lechnickiy, 1977, Lomakin, 1976, Prager, 1955, Sokolovskii, 1969, Timoshenko and Goodier, 1979, Tolokonnikov, 1979).

## NOTES

1. *Dirac, Paul Adrien Maurice* (1902–1984)—English physicist, one of the creators of quantum mechanics. Nobel Prize in 1933.

2. *Mindlin, R.D.* (1906–1987), American scientist and mechanical engineer.
3. *Cerrutti, V.* (1850–1909), Italian scientist and mechanical engineer.
4. *Fourier, Jean Baptiste Joseph* (1768–1830), French mathematician and physics, works in algebra, differential equations, and mathematical physics.
5. *Prandtl, Ludwig* (1875–1953) German mechanic; works in hydro and aerodynamics.

## KEYWORDS

- **Contact surfaces**
- **Fourier's transform**
- **Prandtl equation**
- **Stress–strain state**

# Chapter 8

## Foundations of the Theory of Plasticity

Deformability of solids under the effect of external forces and their capability of perceiving constant or *residual* (*plastic*) *strains* at unloading is called *plasticity*. There is, however, no unequivocal dependence between stresses and strains arising in the body, which means that it is impossible to find strains in terms of stresses, and vice versa, one cannot determine stresses proceeding from known strains.

The theory of plasticity deals with the laws interrelating stresses with elastoplastic deformations and the development of problem-solving methods on equilibrium and motion of deformed solid bodies. The theory of plasticity, forms the basis of today's calculations of different structures, forging processes, rolling, punching, and so forth, as well as natural processes (e.g., orogenesis). This allows to reveal strength and deformation potential of materials. Plastic deformation till fracture may reach 10–20%, while elastic deformations—only 0.3–0.5%. That is why strength calculations based on the assumption of only elastic deformations are often inexpedient both technically and economically. By taking plastic deformations into account, we may reduce stress concentration in structures, increase resistance of bodies to impact loads, define safety margins, rigidity and stability, ensuring thereby most efficient functioning, reliability and safety of structures.

### PLASTICITY OF MATERIALS AT TENSION AND COMPRESSION

The phenomena of elasticity and plasticity are displayed at sufficiently slow so-called *static* or *quasi-static* application of external forces. In this case, the phenomenon of deformability does not in fact depend on time, loading rate, and duration of external forces.

Let us consider the basic phenomena of plasticity by a simplest example of tension and compression of a cylindrical sample. Let us assume for definiteness that the sample material has the same elastoplastic properties as steel, aluminum, copper, nickel, and other metals. Besides, it was isotropic prior to the experiment and had equal yield strength values at tension and compression. Let us denote tensile stress in the sample by $\sigma$, and relative elongation by $\varepsilon$.

The dependence $\sigma \sim \varepsilon$ at tension (Fig. 8.1) for most materials has a straight initial portion until the stress exceeds the proportionality limit $\sigma_{pr}$. Hooke's law is valid in this area. The straight portion is followed by a curvilinear portion on which the stress to strain relationship is violated. Notice that some materials have a well-defined yield segment (Fig. 8.1a), while others (Fig. 8.1b) do not. For this case, a *conventional yield strength* is accepted, that is, the stress at which plastic deformation remains after unloading is 0.2%, or 0.002. It is denoted by $\sigma_{0.2}$.

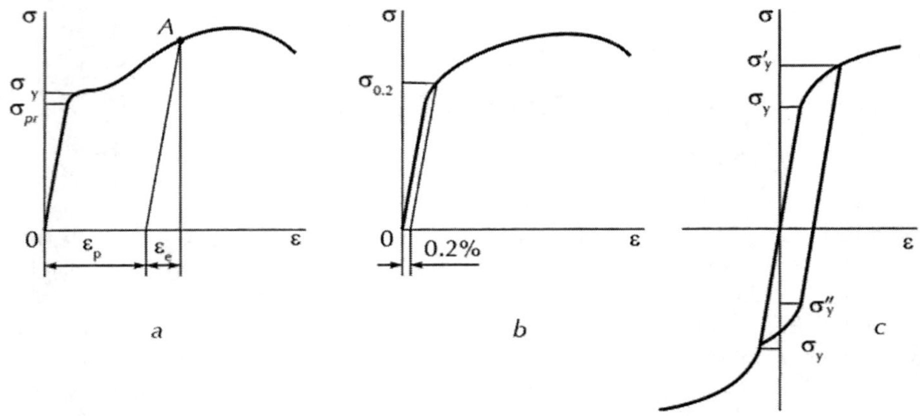


**Figure 8.1.**

We shall further denote yield strength by $\sigma_y$. Point *A* on the stress–strain diagram of the sample (Fig. 8.1a) corresponds to stress $\sigma$ and the total deformation $\varepsilon$, which, in its turn, may be divided into elastic deformation $\varepsilon_e$ and plastic deformation $\varepsilon_p$. After unloading the elastic deformation vanishes, but the plastic one remains.

Upon unloading the stresses–strain dependence is of a linear character, where the slope of the corresponding straight line is similar to the initial portion. Thus, the sample material behaves at unloading as the elastic one.

If to stretch the sample upon the complete unloading once again, as Fig. 8.1c shows the linear section lengthens till greater values $\sigma = \sigma'_y$ as compared to those observed for the first time ($\sigma'_y > \sigma_y$). Therefore, at repeated loading of the plastically deformed sample the limit of proportionality increases. This phenomenon is known as *strain hardening*.

If to repeat the loading by the opposite in sign force, that is compress it, the material achieves plasticity at stresses less than the initial limit of proportionality $\sigma_y$ ($\sigma''_y < \sigma_y$). This phenomenon has been investigated thoroughly by *Bauschinger*[1] and bears his name—*Bauschinger's effect*.

If the equality $\sigma'_y + \sigma''_y = 2\sigma_y$ is met (the more the yield point increases at tension, the more it decreases at compression), the material is called cyclically perfect.

Thus, the sample prestretched till the appearance of plastic deformations, has an increased limit of proportionality due to strain hardening at tension and decreased at compression. We should note that the study of the type requires a separate experience, since long samples used at tension lose stability at compression. Usually, a short cylindrical sample is cut out from the stretched sample and is subjected to compression. We may, however, plot the results of testing on the previous diagram and make reasoning as if the sample remained the same.

In general, keeping to Bell (1984), each aspect of this phenomenon systematically studied by Bauschinger has been discovered by other researchers. In 1858 residual deformations at twisting were studied by *Viedemann*[2] as a functions of the opposite in sign load. *Thurston*[3] was the first to publish the work on the elastic limit increase of a plastically deformed sample as early as in 1874. Subsequently, in 1876 it was also he who supposed that plasticity might include viscosity.

As the experiments have shown, the *rate of loading*, or the *strain rate* exerts certain influence on the sample deformation curve. The strain rate in conventional test machines varies within $\dot{\varepsilon} = 10^{-5} - 10^{-2}\ s^{-1}$. This deformation mode is called *static*. The diagram of static tests is independent of the strain rate. This dependence becomes distinct beginning with the rates of the order of $10^{-1}\ s^{-1}$. Figure 8.2 shows schematic test diagrams for a low-carbon steel obtained at three different levels of deformation rates: curve *1* corresponds to $\dot{\varepsilon} = 10^{-4}\ s^{-1}$, *2*—0,5 $s^{-1}$, *3*—$10^2\ s^{-1}$.

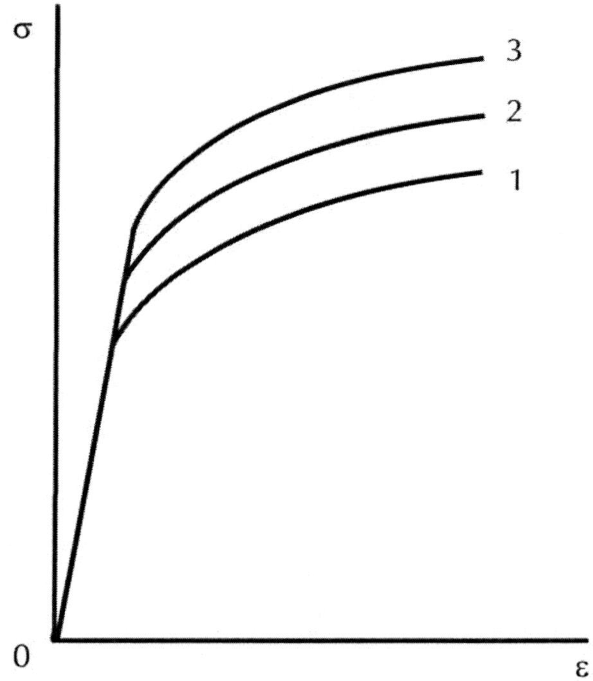


**Figure 8.2.**

Their comparison shows that the elasticity modulus at uniaxial tension does not virtually change. The stress–strain yield points increase with increasing tension rate. This expansion of the elastic deformation range is related to inertia of the plastic deformation mechanism. Analytical presentation of the yield point dependence on the strain rate may be assumed as linear:

$$\sigma_y = \sigma_{y0} + \mu_0 \dot{\varepsilon},$$

where $\sigma_{y0}$ is a static elastic limit, $\mu_0$—material constant to be defined experimentally.

At increasing strain rate the material *ultimate strength* $\sigma_b$ (the stress equal to material fracture) grows noticeably, and deformation $\varepsilon_b$ (*ultimate plastic extension*) related to it decreases. This phenomenon is known as *a material embrittlement*. As a matter of fact, a brittle material fails without any evident plastic deformations prior to rupture.

For high-strength steels the range of plastic deformations occurring before fracture is not wide, and shortens sharply at dynamic loading. Embrittlement of the carbon steel SAE1020, and abrupt reduction of the ultimate plastic elongation starts with deformation rates $2\cdot10^2$ s$^{-1}$. At the rates till $3\cdot10^2$ s$^{-1}$ the ultimate deformation decreases almost twice as much.

High-strength titanium alloys demonstrate a clear-cut dependence on the strain rate beginning from $10^{-1}$ s$^{-1}$. Further increase results in decreasing ultimate plastic elongation and increasing ultimate strength of the material.

It should be noted, that until now there are no so far any reliable systematic experimental studies on the influence of deformation rates for a number of materials. This is because ultimate rate at which above-effects may be observed, are not widely used in engineering. The static diagrams for tensile testing of materials turn to be sufficient in majority of cases for characterizing their mechanical properties.

The dependence of material properties on testing velocity of metals is more apparent under high temperatures. In this case, not only the value of ultimate deformations, but also the deformation curve slope at the initial portion changes essentially.

## PLASTICITY CONDITIONS

The theory of plasticity establishes the relationship between stresses and strains (*deformation theories*) or between the strain rates (*plastic flow theories*) in the plastic range of materials. Stresses often depend not only on the current deformations, but on the history (process) of deformation as well. It is natural that the corresponding problems in view of the nonlinear relationship between stresses and strains turn to be more complex than the elasticity problems. For simplicity, the dependence $\sigma \sim \varepsilon$ for a real material is often approximated as the piecewise polygonal straight lines shown in Fig. 8.3a–c. *Prandl's diagram*, shown in Fig. 8.3a, is the simplest one for the perfectly plastic material (mild steels, titanium alloys).

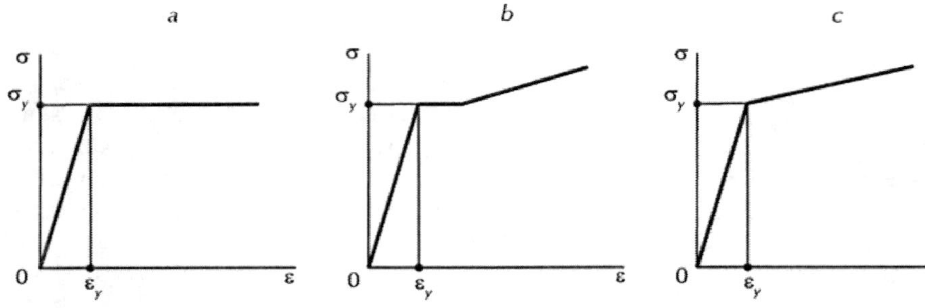


**Figure 8.3.**

A bilinear curve for elastic strain-hardening is shown in Fig. 8.2c. These two approximations are commonly used to solve plasticity problems. For example, the theory of perfect plasticity is used to calculate bearing capacity of construction elements.

Curves $\sigma \sim \varepsilon$ are very close to each other at tension and compression for most materials, so hereafter we shall consider them as coincident. The comprehensive experiments show that the law of unloading is not always linear. The existing plasticity theories ignore these insignificant deviations from Hooke's law at unloading, and the difference between the proportionality limit and the yield point. The constitutive equation for the stress–strain relationship does not include time explicitly.

Let us consider an arbitrary shaped body, assuming that the initial stresses and strains are not present in it. At the initial stage of loading of the body there occur only elastic strain; consequently, the appearance of plastic deformations is unambiguously defined by the acting stresses. In this connection, the *condition of plasticity* may be written as a certain function of the stress tensor components. For an isotropic material the condition, under which plastic deformations may appear, cannot depend on selection of the coordinate system. This is why the corresponding function will be defined by three invariants of the stress tensor, which may be presented, for example, by three principal stresses:

$$f(\sigma_1, \sigma_2, \sigma_3) = 0 \tag{8.1}$$

As a rule, some constant (*yield point*) characteristic for a given material is included in the relation. At a uniaxial tension, we have $\sigma_1 = \sigma_y$.

As the experiments show plastic deformations are associated with the phenomena of shear. So, it is quite natural that widely-spread plasticity theories fundamentally compare some tangential stresses with ultimate ones, which cause flow. Based on *Tresca*'s[4] experiments for metal outflow through holes, Saint-Venant[5] supposed that the transition from the elastic state into plastic happens if the maximum tangential stress reaches a certain limiting value for a given material:

$$|\tau_{\max}| = \frac{|\sigma_1 - \sigma_3|}{2} = \frac{\sigma_y}{2},$$

or

$$\sigma_1 - \sigma_3 = \sigma_y, \tag{8.2}$$

where the principal stresses similarly to previous ones are numbered in the decreasing order $\sigma_1 \geq \sigma_2 \geq \sigma_3$; $\sigma_y$ is the yield point of material at tension. The body material is assumed initially isotropic. It should be noted that Tresca–Saint-Venant's criterion is also known as the *strength theory of the maximum shearing stress* ($\sigma_{red} = \sigma_1 - \sigma_3 \leq [\sigma]$) in the textbooks on strength of materials.

*Another plasticity criterion* defines the transition of initially isotropic material from the elastic into the plastic state, if

$$\sigma_u = \sigma_y. \tag{8.3}$$

Here $\sigma_u$ is the stress intensity (1.24) proportional to the octahedral tangential stress, and its square to the second invariant of the stress deviator:

$$\sigma_u^2 = -3J_{2d} = \tfrac{1}{2}\left[(\sigma_1 - \sigma_2)^2 + (\sigma_2 - \sigma_3)^2 + (\sigma_3 - \sigma_1)^2\right].$$

In this case, it is the octahedral tangential stress, but not the maximal one that reaches a certain limiting value for a given material. This criterion corresponds to the known condition of the maximum strain energy of failure and is called *Mises condition of plasticity* (Hyber[6]—Mises[7]—Hencky[8]).

The results obtained on the basis of the plasticity criteria (8.2), (8.3) are essentially similar. It should be noted, however, that most of experiments conducted for a plane stress condition show better agreement with the Mises criterion. At solving specific problems we use, as a rule, the criterion that simplifies the solution.

Numerous experiments testify that the material is deformed elastically at a uniform tension or compression. Then, we may assume that the plasticity condition in a general case is defined not by the entire stress tensor, like in (8.1), but only by its deviator part. As we have already noted, a transition into the plastic state cannot depend on selection of the coordinate system, so the plasticity condition is a certain function of the stress deviator invariants. The first invariant is equal to zero $J_{1d} = 0$, that is why in a general case the condition, under which plastic deformations appear, is defined by the second and third invariants of the stress deviator:

$$f_1(J_{2d}, J_{3d}) = 0 \tag{8.4}$$

This equation in the coordinate system $\sigma_1$, $\sigma_2$, $\sigma_3$ describes a certain surface, which is called *yield surface*.

Above-mentioned plasticity criteria enable us to fix the moment of the appearance of the first plastic deformations. These criteria are suffice for solving problems of plasticity in the case, the material deformation under a uniaxial stress state obeys Prandtl's diagram (Fig. 8.3a). This is due to the fact that the condition under which the plastic deformation appears, does not change at repeated loading of such materials.

This situation changes if the material under consideration is *hardened* (Fig. 8.3b, c). Any increase in the yield point, which value depends on the accumulated plastic strains, is typical for such materials at repeated loading. We shall introduce a condition of hardening, which looks like the condition of plasticity by its form (8.4):

$$f_1(J_{2d}, J_{3d}) = \Phi(\eta) \tag{8.5}$$

Condition (8.4) contains the stress deviator invariants and constants of the material, for example, its yield point. Condition (8.5) includes a certain function $\Phi(\eta)$ dependent on the material hardening parameter $\eta$. This equation also defines the yield surface within the space of principal stresses, whose change of position, shape, and dimensions during loading characterizes *strain hardening* of the material. If the hardening parameter coincides with the strain rate, generalization of the Mises condition may be taken as an optional criterion (8.5)

$$\sigma_u = \Phi(\varepsilon_u).$$

Here are some examples of constructing yield surfaces (8.4) for above-considered plasticity criteria. In case of the Mises criterion (8.3), the equation ($\sigma_u = \sigma_y$) for the yield surface may be written as:

$$(\sigma_1-\sigma_2)^2+(\sigma_2-\sigma_3)^2+(\sigma_3-\sigma_1)^2-2\sigma_y^2=0\,.$$

This equation in axes $\sigma_1$, $\sigma_2$, $\sigma_3$ describes a cylinder (Fig. 8.4) whose axis is equally inclined to the coordinate axes. If to dissect the cylinder by plane $\sigma_3 = 0$, the sectional view will be presented by an ellipse, the equation for which is

$$\sigma_1^2-\sigma_1\sigma_2+\sigma_2^2=\sigma_y^2\,.$$

Consequently, the yield surface in the form of a circular cylinder with the radius in the plane perpendicular to the axis equal to $\sigma_y/\sqrt{2}$ corresponds to the Mises plasticity criterion.

This equation in axes $\sigma_1$, $\sigma_2$, $\sigma_3$ describes a cylinder (Fig. 8.4), whose axis is equally inclined to the coordinate axes. If to dissect the cylinder by plane $\sigma_3 = 0$, the sectional view will be presented by an ellipse, the equation for which is

$$\sigma_1^2-\sigma_1\sigma_2+\sigma_2^2=\sigma_y^2\,.$$

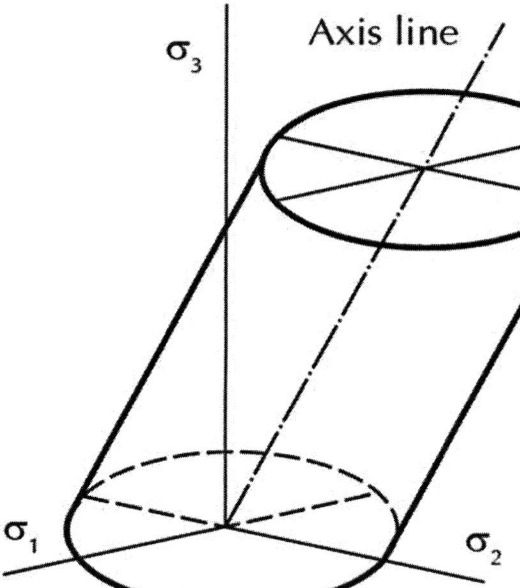


**Figure 8.4.**

Consequently, the yield surface in the form of a circular cylinder with the radius in the plane perpendicular to the axis equal to $\sigma_y/\sqrt{2}$ corresponds to the Mises plasticity criterion.

If to accept Tresca–Saint-Venant's criterion ($\sigma_1 - \sigma_3 = \sigma_y$), bearing in mind that numbering of the principal stresses is not always known in advance, the following equalities may be true:

$$\sigma_1 - \sigma_3 = \pm\sigma_y,\quad \sigma_3 - \sigma_2 = \pm\sigma_y,\quad \sigma_2 - \sigma_1 = \pm\sigma_y .$$

In this case, the yield surface is presented as a hexagonal prism with the axis equally sloping also to axes $\sigma_1$, $\sigma_2$, $\sigma_3$ (Fig. 8.5a). This prism, known as the *Coulomb*[9] *prism*, turns to be inscribed into the *Mises cylinder*. The axes of the prism and the cylinder coincide. The equation for this axis is $\sigma_1 = \sigma_2 = \sigma_3$. Figure 8.5b shows the plane section of the cylinder and the prism $\sigma_3 = 0$ corresponding to the plane stress condition.

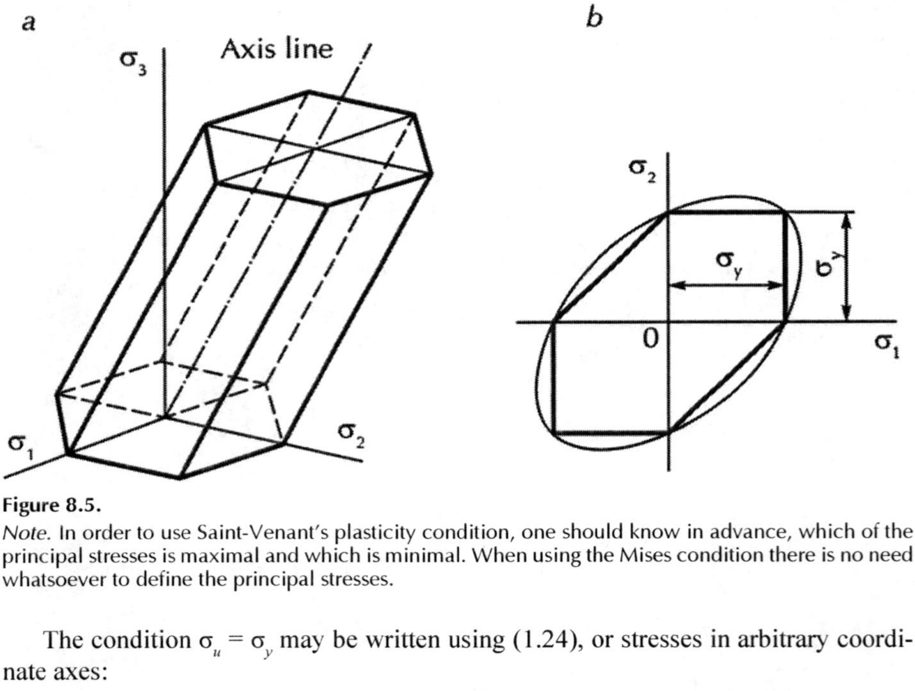


**Figure 8.5.**
*Note.* In order to use Saint-Venant's plasticity condition, one should know in advance, which of the principal stresses is maximal and which is minimal. When using the Mises condition there is no need whatsoever to define the principal stresses.

The condition $\sigma_u = \sigma_y$ may be written using (1.24), or stresses in arbitrary coordinate axes:

$$(\sigma_{11} - \sigma_{22})^2 + (\sigma_{22} - \sigma_{33})^2 + (\sigma_{33} - \sigma_{11})^2 + 6(\sigma_{12}^2 + \sigma_{23}^2 + \sigma_{31}^2) = 2\sigma_y^2 .$$

As it is seen, the use of the Mises plasticity condition for solving these problems turns to be more suitable in many cases as compared to Saint-Venant's conditions. The maximal difference between the calculations according to these criteria does not exceed 13% (the case of pure shear).

## SIMPLE AND COMPLEX LOADING

The definition of the relations of the plasticity theory, namely, stress–strain dependencies, should obviously take into account not only the current values of stress and strain tensor components, but also the ways of achieving them. In the plasticity theory we distinguish between two types of loading a body: simple and combined.

Remember (1.9) that loading is called *simple* if all stress tensor components are increasing proportionally to a single common parameter $t$ (e.g., to time).

In this case, the directional tensor components (1.26) remain invariable, or otherwise the loading is called *combined.*

Let us now consider some examples of simple and combined loading. We shall assume that a cylindrical tube is subjected to a uniform axial tension and torsion (Fig. 8.6). If the wall of this tube is sufficiently thin, its stress state may be considered to be plane. Normal ($\sigma_x$) and tangential ($\tau$) stresses are obtained from the known expressions

$$\sigma_x = \frac{P}{2\pi R\delta}, \tau = \frac{m_{ê}}{\pi R^2 \delta}.$$

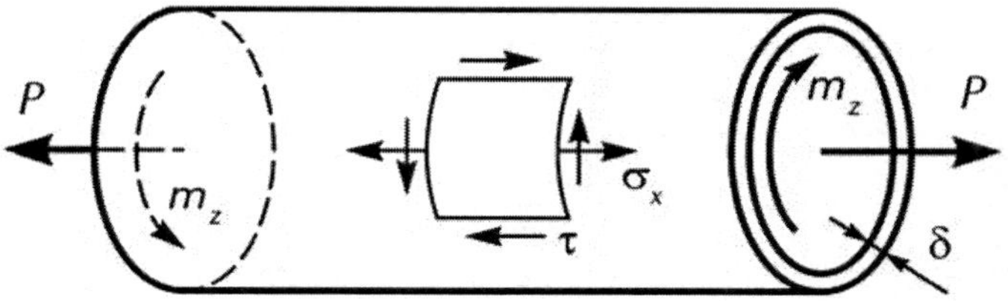


Figure 8.6.

If external effects $P$, $m_z$ vary proportionally to a single parameter $\lambda$, for example, time, then *the simple* loading is performed, since the stress tensor components are varying proportionally to the same parameter. Figure 8.7 shows the loading path in stress axes $\sigma_x$, $\tau$. Simple loading corresponds to the beam *OA*.

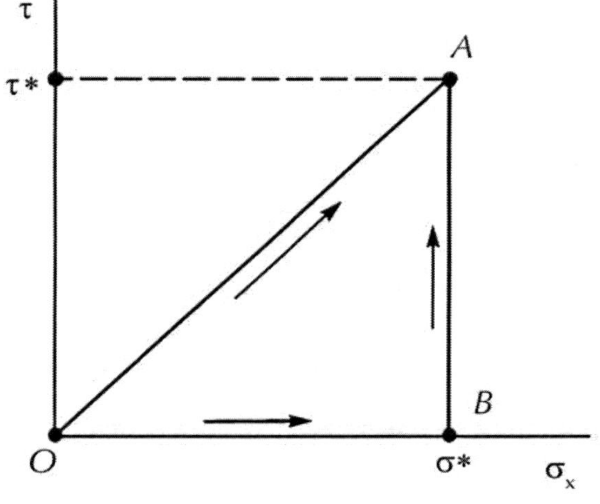


Figure 8.7.

Hence, for a simple loading the following equality is valid:

$$\sigma_{ij} = \lambda \sigma_{ij}^{0},$$

where $\sigma_{ij}^{0}$ are some initial values of the stress tensor at load application. Then, mean stress $\sigma$, stress rate $\sigma_u$, and the stress deviator modulus $s$ in view of formulas (1.24), (1.25) will be also linear with respect to parameter $\lambda$:

$$\sigma = \lambda \sigma^{0}, \sigma_u = \lambda \sigma_u^{0}, s = \sqrt{\tfrac{3}{2}}\sigma_u = \lambda s^{0}.$$

The components of the directional stress tensor (1.26), in their turn, are independent of $\lambda$, because

$$\bar{s}_{ij} = \frac{\sigma_{ij} - \sigma\delta_{ij}}{s} = \frac{\lambda\sigma_{ij}^{0} - \lambda\sigma^{0}\delta_{ij}}{\lambda s^{0}} = \bar{s}_{ij}^{0}.$$

Let us now consider another loading (Fig. 8.7), under which axial load $P$ was applied first to the tube to create the normal stress. After the normal stress value has reached $\sigma^*$, we apply torque $m_z$. Normal stress $\sigma^*$ remained invariable during torque application, and the tangential one was growing from zero till $\tau^*$value. As a result, we arrive again in point $A$, but along the polygonal trajectory. Such loading is *combined.* We would come to a similar result if to subject the tube first to torsion till the value $\tau^*$, and then to tension till stress $\sigma^*$.

For an elastic body the sequence of loading plays no part since there is an unequivocal agreement between stress and strain states, no matter how they have been created.

In elastoplastic bodies the situation is different in principle: not only the nature of the stress state in the points, but also the trajectory it has been created is of great significance. So the strain state may vary considerably in the same points of the body depending on these phenomena.

Ilyushin has proved a *theorem on conditions sufficient for simple loading.* For this purpose, proportionality is to be provided for the external stress to some common parameter, as well as exponential dependence of the stress rate versus strain rate:

$$\sigma_u = A\varepsilon_u^{\alpha},$$

where $A$ and $\alpha$ are constants. When $\alpha = 0$ the equation agrees with the Mises plasticity condition $\sigma_u$ = const. At small $\alpha$, it produces the curves with small hardening $d\sigma_u/d\varepsilon_u$, and if $\alpha = 1$, it follows Hooke's law.

It should be noted, that this condition is sufficient, but it is not obligatory. Simple loading may also occur sometimes, when above-mentioned condition is violated.

When studying plasticity it is essential to distinguish between the active and passive processes of deformation. During the *active* process, also called *load,* the increment of the work induced by external stresses acting on a body (or by the interaction forces between the considered material element and surrounding parts of the medium) is positive. The negative increment of this work corresponds to the *passive* process or *unloading*. Considering an elementary volume of the material, the *active process condition* may be written in the form of the next equation:

$$\sigma_{ij}\varepsilon_{ij} > 0,$$

the *condition for a passive process* may be given by

$$\sigma_{ij}\varepsilon_{ij} < 0.$$

The active process is sometimes recognized as the one at which plastic deformation increases, while the passive one as the process at which plastic deformation remains invariable. It is assumed in the theory of plasticity that the tensor of the total deformation is presented at any moment of deformation as a sum of *tensors of elastic and plastic deformations*:

$$\varepsilon_{ij} = \varepsilon_{ij}^{e} + \varepsilon_{ij}^{p},$$

where the first tensor varies during both active and passive processes, while the plastic strain tensor varies only during the active process.

*Plastic deformation* is defined as a set of components of a strain tensor preserved in a given point of a medium when all components of the stress tensor in this point turn to zero. Hence, plastic deformations are identified as residual deformations.

## HYPOTHESES OF THE THEORY OF SMALL ELASTOPLASTIC DEFORMATIONS

The hypotheses proposed by Hyber, Mises, and Hencky and generalized for the material with Nadai's hardening lie in the base of this deformation theory. It presumes that one can establish stress–strain dependencies for elastoplastic bodies similarly to Hooke's law for elastic bodies. The development and substantiation of the theory of small elastoplastic deformations are closely connected with Ilyushin's works. Therefore, the theory of small elastoplastic deformations is often called the *Ilyushin theory of plasticity*. It assumes that under the conditions of a simple active deformation of initially isotropic material whose properties are independent of the third invariant of the stress tensor, the following three hypotheses are valid.

**1. Elasticity of volume deformation.** Volume deformation θ is considered to be elastic and directly proportional to the mean normal stress σ, for which Hooke's law is valid

$$\sigma = K\theta = 3K\varepsilon.$$

This means that the volume varies owing to elastic deformations only, and in the process of plastic deformation the material behaves as an incompressible one. Therefore, this hypothesis may be formulated in a different way, namely: plastic deformation does not vary the volume of a body.

**2. Coaxiality of deviators.** The components of strain deviator $э_{ij}$ are proportional to the components of stress deviator $s_{ij}$. Their relation may be written in the form proposed by Ilyushin:

$$s_{ij} = \frac{2}{3}\frac{\sigma_u}{\varepsilon_u} \text{э}_{ij}, \tag{8.6}$$

where $\sigma_u$, $\varepsilon_u$ are the rates of stress and strain tensors.

In the stress and strain components, expression (8.6) looks like:

$$\sigma_{ij} - \sigma\delta_{ij} = \frac{2}{3}\frac{\sigma_u}{\varepsilon_u}(\varepsilon_{ij} - \varepsilon\delta_{ij}) \ (i, j = 1, 2, 3).$$

**3. The hypothesis of hardening**. Irrespective of the type of stress state, there is a universal dependence between the stress and strain rates for each material:

$$\sigma_u = \Phi(\varepsilon_u). \tag{8.7}$$

For an elastic material this relationship is expressed by a linear dependence $\sigma_u = 3G\varepsilon_u$.

The experimental check of above-cited hypotheses has produced rather good results for both simple loading and closely approximated to it one. The deformation process should be active without unloading in this case.

Although the condition of proportional loading is rarely fulfilled in reality, as the experiments have proved, the considered theory of deformation is true also in the cases when some departure from the law of proportional loading takes place.

## FORMULATION OF THE PROBLEM OF SMALL ELASTOPLASTIC DEFORMATIONS

Let us consider an elastoplastic body dependent upon body forces $\rho F_i$ and surface loads $R_i$. To solve the problems of the theory of small elastoplastic deformations, that is, for defining unknown displacements, strains and stresses ($u_i$, $\sigma_{ij}$, $\varepsilon_{ij}$; $i, j = 1, 2, 3$), we have equilibrium equations, Cauchy relations, deformation compatibility equations, and boundary conditions:

$$\sigma_{ij},_j + \rho F_i = 0;$$

$$\varepsilon_{ij} = \tfrac{1}{2}(u_i,_j + u_j,_i);$$

$$\mathring{a}_{\alpha\alpha},_{\beta\beta} + \mathring{a}_{\beta\beta},_{\alpha\alpha} - 2\varepsilon_{\alpha\beta},_{\alpha\beta} = 0,$$

$$(\mathring{a}_{\alpha\beta},_\gamma - \mathring{a}_{\beta\gamma},_\alpha + \mathring{a}_{\gamma\alpha},_\beta),_\alpha - \varepsilon_{\alpha\alpha},_{\beta\gamma} = 0;$$

$$u_i = u_{i0}(x) \text{ on } S_u, \sigma_{ij} l_j = R_i \text{ on } S_\sigma. \tag{8.8}$$

It is convenient to write the constitutive equations as follows:

$$s_{ij} = 2G\varphi(\varepsilon_u)\text{э}_{ij}, \sigma = 3K\varepsilon. \tag{8.9}$$

where, $s_{ij} = \sigma_{ij} - \sigma\delta_{ij}$; $\text{э}_{ij} = \varepsilon_{ij} - \varepsilon_{ij}\delta_{ij}$; $i, j = 1, 2, 3$.

By comparing (8.7) and (8.9), we can find an expression for plasticity function $\varphi(\varepsilon_u)$ in terms of a general function $\Phi(\varepsilon_u)$ introduced earlier:

$$\varphi(\varepsilon_u) \equiv \frac{\Phi(\varepsilon_u)}{3G\varepsilon_u}.$$

Equations (8.8) are valid only under loading. In the case of elastic unloading the following relations result from the generalized Hooke's law (2.6)

$$\sigma_{ij} - \sigma'_{ij} = 2\mu(\varepsilon_{ij} - \varepsilon'_{ij}) + \lambda(\theta - \theta')\delta_{ij}, \tag{8.10}$$

where $\sigma_{ij}'$ and $\varepsilon_{ij}'$ are stresses and strains, which existed prior to unloading onset; $\lambda$ and $\mu$—Lame's constants.

The equations for unloading in the form of (8.10) are preserved until new (secondary) plastic deformations appear in the process of unloading.

When solving the problems of elastoplasticity, the sought functions are either displacements or stresses, depending on the statement. The obtained solution should meet not only the force and kinematic boundary conditions, but also auxiliary conditions at the interface separating the zones of elastic and plastic deformations.

Since the problem of the plasticity theory is linear, it poses a question on its correctness and uniqueness. It was proved in (Lenskiy, 1969) that under the condition

$$3G \geq \frac{\sigma_u}{\varepsilon_u} \geq \frac{d\sigma_u}{d\varepsilon_u} > 0 \tag{8.11}$$

the series of equations (8.8), (8.9) is elliptical, this boundary problem can be solved if there exists a solution of the corresponding problem of the linear theory of elasticity.

As for its uniqueness, Ilyushin has proved a theorem (Ilyushin, 1948, 1963) that under given volume forces $\rho F_i$ and surface forces $R_i$ on a part of the boundary surface $S_\sigma$ and displacements $u_i$ on a part of the boundary surface $S_u$, the stress–strain state (SSS) of the body, that is, $u_i$, $\sigma_{ij}$, $\varepsilon_{ij}$, are defined uniquely if the loading is simple.

## A METHOD OF ELASTIC SOLUTIONS

The plastic problem solution is connected with calculation of a series of nonlinear partial differential equations. This presents an intricate mathematical problem, which is solved in a closed-form in exceptional cases. Therefore, we use approximate methods more often. One of them is the method of successive approximations proposed by Ilyushin to solve the problems of the theory of small elastoplastic deformations at active loading. It is known in the theory of plasticity as the *method of linear approximations*. Its essence consists in considering a sequence of linear elastic problems, whose solution is brought to solving the elasticity theory problem as their number in a series ascends.

To develop an algorithm for this method let us present function $\varphi(\varepsilon_u)$ by

$$\varphi(\varepsilon_u) = 1 - \omega(\varepsilon_u). \tag{8.12}$$

Let us substitute (8.12) into (8.9), then the physical relations become:

$$s_{ij} = 2G(1 - \omega(\varepsilon_u))\,Э_{ij}, \sigma = 3K\varepsilon, \tag{8.13}$$

where $0 \leq \omega < 1$, so $\omega(\varepsilon_u) = 0$, if $\varepsilon_u \leq \varepsilon_y$.

Thus, at ω = 0, equations (8.13) coincide with the relations of the linear elasticity theory. We shall present the differential equilibrium equation and boundary conditions by expanding the stress tensor into the deviator and spherical parts:

$$s_{ij},_j + \sigma,_i + \rho F_i = 0\,,$$

$$s_{ij} l_j + \sigma l_i = R_i \quad \text{on} \quad S_\sigma,\ u_i = u_{i0}(x) \quad \text{on} \quad S_u\,. \tag{8.14}$$

To solve the elastoplastic problem in terms of displacements let us substitute components (8.13) into the equilibrium equations and natural boundary conditions (8.14), and consider Cauchy's relations. As a result, we obtain generalization of Lame's equations (3.6):

$$(\lambda + \mu)\theta,_i + \mu \Delta u_i + \rho F_i - \rho F_{\omega i} = 0\,,$$

$$\lambda \theta l_i + \mu (u_i,_j + u_j,_i) l_j = R_i + R_{\omega i}$$

where

$$F_{\omega i} = 2G(\omega\, э_{ij}),_j\,,\ R_{\omega i} = 2G\, э_{ij}\, l_j \omega\,.$$

If we assume $\omega^{(0)} = 0$ as a zeroth approximation, then loads $F_{\omega i} = R_{\omega i} = 0$ will be fictitious, and to define the first approximation $u_i^{(1)}$ we have an ordinary problem of the linear elasticity. From obtained displacements $u_i^{(1)}$ we shall find the following values:

$$э_{ij}^{(1)},\ \varepsilon_u^{(1)},\ \omega^{(1)} \equiv \omega(\varepsilon_u^{(1)}),\ F_{\omega i}^{(1)},\ R_{\omega i}^{(1)}\,.$$

The equilibrium equations and boundary conditions take place for any $k$-th approximation

$$(\lambda + \mu)\theta,_i^{(k)} + \mu \Delta u_i^{(k)} + \rho F_i - \rho F_{\omega i}^{(k-1)} = 0\,,$$

$$\lambda \theta^{(k)} l_i + \mu (u_i^{(k)},_j - u_j^{(k)},_i) l_j = R_i + R_{\omega i}^{(k-1)}\,, \tag{8.15}$$

where $F_{\omega i}^{(k\text{-}1)}$, $R_{\omega i}^{(k\text{-}1)}$ are found by preceding $(k - 1)$-th approximation.

Modified Lame's equations and boundary conditions (8.15) are linear with respect to unknown displacements $u_i^{(k)}$. They differ from the corresponding equations of elasticity in that the fictitious forces $F_{\omega i}^{(k\ 1)}$, $R_{\omega i}^{(k\ 1)}$ are added to external forces $\rho F_i$, $R_i$. This allows us to construct a solution of the elastoplastic problem recursively in terms of the corresponding known solution of the elastic problem. An example of such iterative solution is disclosed in 8.11.

To provide the convergence of the method of linear approximation it is necessary that parameter ω related to function $\varphi(\varepsilon_u)$ by equation (8.11) be small as compared to a unity. For this aim, the following condition (Moskvitin, 1981) shall be met:

$$1 > \omega + \varepsilon_u \frac{d\omega}{d\varepsilon_u} \geq \omega \geq 0\,.$$

The convergence of this method has been studied by various authors. A detailed enough bibliography on this matter is contained in the works of Pobedrya. It has been established in practice at solving specific problems that the rate of convergence of the method of linear approximation is quite high, so only few approximations are suffice to obtain a required accuracy. For example, for solving the problem on variable elastoplastic bending of a circular sandwich plate (refer to 8.13) five iterations were necessary.

In conclusion, we give a plasticity function used to describe the behavior of duralumin in the plastic domain (Starovoitov, 2004). Its closed form may look like:

$$\omega(\varepsilon_u) = \begin{cases} 0, & \varepsilon_u \le \varepsilon_y, \\ A(1-\varepsilon_y/\varepsilon_u)^{\alpha}, \varepsilon_u > \varepsilon_y, \end{cases}$$

where $A = 096$, $\alpha = 2.34$, $\varepsilon_y = 0.735\%$—deformation corresponding to the yield point of Д16Т alloy.

## GEOMETRIC INTERPRETATION OF LOADING PROCESS

When we consider the theory of plastic flow, it is convenient to proceed from the geometric interpretation of the loading process, namely of the yield surface mentioned above (8.2). Let us single out an element in the form of a parallelepiped in the body under consideration (Fig. 8.8), being so small that its stress state can be assume a homogeneous. Let us relate this element to axes $x_1$, $x_2$, $x_3$ and denote stress components acting on its faces by $\sigma_{ij}$ $(i, j = 1, 2, 3)$.

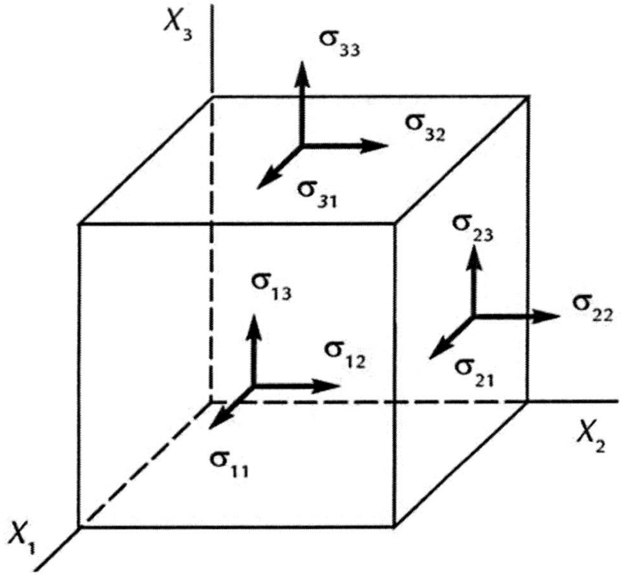


**Figure 8.8.**

The stress tensor is symmetric, therefore six values $\sigma_{ij}$ would be suffice to characterize the stress state. Let us now take a *six-dimensional stress space* where we shall plot one stress component along each axis. Let us compare the stress state of the element to a point with Cartesian coordinates $\sigma_{ij}$ in the stress space.

Here, the origin of coordinates complies with the unloaded condition. Loading of the specimen is accompanied by variations in $\sigma_{ij}$ values and, hence, the point in the stress space describing the stress state of the considered element draws a certain trajectory, that is, a *load path*.

At a uniaxial stress state all components $\sigma_{ij}$, except one, for example, $\sigma_{11}$, are equal to zero. In this case, the load path coincides with axis $\sigma_{11}$. The appearance of plastic deformation, according to the modules of the preceding paragraph, is because $\sigma_{11}$ has reached value $\sigma_y$, which is typical for the given material. Thus, we may single out such domain on axis $\sigma_{11}$, which contains the origin of coordinates, inside which the state of the material is elastic at the initial loading. In Figs. 8.9 and 8.10 this domain is denoted by $\Omega$. Its boundaries are the points with coordinates $\pm\sigma_y$ corresponding to the case of equal yield limits at tension and compression.

If a body is *perfectly plastic* (Figs. 8.3a, 8.9), the points situated outside domain $\Omega$, cannot be revealed, as the stresses greater than $\sigma_y$ are inaccessible according to the model of the perfectly plastic body. The emergence of a representative point on the boundary $\Omega$ means the transition of the body element into the state of flow. The deformation in this case is indefinite. Displacements of the point from the boundary inside domain $\Omega$ (unloading) corresponds to a variation of the elastic stage only of deformation.

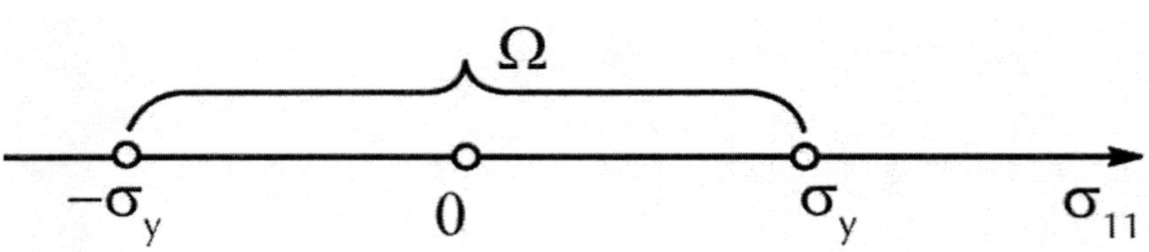


Figure 8.9.

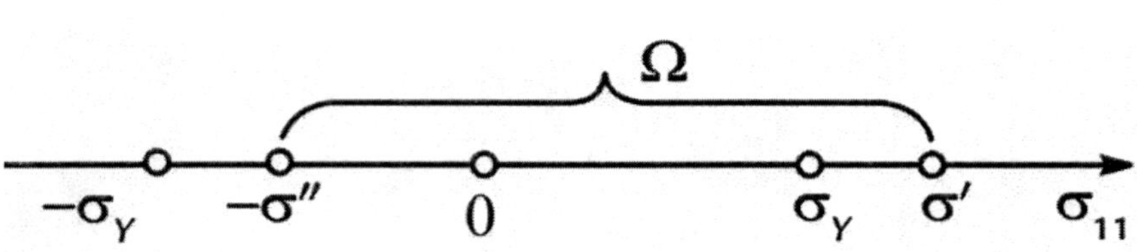


Figure 8.10.

Let us now consider a *work-hardened body* (Fig. 8.10). Its model can be presented by a diagram, for example, like in Fig. 8.3c. When the representative point shifts outside domain $\Omega$, there appears a residual deformation. If tensile stress $\sigma_{11} = \sigma' > \sigma_y$ is reached, the elasticity limit at this moment equals to σ'. This means that the upper boundary of domain Ω has shifted to the right along $\sigma_{11}$ axis from value $\sigma_y$ till σ'. The lower boundary of this domain, due to Bauschinger's effect for a cyclically perfect work-hardened body, attains a similar right shift so that the representative point entrains domain Ω as a rigid whole. As the point travels from the boundary inside domain Ω, only the elastic component of deformation $\varepsilon_{ij}^{e}$ is varying, while its plastic share $\varepsilon_{ij}^{p}$ remains unchanged.

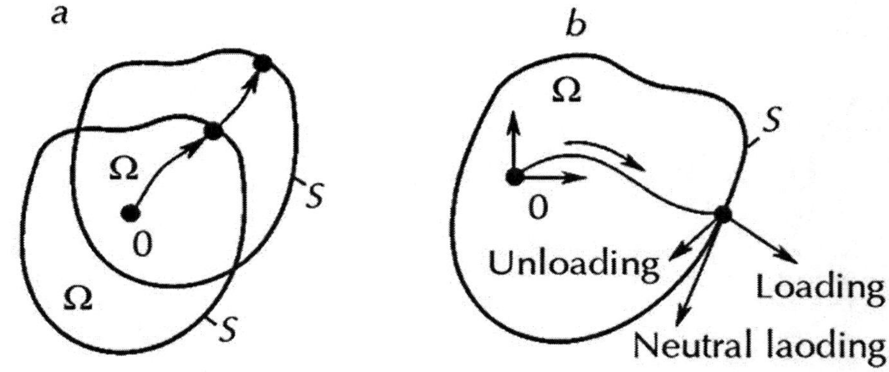


**Figure 8.11.**

Figure 8.11 shows a natural generalization of the described picture for a general case of the stress state. The key factor here is that there is such domain Ω in the stress space containing the origin of coordinates that the deformation of the element remains elastic at any loading path situated fully inside domain Ω.

If a body is *perfectly plastic*, the emergence of the point on boundary *S* of domain Ω means transition of the body into the state of flow under which the deformation becomes indefinite. Thus, boundary *S* presents the locus of the yield points at various load paths. For a perfectly plastic body the points outside domain Ω are not realized. The transition of the point from boundary *S* inside domain Ω is accompanied by a change of only the elastic component of deformation, meaning that unloading takes place, although some stress components $\sigma_{ij}$ may increase.

In case of a *work-hardened body* the process of loading that carries the representative point outside domain Ω shall be accompanied by a displacement of this domain. The analog of the corresponding unidimensional model is the one (Fig. 8.11a), in which domain Ω is displaced translationally as a rigid whole called a *translational hardening*. When the point moves inside domain Ω only the elastic component of deformation varies (unloading). If the point shifts outside domain Ω (Fig. 8.11b), further increase of the plastic component of deformation takes place (active process, additional loading). There may be also such changes in the stresses at which the load path is tangential to the boundary of domain Ω (neutral loading). It may be shown that in a work-hardening body only elastic deformations shall vary.

In the case of *a simple loading*, when all components of the stress tensor vary proportionally to a common parameter (see 8.3), the process of loading in the stress space is presented by a beam emanating from the origin of coordinates (Fig. 8.12). It is possible to show that this beam intersects the yield point not more than once. Let us assume that unloading takes place after reaching the plastic domain in point $A$ followed by a subsequent proportional loading by stresses of the same value but opposite in sign, so that the process keeps to a straight line $AB$. We shall further call such type of deformation an *alternate simple (cyclic) loading.* Yield surface $S$ that confines domain $\Omega$ transfers into $S'$ at loading from the natural state at the end of the first half-cycle and into $S''$ after application of the alternate load.

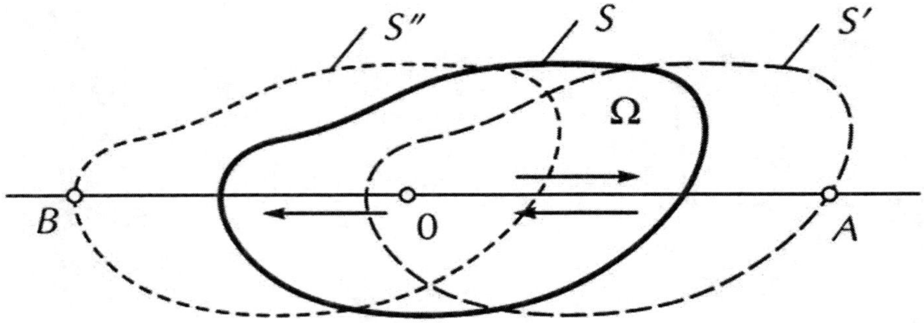


**Figure 8.12.**

Boundary $S$ of domain $\Omega$ is called a *yield surface* or *loading surface*. If we have a perfectly plastic material, this surface is fixed. For a work-hardening body the loading surface varies as the plastic deformation builds up. In a six-dimensional stress space it divides the domains of elastic and plastic deformation at each given moment of time. At translational hardening the yield surface is displaced translationally as a rigid whole. Other types of hardening are also possible, for example (8.5), at which not only the position of this surface in the stress space varies, but its shape and dimensions as well.

## THE THEORY OF PLASTIC FLOW

For making calculations in the field of large plastic strains the *theory of the plastic flow (yielding)* is used. Its principal difference from the *deformation theory* lies in the assumption that no unambiguous relationship between stresses and plastic deformations both at simple and combined loading is accepted. In practice, analogous problems are encountered at metal forming, metal cutting, and in calculations of marginal states of construction foundations.

The plastic deformation of materials within the theory of flow is identified with a viscous fluid flow. As it has been mentioned earlier, transition into the plastic state about the vicinity of a body point is defined by equation $f(\sigma_{ij}) = 0$, which describes the yield surface (loading surface) in a six-dimensional stress space. In the case of material hardening the yielding surface varies as the plastic deformation increases. Its equation contains a certain hardening parameter $\eta$. The following hypotheses underlie the plastic theory of flow:

**1. Elasticity of volume deformation.** Similarly to the theory of small elastoplastic deformations (8.4), the following relation is met:

$$\sigma = 3K\varepsilon.$$

Meaning that the material is incompressible in the plastic state:

$$\varepsilon^p = \tfrac{1}{3}(\varepsilon_{11}^p + \varepsilon_{22}^p + \varepsilon_{33}^p) = 0.$$

This *incompressibility condition* may be written also as an equality to zeroth velocity of the volume plastic deformation $\mathrm{d}\varepsilon^p/\mathrm{d}t = 0$.

**2. The hypothesis of gradientality.** The vector of deformation increment is assumed to be directed perpendicular to the yield surface. It is equivalent to the assumption of proportionality of the components of deformation increment and those of the gradient-vector to the yield surface (partial derivatives of the equation of the surface with respect to the corresponding stress components):

$$\mathrm{d}\varepsilon_{11}^p = \mathrm{d}\lambda \frac{\partial f}{\partial \sigma_{11}},\ 2\mathrm{d}\varepsilon_{13}^p = \mathrm{d}\lambda \frac{\partial f}{\partial \sigma_{13}}. \tag{8.16}$$

Relations (8.16) express the *rule of flow associated with the accepted state of plasticity* $f = 0$, where $d\lambda$ is a differentially small multiplier, which mechanical value is found when considering an elementary work of internal forces on plastic deformations. According to the definition, we have

$$\mathrm{d}U = \sigma_{ij}\,\mathrm{d}\varepsilon_{ij}^p.$$

Using relation (8.16), we come to

$$dU = d\lambda \sigma_{ij} \frac{\partial f}{\partial \sigma_{ij}} = d\lambda \sigma\, grad f. \tag{8.17}$$

Where $\sigma$ is a radius vector of the point in six-dimensional stress space obeying components $\sigma_{ij}$.

Figure 8.13 shows the yield subsurface and the vectors comprising a scalar product in formula (8.17). Consequently, multiplier $d\lambda$ is proportional to the work of stress for plastic deformations, where the proportionality coefficient is nonzero.

The definition of the plastic deformation increments by derivatives of function $f$ with respect to corresponding arguments (8.16) provides the basis for defining $f$ as a *plastic potential*. The yield surface or the plastic potential is found experimentally in agreement with some general physical notions. One of general requirements to construction of the plastic potential is the criterion of hardening stated by Drucker (1951), which is the next.

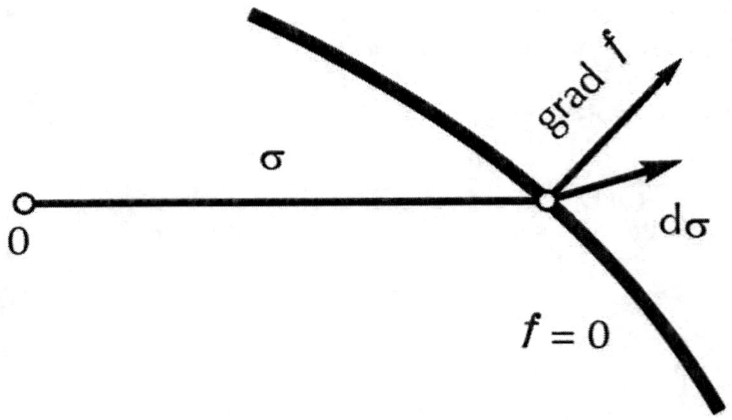


**Figure 8.13.**

Let us imagine (Fig. 8.14) some stress state $\sigma^*$ corresponding to a yield surface $S^*$. The yield surface divides the six-dimensional stress space into two regions: transition from point $\sigma^*$ into the region of its positive values is accompanied by the active plastic deformation with an increment of the plastic deformation vector. The transition from terminus $\sigma^*$ into the region of negative values corresponds to unloading that observes the elastic laws, where plastic deformation remains invariable.

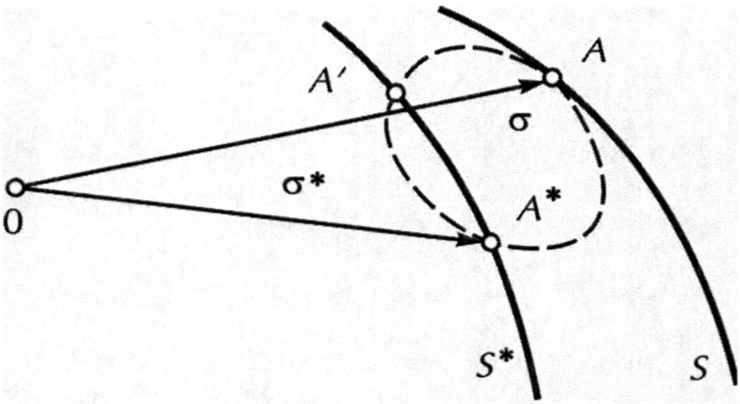


**Figure 8.14.**

Let us now consider a state defined by vector σ whose terminus lies in the positive region to which yield surface $S$ corresponds. *Drucker's postulate* states nonnegativity of the work of stress increments on actual displacements of deformations per cycle, when the state following a certain path transforms from point $\sigma^*$ into σ, and then returns into $\sigma^*$ again

$$\int (\sigma_{ij} - \sigma_{ij}^{*}) \mathrm{d}\,\varepsilon_{ij} \geq 0. \tag{8.18}$$

By assuming that the closed path of integration consists of three portions we may obtain an important conclusion from this postulate. There is only the elastic deformation on sections *A***A'* and *AA**, but on section *A'A* there is an elastoplastic one:

$$\mathrm{d}\,\varepsilon_{ij} = \mathrm{d}\,\varepsilon_{ij}^{e} + \mathrm{d}\,\varepsilon_{ij}^{p}.$$

Due to reversibility of elastic deformations, the stage corresponding to them in the integral (8.18) turns to zero and remains as follows:

$$\int_{A'}^{A} (\sigma_{ij} - \sigma_{ij}^{*}) \mathrm{d}\,\varepsilon_{ij}^{p} \geq 0.$$

Point *A'* may be assumed in the limiting case as coincident with *A*, while conditions $\sigma^*$ and σ—as infinitely close elements, therefore the left-hand side of the preceding equality is reduced to a scalar product

$$d\sigma_{ij} \times d\varepsilon_{ij}^{p} \; ^{3}\, 0\cdot$$

This means that the stress increment vector and the plastic deformation increment vector form an acute angle.

In other words, the yield surface is convex toward the active plastic deformations, as shown in Figs. 8.11, 8.12. This requirement is sometimes called the *condition of stability of plastic deformation,* which is shown in a simplest case of uniaxial tension of a specimen.

Figure 8.15 shows a diagram of a corresponding deformation. Transition from point $A^*$ into state *A* is accompanied by the work of stress increment 0,5*d*σ*d*ε.

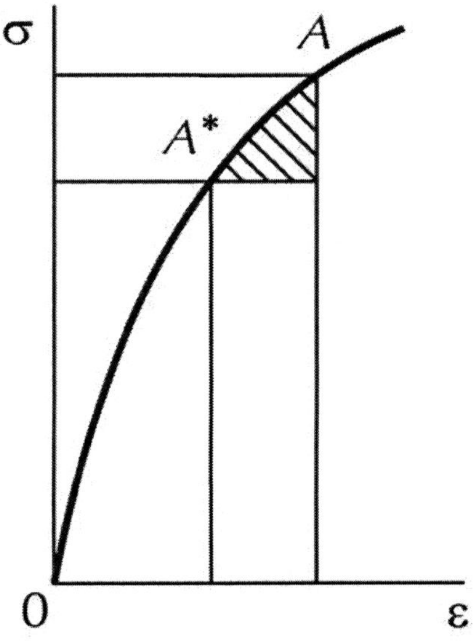


**Figure 8.15.**

This work is positive when the stress augments with increasing strain.

This process is absent if a neck is formed on the stretched sample when the resultant defects of the material lead to the intensively growing strains even at decreasing stresses. This process is unstable and the corresponding diagram is declining, as we have mentioned above (see 3.5).

The equilibrium equations, Cauchy's relations, deformation compatibility equations, and boundary conditions are valid for the problems of the theory of plastic flow

$$\sigma_{ij},_j + \rho F_i = 0;\ \varepsilon_{ij} = \tfrac{1}{2}(u_i,_j + u_j,_i);$$

$$\mathring{a}_{\alpha\alpha},_{\beta\beta} + \mathring{a}_{\beta\beta},_{\alpha\alpha} - 2\varepsilon_{\alpha\beta},_{\alpha\beta} = 0,\ (\mathring{a}_{\alpha\beta},_{\gamma} - \mathring{a}_{\beta\gamma},_{\alpha} + \mathring{a}_{\gamma\alpha},_{\beta}),_{\alpha} - \varepsilon_{\alpha\alpha},_{\beta\gamma} = 0;$$

$$u_i = u_{i0}(x) \quad \text{on} \quad S_u, \sigma_{ij} l_j = R_i \quad \text{on} \quad S_\sigma .$$

The theorem on uniqueness of incremental fields of stresses, strains, and displacements has been proved for a work-hardened body in the theory of plastic flow. One cannot guarantee the uniqueness of strain and stress increments for a nonhardening material, although the uniqueness of specified increments may be proved in partial problems for a perfectly plastic material.

As we see, the equations of the flow theory turn to be more complex than in the theory of small elastoplastic deformations. At a simple loading both theories have proved to give the same solution. In the case of a combined loading the results obtained via the plastic flow theory agree with the experimental data best of all. For further details the theory and problems of the plastic yielding refer to monographs (Ishlinskii and Ivlev, 2001, Kiiko, 1978, Klyushnikov, 1979, Malinin, 1968, Rabotnov, 1966, Starovoitov, 1987).

## AN EXAMPLE OF A LIMITING SURFACE

Let us study what stress–strain relations may be derived if to accept equation $f = 0$ as the yield surface according to Mises condition

$$f = (\sigma_{11} - \sigma_{22})^2 + (\sigma_{22} - \sigma_{33})^2 + (\sigma_{33} - \sigma_{11})^2$$
$$+ 6(\sigma_{12}^2 + \sigma_{23}^2 + \sigma_{31}^2) - 2\sigma_y^2 = 0 .$$

Let us consider partial derivatives (8.16) of function $f$. Then,

$$\frac{\partial f}{\partial \sigma_{11}} = 4\sigma_{11} - 2\sigma_{22} - 2\sigma_{33} = 6(\sigma_{11} - \sigma), \frac{\partial f}{\partial \sigma_{13}} = 12\sigma_{13} ,$$

where $\sigma = \frac{1}{3}(\sigma_{11} + \sigma_{22} + \sigma_{33})$.

We shall assume that coefficient dλ includes also a numerical factor 6. Then, the preceding relations may be generalized and rewritten with account of (8.16) as follows:

$$d\varepsilon_{ii}^{p} = d\lambda(\sigma_{ii} - \sigma),\, d\varepsilon_{ij}^{p} = d\lambda\sigma_{ij} \;\; (i, j = 1, 2, 3;\; i \neq j). \tag{8.19}$$

Let us represent the incremental plastic deformation components as an expansion into the deviatoric and spherical parts:

$$d\varepsilon_{ij}^{p} = d\,э_{ij}^{p} + d\varepsilon^{p}\delta_{ij}\,.$$

Since the volume plastic deformation is absent, then

$$d\varepsilon^{p} = \tfrac{1}{3}(d\,э_{11}^{p} + d\varepsilon_{22}^{p} + d\varepsilon_{33}^{p}) = 0\,,$$

and values $d\,э_{ij}^{p}$ are the incremental deviator components of the plastic deformation $d\,э_{ij}^{p}$. The components of stress deviator $s_{ij} = \sigma_{ij} - \sigma\delta_{ij}$ are formed in the right-hand side of relations (8.19) therefore,

$$d\,э_{ij}^{p} = d\lambda s_{ij}\,.$$

Thus, for a plastic yield of a metal it is assumed that there is a linear relationship between the deviator increments of plastic deformations $dэ_{ij}^{p}$ and the components of stress deviator $s_{ij}$. This conclusion may be interpreted also as a linear dependence between the components of the velocities of plastic deformations and stress components.

By analogy with deformations (1.31) we shall introduce a concept of the *increment intensity of plastic deformations*:

$$d\varepsilon_{u}^{p} = \tfrac{\sqrt{2}}{3}\Big\{(d\varepsilon_{11}^{p} - d\varepsilon_{22}^{p})^2 + (d\varepsilon_{22}^{p} - d\varepsilon_{33}^{p})^2 + (d\varepsilon_{33}^{p} - d\varepsilon_{11}^{p})^2$$

$$6\Big[(d\varepsilon_{12}^{p})^2 + (d\varepsilon_{23}^{p})^2 + (d\varepsilon_{31}^{p})^2\Big]\Big\}^{1/2}.$$

It should be noted, that the increment intensity of plastic deformations (8.19) is not equal to the increment of intensity of plastic deformations.

After substitution of (8.19) into the previous relation and some transformations, we obtain the following expression for $d\lambda$:

$$d\lambda = \frac{3}{2}\frac{d\varepsilon_{u}^{p}}{\sigma_{u}}\,.$$

Based on obtained expression $d\lambda$ in the formula for the increment of deviator components of plastic deformation $dэ_{ij}^{p}$, the stress deviator components will be

$$s_{ij} = \frac{2}{3}\frac{\sigma_{u}}{d\varepsilon_{u}^{p}}\,d\,э_{ij}^{p}. \tag{8.20}$$

By assuming elastic deformations small in contrast to plastic ones, we may ignore them and withdraw superscript $p$ throughout equations (8.20). It is assumed also that the total deformations are equal to the plastic ones. Let us in addition divide deformation components by $dt$ in the numerator and denominator of the right-hand side. We obtain the components of *deformation rates* $\xi_{ij} = d\varepsilon_{ij}/dt$, and the *deformation rate intensity* $\xi_{u} = d\varepsilon_{u}/dt$.

As a result, expressions (8.20) take the form:

$$s_{ij} = \frac{2}{3}\frac{\sigma_u}{\xi_u}\xi_{ij}\,.$$

That is, the stress deviator components are proportional to the deformation rate of components.

## FOUNDATIONS OF THE GENERAL MATHEMATICAL PLASTICITY THEORY BY A.A. ILYUSHIN

The laws of stresses and strains in a general case of a combined loading are based on the condition of ambiguity, isotropy postulate, principle of delay of the vector and scalar properties, the hypothesis of unloading, and plasticity postulate. They are formulated with regard to the following assumptions:

1. In the initial state a body is isotropic.
2. It is assumed that in a narrow enough neighborhood of any point of a deformed body the state is homogeneous; the variations in homogeneous deformation of the neighborhood of a point of an inhomogeneously deformed body in time and homogeneous deformation of a finite in size specimen at similar stresses and environment conditions are alike.
3. Deformations are assumed to be so small that their squares as compared to eigen-values may be ignored.
4. Rheonomic properties of the material are excluded from consideration.

Let us consider in brief the basic provisions of the general mathematical theory of plasticity formulated by A.A. Ilyushin (1963).

**1. Single-valuedness condition**. At a given initial state, known variation of external parameters in time (temperature, radiation dose, etc.) and preset deformation process (i.e., variation of deformation tensor in time) the stress tensor is defined unambiguously at each moment of time.

The necessity of introducing this condition is connected with impossibility to accomplish absolutely equal variations of external parameters and identical deformation processes in two experiments. This, in turn, leads to deviations in the values of the stress tensor. But the condition of single-valuedness actually postulates that deviations in the stress tensor values are small if deviations in the values of external parameters and the stress tensor are small too.

**2. Presentation of the deformation process vector and isotropy postulate.** It is recognized that stress tensor $\varepsilon_{ij}$, related to a fixed Cartesian system $x = (x_1, x_2, x_3)$ may be presented as a sum of the deviator and spherical part

$$\varepsilon_{ij} = \text{э}_{ij} + \varepsilon\delta_{ij}\,,$$

only five out of $\text{э}_{ij}$ values are independent since the first invariant of the deviator equals to zero:

$$э_{11} + э_{22} + э_{33} = 0,$$

The stress tensor is presented in a similar way

$$\sigma_{ij} = s_{ij} + \sigma\delta_{ij}.$$

Since the spherical parts are interrelated via a simplest relation

$$\sigma = 3K\varepsilon.$$

We shall assume further that the process in stresses and strains is set if, respectively, $s_{ij}(t)$ and $э_{ij}(t)$ are set at any moment of time $t$. It should be noted, that pressure $p = -\sigma$ in the experimental studies $\sigma_{ij} \sim \varepsilon_{ij}$ may be related to external parameters, for example, temperature.

Since only five of $э_{ij}$ values are independent, it is convenient for vector presentation of the deformation deviator by introduction of a space of five variables $э_k$, in which a metrics (five-dimensional Euclidean space $Э_5$) is introduced in a certain manner. Let us choose in this space a unit orthogonal frame $e_k$ ($e_k e_i = \delta_{ki}$), in which we set a five-dimensional deformation vector

$$э = э_k e_k, \tag{8.21}$$

which is identical to deviator $э_{ij}$, that is, the following condition should be met:

$$э_{ij} э_{ij} = э^2 = э_k э_k. \tag{8.22}$$

From relation (8.22) it follows that five components $э_k$ are defined ambiguously by six components $э_{ij}$. Let us consider one of possible definitions. It is suitable to take values $э_3$, $э_4$, $э_5$ in the form

$$э_3 = \sqrt{2}\, э_{12},\ э_4 = \sqrt{2}\, э_{23},\ э_5 = \sqrt{2}\, э_{31}.$$

While from (8.21) and (8.22) it follows

$$э_1^2 + э_2^2 = э_{11}^2 + э_{22}^2 + э_{33}^2,\ э_{11} + э_{22} + э_{33} = 0.$$

It is not easy to verify that these equations will be met if to assume that:

$$\sqrt{\tfrac{3}{2}}\, э_{11} = э_1 \cos\beta + э_2 \sin\beta,\ \sqrt{\tfrac{3}{2}}\, э_{22} = -э_1 \sin(\beta + \pi/6) + э_2 \cos(\beta + \pi/6),$$

$$\sqrt{\tfrac{3}{2}}\, э_{33} = э_1 \sin(\beta - \pi/6) - э_2 \cos(\beta - \pi/6). \tag{8.23}$$

In turn,

$$э_1/\sqrt{2} = э_{11} \cos(\beta + \pi/6) - э_{22} \sin\beta,\ э_2/\sqrt{2} = э_{11} \sin(\beta + \pi/6) + э_{22} \cos\beta,$$

$$э_3 = \sqrt{2}\, э_{12},\ э_4 = \sqrt{2}\, э_{23},\ э_5 = \sqrt{2}\, э_{31}. \tag{8.24}$$

Let us accept that value β, included in these relations is constant since rotation of vector э(э$_k$) is observed in the plane of coordinate vectors ($e_1$, $e_2$) when parameter β varies.

When values $э_{ij}$ are varying during loading, values $э_k$ vary with time as well. So the terminus of the deformation vector э describes in the deformation space a curve called hereunder a strain *path*. The square of the arc element of this path is

$$ds^2 = d\,э_k\, d\,э_k = d\,э_{ij}\, d\,э_{ij},$$

hence,

$$s(t) = \int_0^t e\,dt,\ e^2 = e_{ij}e_{ij}.$$

We can now introduce arc length $s$ instead of time $t$ and assume vector э as a function of $s$.

The inner geometry of the strain path is defined by a motion of a so-called Frenet pentahedron over it. To plot it, in each point of the strain path we introduce its own nonorthogonal frame

$$э^k = \frac{d^k\,э}{d\,s^k}\ (k = 1, 2, 5),$$

assuming э($s$) as a differentiated sufficient number of times function. These five vectors are, generally speaking, linearly independent. We shall plot against them a local orthogonal frame $q_k$.

Let us give $q_1$ as

$$q_1 = \frac{d\,э}{ds},$$

that is, as a unit vector directed at a tangent to the strain path. The next vector will be

$$q_2 = \frac{d^2\,э}{ds^2} + \alpha_1 q_1,$$

where $\alpha_1$ will be obtained from the orthogonality condition $q_1$ and $q_2$. If we continue in a similar manner, we may find $q_3$, $q_4$, $q_5$. By normalizing $q_k$, we may find unit vectors of the sought accompanying Frenet's pentahedron

$$p_1 = q_1,\ p_2 = \frac{q_2}{q_2},\ p_5 = \frac{q_5}{q_5}.$$

Above-considered construction shows that $p_k$ are linear combinations of vectors $э^k$. In their turn, $э^k$ are linearly expressed by $p_k$, where coefficients of these relations depend on four scalar values

$$\kappa_{kk} = \frac{d\,э^k}{ds^k}\frac{d\,э^k}{ds^k}. \tag{8.25}$$

The deduced generalized Frenet formulas expressing derivatives $dp_k / ds$ in terms of $p_k$ are

$$\frac{dp_k}{ds} = -\kappa_{k-1} p_{k-1} + \kappa_k p_{k+1} \ (k = 1, 2, \ldots, 5), \ \kappa_0 = \kappa_5 = 0, \tag{8.26}$$

where $\kappa_1(s), \ldots, \kappa_4(s)$ are the parameters of curvature and torsion expressed in terms of scalar values $\kappa_{kk}$ (8.25). Values $\kappa_1(s), \ldots, \kappa_4(s)$ are characteristic of the inner geometry of the strain path. In other words, the strain path is explicitly defined by setting $\kappa_1(s)$, $\kappa_2(s)$, $\kappa_3(s)$, $\kappa_4(s)$ within the position of strains $Э_5$ in space.

Using the generalized Frenet formulas (8.26), one may express a derivative of any order for vector э($s$) with respect to $s$ and an integral of any multiplicity for э($s$) over $s$ (the paths with an infinite number of derivatives э with respect to $s$) in terms of $p_1$, ..., $p_5$.

Therefore, any vector-linear operator L(э) on э with respect to $s$ with coefficients depending on $\kappa_k$ and $s$, may be in the form

$$L(э) = \lambda_k p_k \ (k = 1, 2, \ldots, 5), \tag{8.27}$$

where $\lambda_k$ depends functionally on $\kappa_1, \ldots, \kappa_4$, and $s$. Therefore, L(э) is defined only by the inner geometry of the strain path, that is, operator L(э) is an invariant of rotation transformation and space reflection of deformations $Э_5$.

In view of generality of operator L(э), any physical vector related to the strain path and presented as (8.27) will be defined by the inner geometry of the strain path as well.

In the physical values we are interested first of all in stress vector σ, which may be constructed fully similarly to deformation vector **ε**. At a given pressure $p = -\sigma$ the stresses are defined by five independent components of the stress deviator tensor $s_{ij}$. Let us introduce a stress space, in which stress vector σ with coordinates $\sigma_1, \ldots, \sigma_5$ is set, where the latter's are related to $s_{ij}$ via equations similar to (8.24):

$$\sigma_1/\sqrt{2} = s_{11}\cos(\beta + \pi/6) - s_{22}\sin\beta,$$

$$\sigma_2/\sqrt{2} = s_{11}\sin(\beta + \pi/6) + s_{22}\cos\beta,$$

$$\sigma_3 = \sqrt{2}s_{12}, \sigma_4 = \sqrt{2}s_{23}, \sigma_5 = \sqrt{2}s_{31}.$$

Deviator $s_{ij}$, in its turn, is expressed in terms of $\sigma_k$ according to the formulas like (8.23).

The terminus of vector σ in the stress space will describe a trajectory called hereinafter a *loading path*.

In view of aforesaid, stress vector σ may be presented in the form of (8.26) at each point of the strain path

$$\sigma = \lambda_k p_k, \tag{8.28}$$

where $\lambda_k$, in a general case, are scalar-valued functionals of the curvature parameters and those of torsion along the length of arc $s$:

$$\lambda_k = \lambda_k \{\kappa_m(\xi), \xi\}_{\xi=0}^{\xi=s} \ (k = 1, \ldots, 5; m = 1, \ldots, 4).$$

In practice, it is convenient to use instead of the vector notation the relations between stress tensor $\sigma_{ij}$ and strain tensor $\varepsilon_{ij}$:

$$\sigma_{ij} = A_k \varepsilon_{ij}^k,$$

where $\varepsilon_{ij}^k$ is the basis related to $\varepsilon_{ij}$, for example, by equations

$$\varepsilon_{ij}^k = \int_0^t f_k(t,\tau)\varepsilon_{ij}(\tau)\,d\tau \cdot$$

Values $f_k$ are the linearly defined independent functions. Values $A_k$ in (8.28) are the invariant functionals of strain tensor $\varepsilon_{ij}$.

If to set stress vector σ by its modulus σ and by angles $\alpha_1, \ldots, \alpha_5$ of orientation σ in earlier introduced natural orthogonal frame $p_k$ of the strain path, we may write the following:

$$\sigma = \hat{O}\{\kappa_m(\xi), \xi\}_{\xi=0}^{\xi=s}, \alpha_i = \alpha_i\{\kappa_m(\xi), \xi\}_{\xi=0}^{\xi=s},$$

where four out of $\alpha_i$ values are independent.

A combination of the strain path, physical vectors connected with the strain path in its every point and scalar values (temperature $T$, pressure $p$, etc.) is known as *an image of the process body loading in the deformation space*. Using the concept of the process image the isotropy postulate is formulated, which forms the basis of the presentation (8.28): the image of the loading process in five-dimensional deformation space $Э_5$ is defined by the inner geometry of the strain path and scalar values of pressure $p(s)$, temperature $T(s)$, and so forth.

In view of the isotropy postulate, the image of the loading process is invariant with respect to rotation transformations and space reflection of deformations $Э_5$. This circumstance allows us to reduce the number of experiments for studying elastoplastic properties of materials at applying arbitrarily combined loads, including fluctuating ones.

The isotropy postulate has been checked experimentally being so far convincingly justified. We should name in this respect the works by Lenskiy[10] and some other authors first of all. For the most part, these were the experiments with thin-walled tubes subjected to tension and torsion (with varying torsion direction included) in different sequences.

If to select an appropriate coordinate system for the considered case, we have the incompressible material: $э_{11} = \varepsilon_{11}$, $э_{22} = э_{33} = -\varepsilon_{11}/2$, $э_{12} = \gamma_{12}/2$, where $\gamma_{12}$ is the deformation of shear. Assuming that $\beta = 0$ in formulas (8.24), we obtain

$$э_1 = \sqrt{\tfrac{3}{2}}\, э_{11} = \sqrt{\tfrac{3}{2}}\varepsilon_{11},\ э_3 = \sqrt{2}\, э_{12} = \tfrac{\sqrt{2}}{2}\varepsilon_{12},\ э_2 = э_4 = э_5 = 0.$$

Thus, only two components of the deformation vector are nonzero in this case. They are used to preset the deformation process in plane $э_1 \sim э_3$ according to an arbitrary program.

Let us consider a deformation process shown in Fig. 8.16 as an example. We shall first stretch the specimen till some deformation $э_1'$, then, keeping this axial deformation unchangeable we shall twist the sample till deformation $э_3'$ appears.

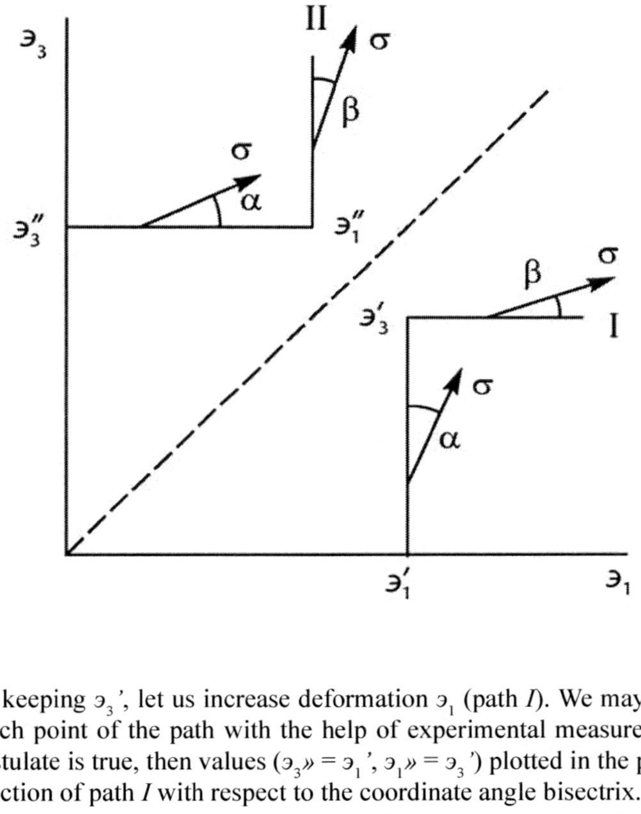


**Figure 8.16.**

Further, keeping $э_3'$, let us increase deformation $э_1$ (path *I*). We may plot a stress vector in each point of the path with the help of experimental measurements. If the isotropy postulate is true, then values ($э_3$» = $э_1'$, $э_1$» = $э_3'$) plotted in the points of path *II* are a reflection of path *I* with respect to the coordinate angle bisectrix.

The stress vectors shall be a reflection of a corresponding vector of path *I*. The experiments have revealed that stress vectors in the corresponding points of paths *I* and *II* are in fact closely orientated toward the paths and coincide in modulus. A variety of strain paths studied at experimental checking of the isotropy postulate and a set of pure metals and alloys used make grounds to affirm that the isotropy postulate is a general law describing the behavior of initially isotropic material at arbitrary loading.

**3. The principle of delay of vector and scalar properties.** The orientation and modulus of the stress vector with respect to the strain path are rather dependent on a certain final section of the strain path (*delay trace*) immediately preceding the considered moment than on the whole history of the deformation process beginning from the initial state.

Let us denote the value of the delay trace by $h$, which varies within (3–10)$\varepsilon_y$ depending on the material, where $\varepsilon_y$ is a tensile yield point at pure tension. It is recommended to find $h$ based on the experiments of vector properties since the parameter of limited "memory" of scalar properties becomes apparent on smaller sections of the strain path.

With account of the principle of delay of the vector and scalar properties, functionals $\lambda_k$ in (8.28) are given by:

$$\lambda_k = \lambda_k \{\kappa_m(\xi)\}_{\xi=s-h}^{\xi=s}.$$

The principle of delay of the vector and scalar properties is especially important for studying cyclic loading, because in this case the arc length of the strain path, unlike deformations themselves, may be considerable at a greater number of cycles. In this connection, it is extremely complicated in practice to account the prehistory as a whole.

We shall call the *paths of simple loading (deformation)* the paths in a deformation space, for which a unit vector э/э is constant (invariable during loading). As a matter of fact these are straight lines passing through the origin of coordinates. All other paths will be called the *combined loading (deformation) paths.*

It should be noted, that in the stress space the path will be a straight line too passing through the coordinate origin in the case of a simple loading. Stricter formulations introduced here are in a full agreement with those accepted earlier.

**4. The hypothesis of unloading and deformation-induced anisotropy.** It is possible to present the stress tensor as a sum of elastic and plastic stages.

$$\varepsilon_{ij} = \varepsilon_{ij}^e + \varepsilon_{ij}^p. \tag{8.29}$$

In this case, the tensor of elastic deformations is related to the stress tensor via a generalized Hooke's law, which may be written for the isotropic material as follows:

$$э_{ij}^e = \varepsilon_{ij}^e - \varepsilon^e \delta_{ij} = \frac{S_{ij}}{2G}, \ \varepsilon^e = \frac{\sigma}{3K}, \tag{8.30}$$

where $\varepsilon^e = (\varepsilon_{11}^e + \varepsilon_{22}^e + \varepsilon_{33}^e)/3$, $G$ is shear modulus, $K$—modulus of volume elasticity.

Let us assume that the mean stress $\sigma$ is proportional to a relative volume variation:

$$\sigma = 3K\varepsilon. \tag{8.31}$$

By comparing (8.30) and (8.31) we have come to a conclusion that $\varepsilon = \varepsilon^p$, meaning that the relative variation of volume $\varepsilon^p$ due to the plastic stage of deformation is equal in this case to zero. Hence, relation (8.29) may be also written for deviators like:

$$э_{ij} = э_{ij}^e + э_{ij}^p, \ э_{ij}^p = \varepsilon_{ij}^p.$$

This implies a relation for the deformation vector

$$э = э^e + э^p.$$

Let us accept that loading is taking place in some segment of the strain path if vector $э^p$ remains constant in this segment. We may conclude that if the process of loading runs from a certain point $M$ of the strain path (such that $\mathrm{d}\,э^p/\mathrm{d}s \neq 0$ in this point), the paths of all kinds will be divided into two series.

The paths of one series possess a property that the vector of plastic deformation remains invariable while moving over them, that is, we observe unloading. When another series is moving over the path, the vector of plastic deformation $э^p$ varies too. In other words, an active loading (active deformation) takes place.

The paths of active deformation and unloading in a 5D space $Э_5$ are separated by some surface $F(э) = 0$ passing through point $M$. We shall call this surface a *yield surface*, and a corresponding surface in the stress space a *loading surface*. The appearance of the yield surface and the loading surface depends on the preceding (till point $M$) segment of active loading.

Variations in elasticity coefficients due to plastic deformation (deformation anisotropy) have been previously observed by Bauschinger. A number of scientists were engaged in studying the influence of the active loading on their value (refer to Moskvitin, 1981). Ilyushin (1963) has pointed to the importance of variations of elasticity coefficients in the plasticity theory. The variation of elasticity coefficients at plastic deformation may reach quantitatively as much as 15–20%.

**5. Plasticity postulate.** The work of stress vector $W_э$ in any path closed in the deformation space is equal to zero if no variation of the plastic deformation vector $э^p$ is observed along the entire path; and it is positive if the vector of plastic deformation does not remain constant at least in some segments of the path, or otherwise the work of the stress vector

$$W_э = \oint \sigma_k \, \mathrm{d}\varepsilon_k = \oint s_{ij} \, \mathrm{d}\, э_{ij} \geq 0,$$

in any closed in $Э_5$ process of deformations is nonnegative.

This postulate answers the question: is there any segment of active deformation on this path? If to review the stress-closed process of loading in contrast to the strain-closed one and define a corresponding work $W_э'$, it turns out (Ilyushin, 1963) that $W_э' \geq W_э$ all the time, that is, from condition $W_э \geq 0$, it follows that $W_э' \geq 0$.

The complete work of stresses is known to be presented as:

$$W = \int \sigma_{ij} \, \mathrm{d}\varepsilon_{ij} = \int s_{ij} \, \mathrm{d}\, э_{ij} + 3\int \sigma \, \mathrm{d}\varepsilon .$$

If $\sigma = 3K\varepsilon$ and the process is deformation-closed, the second integral is equal to zero (no increments in volume plastic deformations), that is why $W = W_э$, and from condition $W_э \geq 0$ it follows that in this case $W \geq 0$.

Condition $W \geq 0$ implies that if $F(э) = 0$—the equation of the yield point in any point on the surface

$$\mathrm{d}\, э^p = D \,\mathrm{grad}(\mathrm{d}\, F(э))\, \mathrm{d}\lambda ,$$

that is, increment of plastic deformation is directed normal to surface $F$. D is here a certain functional defined by deformation prehistory, $\lambda$—loading parameter. This relation is usually called an associated (with the yield surface $F(э) = 0$) flow rule. It serves the base for constructing various options of the flow theory.

## ELASTOPLASTIC BENDING OF A CIRCULAR THREE-LAYER PLATE

Let us consider an elastoplastic bending of a circular sandwich plate. We shall assume that the materials of the layers of a circular sandwich plate (Fig. 6.11), considered earlier in 6.14, may display elastoplastic properties when experiencing deformation. To describe them we shall use the relations of the theory of small elastoplastic deformations (8.9). Let us present the components of the stress tensor in terms of the deviator and spherical part of the strain tensor as follows:

$$\sigma_{\alpha}^{(k)} = 2G_k \,э_{\alpha}^{(k)} + K_k\theta^{(k)} - 2G_k\omega_k(\varepsilon_u^{(k)})\,э_{\alpha}^{(k)},$$

$$\sigma_{rz}^{(3)} = 2G_3 \,э_{rz}^{(3)} - 2G_3\omega_3(\varepsilon_u^{(3)})\,э_{rz}^{(3)}. \tag{8.32}$$

Here, as earlier, $k$ is number of the layer; $\omega_k(\varepsilon_u^k)$—plasticity function of the $k$-th layer material; $\varepsilon_u^k$—strain intensity. Volume deformation is assumed to be elastic.

The linear and nonlinear components in stresses (8.32) will be:

$$\sigma_{\alpha}^{(k)} = \sigma_{\text{åe}}^{(k)} - \sigma_{\text{å}\omega}^{(k)},\ \sigma_{rz}^{(k)} = \sigma_{rze}^{(k)} - \sigma_{rz\omega}^{(k)},$$

$$\sigma_{\text{åe}}^{(k)} = 2G_k \,э_{\alpha}^{(k)} + K_k\theta^{(k)},\ \sigma_{\text{å}\omega}^{(k)} = 2G_k\omega_k(\varepsilon_u^{(k)})\,э_{\alpha}^{(k)},$$

$$\sigma_{rze}^{(3)} = 2G_3 \,э_{rz}^{(3)},\ \sigma_{rz\omega}^{(3)} = 2G_3\omega_3(\varepsilon_u^{(3)})\,э_{rz}^{(3)}. \tag{8.33}$$

The internal forces and moments in the plate layers will be presented as a difference between the linear and nonlinear components:

$$T_{\alpha}^{(k)} = T_{\alpha e}^{(k)} - T_{\alpha\omega}^{(k)},\ M_{\alpha}^{(k)} = M_{\alpha e}^{(k)} - M_{\alpha\omega}^{(k)},$$

$$Q = Q_e - Q_\omega\ (\alpha = r, \varphi).$$

Values $T_{\text{åe}}^{(k)}, T_{\text{å}\omega}^{(k)}, M_{\text{åe}}^{(k)}, M_{\alpha\omega}^{(k)}, Q_e, Q_\omega$ are calculated by formulas (6.50), in which values $\sigma_{\alpha}^{(k)}$ shall be replaced by stresses $\sigma_{\alpha e}^{(k)}, \sigma_{\alpha\omega}^{(k)}$ (8.33), respectively. As a result, the generalized forces will be the next:

$$T_{\alpha} = T_{\alpha e} - T_{\alpha\omega} = \sum_{k=1}^{3} T_{\alpha e}^{(k)} - \sum_{k=1}^{3} T_{\alpha\omega}^{(k)},$$

$$M_{\alpha} = M_{\alpha e} - M_{\alpha\omega} = \sum_{k=1}^{3} M_{\alpha e}^{(k)} - \sum_{k=1}^{3} M_{\alpha\omega}^{(k)},$$

$$H_{\alpha e} = M_{\alpha e}^{(3)} + c\left(T_{\alpha e}^{(1)} - T_{\alpha e}^{(2)}\right),\ H_{\alpha\omega} = M_{\alpha\omega}^{(3)} + c\left(T_{\alpha\omega}^{(1)} - T_{\alpha\omega}^{(2)}\right). \tag{8.34}$$

After substitution of expressions (8.34) in a series of equilibrium equations of the forces (6.55) describing deformation of a circular sandwich plate, we obtain

$$T_{r,r} + \frac{1}{r}(T_r - T_\phi) = -p + p_\omega,$$

$$H_{r,r} + \frac{1}{r}(H_r - H_\varphi) - Q = h_\omega,$$

$$M_{r},_{r}+\frac{1}{r}(2M_{r},_{r}-M_{\varphi},_{r})=-q+q_{\omega}\,. \tag{8.35}$$

In the left-hand side of above equation the linear components of the generalized internal forces are assembled, where index «e» is dropped for convenience. The non-linear additional components are concentrated in the right-hand side and are included into the summands with a subscript "ω":

$$p_{\omega}=T_{r\omega},_{r}+\frac{1}{r}(T_{r\omega}-T_{\varphi\omega}),\; h_{\omega}=H_{r\omega},_{r}+\frac{1}{r}(H_{r\omega}-H_{\varphi\omega})-Q_{\omega},$$

$$q_{\omega}=M_{r\omega},_{rr}+\frac{1}{r}(2M_{r\omega},_{r}-M_{\varphi\omega},_{r})\,. \tag{8.36}$$

The linear generalized internal forces are expressed in terms of displacements by the formulas similar to (6.57). So, a series of differential equations (8.35) for nonlinear displacement takes the form:

$$L_2(a_1u+a_2\psi-a_3w,_{r})=-p+p_{\omega},$$

$$L_2(a_2u+a_4\psi-a_5w,_{r})-2cG_3\psi=h_{\omega},$$

$$L_3(a_3u+a_5\psi-a_6w,_{r})=-q+q_{\omega}\,. \tag{8.37}$$

Coefficients $a_i$ and differential operators $L_2$, $L_3$ are defined by relations (6.59).

According to the method of linear approximation let us rewrite equations (8.37) in the iterative form:

$$L_2(a_1u^n+a_2\psi^n-a_3w,_{r}^{n})=-p+p_{\omega}^{n-1},$$

$$L_2(a_2u^n+a_4\psi^n-a_5w,_{r}^{n})-2cG_3\psi^n=h_{\omega}^{n-1},$$

$$L_3(a_3u^n+a_5\psi^n-a_6w,_{r}^{n})=-q+q_{\omega}^{n-1}\,. \tag{8.38}$$

where $n$ is approximation number; $p_{\omega}^{n-1}, h_{\omega}^{n-1}, q_{\omega}^{n-1}$ are called «additional» external stresses, assumed to be zero at the first step and are further calculated from the results of preceding approximation. For this aim, the formulas similar to (8.36) are used in which all summands have a superscript «$n-1$»:

$$p_{\omega}^{n-1}=T_{r\omega}^{n-1},_{r}+\frac{1}{r}(T_{r\omega}^{n-1}-T_{f\omega}^{n-1}),\; h_{\omega}^{n-1}=H_{r\omega}^{n-1},_{r}+\frac{1}{r}(H_{r\omega}^{n-1}-H_{f\omega}^{n-1})-Q_{\omega}^{n-1},$$

$$q_{\omega}^{n-1}=M_{r\omega}^{n-1},_{rr}+\frac{1}{r}(2M_{r\omega}^{n-1},_{r}-M_{\omega\omega}^{n-1},_{r})\,. \tag{8.39}$$

Hence,

$$T_{\alpha\omega}^{n-1}\circ\sum_{k=1}^{3}\int_{h_k}\sigma_{\alpha\omega}^{(k)n-1}dz=\sum_{k=1}^{3}\int_{h_k}2G_k\omega_k(\varepsilon_u^{(k)n-1})\,\text{э}_{\alpha}^{(k)n-1}\,dz,$$

$$M^{n\text{-}1}_{\alpha\omega} \circ \sum_{k=1}^{3} \int_{h_k} \sigma^{(k)n\text{-}1}_{\alpha\omega} z dz = \sum_{k=1}^{3} \int_{h_k} 2G_k \omega_k (\varepsilon^{(k)n\text{-}1}_u) \ni^{(k)n\text{-}1}_{\alpha} zdz,$$

$$H^{n\text{-}1}_{\alpha\omega} = M^{(3)n\text{-}1}_{\alpha\omega} + c\left(T^{(1)n\text{-}1}_{\alpha\omega} - T^{(2)n\text{-}1}_{\alpha\omega}\right),$$

$$Q^{n\text{-}1}_{\omega} = \int_{-c}^{c} 2G_3\omega_3(\varepsilon^{(3)n-1}_u) \ni^{(3)n-1}_{rz} \, \mathrm{d}z \; (\alpha = r, \varphi). \tag{8.40}$$

Thus, at each step of approximation we have a linear elastic problem with the known additional external stresses, which are calculated by equations (8.39), (8.40). The procedure for obtaining a solution for (8.38) with regard to its smoothness in the center of the plate does not differ from the one for a corresponding elastic problem (6.58)–(6.67).

As a result, we obtain a solution of the elastoplastic problem in a recurrent form:

$$\psi^n = C^n_2 I_1(\beta r) + \psi^n_r,$$

$$u^n = \frac{a_3}{a_1 a_6 - a_3^2}\left[\mathrm{L}_3^{-1}(q - q^{n-1}_{\omega}) - \frac{a_6}{a_3}\mathrm{L}_2^{-1}(p - p^{n-1}_{\omega}) + \left(a_5 - \frac{a_2 a_6}{a_3}\right)\psi^n + C^n_7 r\right],$$

$$w^n = \frac{1}{b_3}\left[b_2\left(\frac{C^n_2}{\beta} I_0(\beta r) + \int \psi^n_r \, \mathrm{d}r\right) - \int\left(\frac{a_3}{a_1}\mathrm{L}_2^{-1}(p - p^{n-1}_{\omega}) - \mathrm{L}_3^{-1}(q - q^{n-1}_{\omega})\right)\mathrm{d}r + \frac{C^n_5 r^2}{4} + C^n_4\right] \tag{8.41}$$

The partial solution of the inhomogeneous modified Bessel's equation is as follows:

$$\psi^n_r = -K_1(\beta r)\int I_1(\beta r) f^n r \, \mathrm{d}r + I_1(\beta r)\int K_1(\beta r) f^n r \, \mathrm{d}r,$$

where

$$f^n(r) = \frac{b_3}{b_1 b_3 - b_2^2}\left[h^{n\text{-}1}_{\omega} + \frac{a_2 b_3 - a_3 b_2}{a_1 b_3}(p - p^{n\text{-}1}_{\omega}) + \frac{b_2}{b_3 r}\int (q - q^{n\text{-}1}_{\omega}) r \mathrm{d}r\right].$$

Under the boundary conditions for edge-fixing of the plate, we obtain the following recurrent formulas for integration constants:

$$C^n_2 = -\frac{\psi^n_r\big|_{r=1}}{I_1(\beta)}, \quad C^n_4 = -b_2\left(\frac{C^n_2 I_0(\beta)}{\beta} + \int \psi^n_r \, \mathrm{d}r\Big|_{r=1}\right),$$

$$+\int\left(\frac{a_3}{a_1}\mathrm{L}_2^{-1}(p - p^{n\text{-}1}_{\omega}) - \mathrm{L}_3^{-1}(q - q^{n\text{-}1}_{\omega})\right)\mathrm{d}r\Bigg|_{r=1} - \frac{C^n_5}{4},$$

$$C^n_5 = 2\left(\frac{a_3}{a_1}\mathrm{L}_2^{-1}(p - p^{n\text{-}1}_{\omega}) - \mathrm{L}_3^{-1}(q - q^{n\text{-}1}_{\omega})\right)\Bigg|_{r=1},$$

$$C^n_7 = \left(\frac{a_6}{a_3}\mathrm{L}_2^{-1}(p - p^{n\text{-}1}_{\omega}) - \mathrm{L}_3^{-1}(q - q^{n\text{-}1}_{\omega})\right)\Bigg|_{r=1}. \tag{8.42}$$

In the case of a hinge-supported contour, integration constants $C_2^n$, $C_4^n$, $C_7^n$ are the same, and $C_5^n$ is defined according to (6.70).

**Numerical results.** It was assumed in calculations that the bearing layers of the plate are made of Д16T alloy, the filler—of fluoroplastic. The corresponding functions and properties of these materials are described in Ch. 11. Computation of (8.41), (8.42) has shown practical convergence of the method of linear approximation. The maximum difference of the results in the fifth approximation assumed as the sought solution from the preceding ones is less than 1%.

So, application of the method of linear approximations for the sandwich elastoplastic plates allows us to obtain accuracy sufficient for engineering purposes.

Figure 8.17 shows a deflection of a circular sandwich symmetric in thickness plate under impacting by a distributed load of intensity $q = 30$ MPa. Relative thicknesses of the layers ($h_1 = h_2 = 0{,}04$, $h_3 = 0{,}2$) was sorted so that the plastic and nonlinear features of the material were revealed adequately enough. This is justified by the graphs: curve *1* corresponds to the linear elastic plate, *2*—to elastoplastic one.

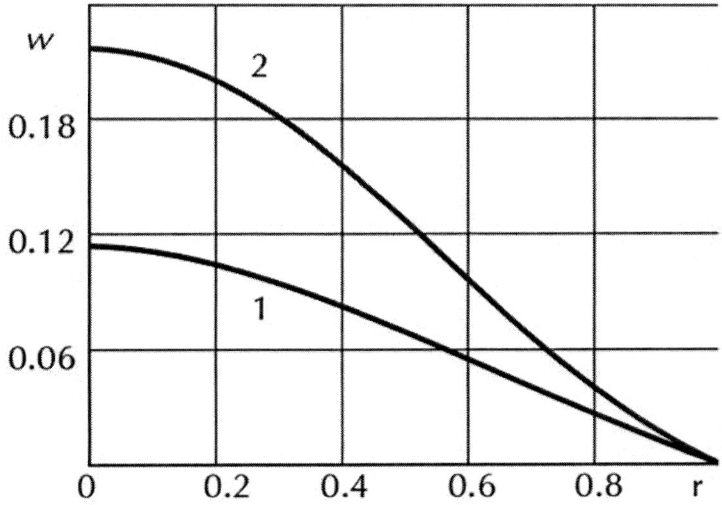


**Figure 8.17.**

Thus, the account of plasticity of the bearing layer, and of physical nonlinearity of the filler leads to increasing of the deflection by about 80%. Similar results are observed for the filler shear as well.

## VARIABLE LOADING OF ELASTOPLASTIC BODIES

When studying the work of the load-bearing structural components experiencing cyclically varying loads under conditions of temperature, radiation, and other physico-mechanical fields, the scientists encounter specific problems. They are related mainly to calculation of the corresponding stress and strain, and determination of conditions at which a threshold is reached: structural failure, inadmissible shear, and so forth.

A distinctive feature of cyclic deformations of elastoplastic bodies, unlike elastic ones, is the influence of the loading prehistory on the deformed state at a given moment. In the present chapter we are going to consider one class of a *simple variable loading* (including temperature and radiation effects) for which a probability is specified of solving the boundary problem of the plasticity theory at any half-cycle, providing a solution at loading from the initial state is known.

To analyze the diagrams of cyclic deformation of elastoplastic materials, various approaches are used nowadays. We shall restrict to *Moskvitin's* theory here (Moskvitin, 1981). Let us preset the conditions used by Moskvitin[11] for describing recurrent and alternate loading for which the equations of small elastoplastic deformations are valid.

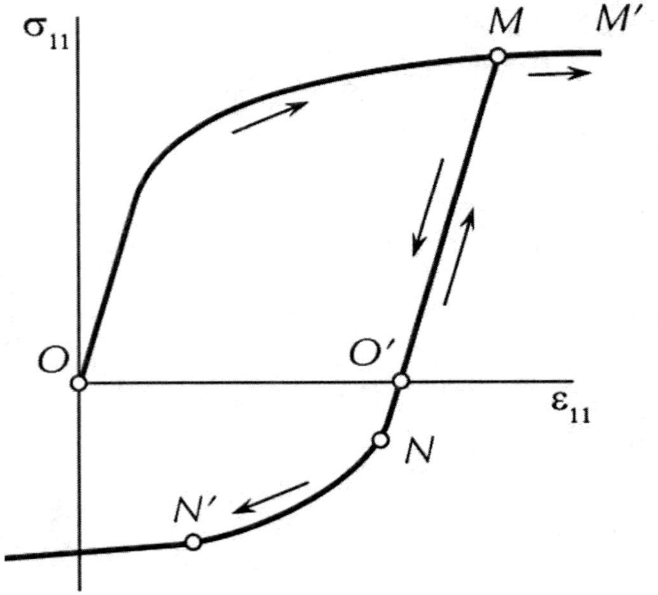


**Figure 8.18.**

1. Unloading and subsequent repeated loading till the point corresponding to unloading start, occur linearly along a straight line *O'M* (Fig. 8.18). The area of the hysteresis loop for most of materials is small and is ignored.
2. At repeated loading toward *O'M* plastic deformations would vary in the case point *M* is reached. Subsequent loading takes place along *M'M*, and curve *OMM'* characterizes elastoplastic properties of the material in its original state. Thus, loading in this case would proceed as if there were no unloading. The yield point is a variable and depends on the preceding SSS.

3. We shall assume the straight line *O'M* to be parallel to that of the elastic loading from the initial state, meaning that plastic deformation does not change the elastic constants characterizing elastic properties of the material.
4. Alternate loading prior to new plastic deformations occurs along a straight line *O'N*, being an extension of the line *O'M*. Inner geometry of curve *NN'* of subsequent plastic deformation depends in general on the position of point *M* on diagram *OMM'*.

These variations of the external force parameters in time at alternating loading/unloading in the body as a whole or on its edges, is called *cyclic* (or *variable*) *loading* of the body. Speaking on the «*body loading from the initial state*», we mean that it was not prestrained, and there are no stresses or strains in it prior to load application.

Let us consider a solid body without any initial stresses and strains. Let volume forces $F_i'$ and surface forces $R_i'$ be acting on the exterior boundary section $S_\sigma$ of the body at the initial moment $t_0 = 0$. These forces and boundary shift $u_{0i}'$ on the boundary section $S_u$ induce stresses $\sigma_{ij}'$, strains $\varepsilon_{ij}'$ and displacements $u_i'$ in the body. As a result, the areas of plastic deformation appear in the body. In elastic domains Hooke's law is valid:

$$s_{ij}' = 2G\,э_{ij}', \sigma' = 3K\varepsilon', \tag{8.43}$$

where $s_{ij}'$, $э_{ij}'$ are deviator components; $\sigma'$, $\varepsilon'$—spherical components of stress and strain tensors; *G*, *K*—modules of shear and volume deformations.

For such areas of a solid body, where plastic deformation has occurred, the relationship between deviators may be written as

$$s_{ij}' = 2G\,э_{ij}'\, f'\left(\varepsilon_u', a_k'\right),$$

where $f'(\varepsilon_u', a_k')$ is a plasticity function introduced by Ilyushin; $\varepsilon_u'$—strain intensity; $a_k'$—approximation parameters defined by inner geometry of the deformation curve.

The plasticity condition $\varepsilon_u' = \varepsilon_y'$ is met on the boundary between elastic and plastic deformations ($\varepsilon_y'$—plastic deformation limit at loading from the initial state).

Thus, the relationship between stresses and strains in the deformed body may be presented as follows:

$$s_{ij}' = 2G\,э_{ij}'\, f'\left(\varepsilon_u', a_k'\right), \sigma' = 3K\varepsilon', \tag{8.44}$$

where the plasticity function is set as $f'(\varepsilon_u', a_k') = 1$ in the domains where $\varepsilon_u' \le \varepsilon_y'$.

Let us add differential equilibrium equations and boundary conditions along with Cauchy relations to equations (8.43), (8.44) in supposition of deformation smallness:

$$\sigma_{ij},_j' + \rho F_i' = 0, 2\varepsilon_{ij}' = u_i',_j + u_j',_i,$$

$$ó_{ij}' l_j = R_i' \text{ on } S_ó, u_i' = u_{0i}' \text{ on } S_u \tag{8.45}$$

where $l_j$ are directional cosines of the normal to a part of the body surface $S_\sigma$.

Now let instantaneous unloading and subsequent loading by forces $F_i''$, $R_i''$ with a boundary shift $u_{0i}''$ to be performed beginning from a certain instant $t = t_1$. In this connection, we suppose that these forces and shifts are of the opposite sign to the forces and shifts occurred prior to unloading. This process is shown schematically in Fig. 8.19. Certain effects induce stresses $\sigma_{ij}''$, strains $\varepsilon_{ij}''$, and displacements $u_i''$ in the body, for which relations (8.45) remain valid:

$$\sigma_{ij}'',_j + \rho F_i'' = 0,\ 2\varepsilon_{ij}'' = u_i'',_j + u_j'',_i,$$

$$\sigma_{ij}'' l_j = R_i'' \text{ on } S_\sigma,\ u_i'' = u_{0i}'' \text{ on } S_u. \tag{8.46}$$

The stress–strain relationship for them is

$$s_{ij}'' = 2g\,\text{э}_{ij}''\, f''\left(\varepsilon_{u1}', \varepsilon_u'', a_k''\right),\ \sigma'' = 3K\varepsilon''. \tag{8.47}$$

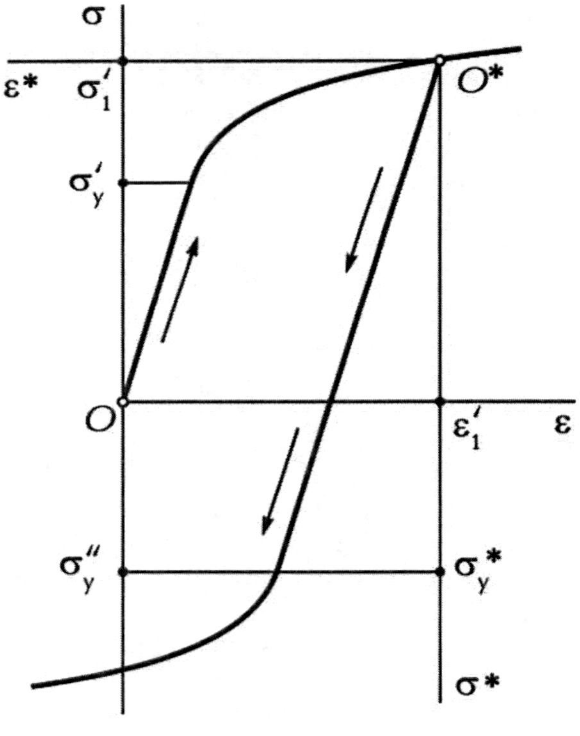


**Figure 8.19.**

Where $f''\left(\varepsilon_{u1}', \varepsilon_u'', a_k''\right)$ is the plasticity function at repeated alternate loading that depends on preceding unloading of strain intensity $\varepsilon_{u1}'$, current strain intensity $\varepsilon_u''$, and approximation parameters $a_k''$ describing the deformation curve of the second half-cycle. The plasticity function shall be $f'' = 1$ in the domains, where no new plastic deformations appear, that is, $\varepsilon_u'' \leq \varepsilon_u''$ by modulus, $\varepsilon_y''$—deformation corresponding to plastic limit $\sigma_y''$ at repeated loading. Characteristics $\varepsilon_y''$, $\sigma_y''$ depend in fact on the values of strain and stress intensity $\varepsilon_{u1}'$ and $\sigma_{u1}'$ available prior to the unloading start, more precisely, on whether the plastic deformation has taken place in a given point of the body or not.

Equations (8.46), (8.47) solve the boundary problem for the double-primed values. Its complexity lies in the dependence of the sought solution on unloading point ($\varepsilon_{u1}'$, $\sigma_{u1}'$). This is because it is necessary to set its own boundary problem in each point and get its own solution.

Let us consider an opportunity to avoid these difficulties by introducing differences for moments $t > t_1$:

$$\sigma_{ij}^{*} = \sigma_{ij}' - \sigma_{ij}'', \varepsilon_{ij}^{*} = \varepsilon_{ij}' - \varepsilon_{ij}''. \tag{8.48}$$

Similar differences are introduced for deviators and spherical parts of stress and strain tensors. Single-primed values correspond to their values before unloading.

In the new domains of plastic deformation, asterisked deviators of the stress and strain tensors are interrelated by:

$$s_{ij}^{*} = 2G\, э_{ij}^{*}\, f^{*}\left(\varepsilon_u^{*}, \varepsilon_{u1}', a_k^{*}\right). \tag{8.49}$$

Where $f^{*}(\varepsilon_u^{*}, \varepsilon_{u1}', a_k^{*})$ is some new universal function describing nonlinearity of the deformation diagram $\sigma^{*} \sim \varepsilon^{*}$ (refer to Fig. 8.18). In the linear segment $f^{*} = 1$.

It should be noted, that the first attempt to compare the curve of the repeated alternate deformation to a corresponding curve of the first loading was made by *Masing*. Moskvitin has generalized Masing's hypothesis by introducing extra scale factors $\alpha_n$, $\beta_n$ in relations

$$\sigma_n^{*} = \alpha_n f'\left(\frac{э_n^{*}}{\beta_n}\right), \quad \alpha_n{}^{1}\beta_n. \tag{8.50}$$

Another Moskvitin's assumption states a possibility of describing curve $f^{*}$ by function of closed form similar to the loading from the initial state, but with different approximation parameters $a_k^{*}$ to be defined experimentally. Therefore, if $f' \equiv f'(\varepsilon_u', a_k')$, then

$$f^{*} \equiv f'(\varepsilon_u^{*}, a_k^{*}) \ (k = 1, \ldots, m), \tag{8.51}$$

where $m$—number of parameters.

By introducing a sufficient number of constants, it is possible to meet assumption (8.51) with a reasonable degree of accuracy, which, in its turn, allows using Moskvitin's hypothesis of variable loading for the arbitrary cyclic deformation curves in practice. Thus, if to use relations (8.50) or (8.51), we can get rid of the dependence of the repeated alternate loading curve upon the loading point.

The rest equations with asterisked values include the relationship for spherical tensors that remains linear throughout the body for being assumed linear at both first (8.44) and second (8.47) loading cycles:

$$\sigma^* = 3K\varepsilon^*. \tag{8.52}$$

The differential equilibrium equations, boundary conditions, and Cauchy relations following from linearity of equations (8.45), (8.46), shall be added to equations (8.49)–(8.52):

$$\sigma^*_{ij},_j + \rho F^*_i = 0, \quad F^*_i = F'_i - F''_i, \ 2\varepsilon^*_{ij} = u^*_i,_j + u^*_j,_i \ ,$$

$$u^*_i = u^*_{0i} = u'_{0i} - u''_{0i} \ \text{on} \ S_u, \ \sigma^*_{ij} l_j = R^*_i, \quad R^*_i = R'_i - R''_i \ \text{on} \ S_\sigma. \tag{8.53}$$

In this connection, we assume that the external boundary sections of the body $S_\sigma$ and $S_u$ where stresses and strains are set remain the same as at preceding loading.

Let us additionally suppose that plastic deformations at variable loading do not cover the areas with only elastic deformations appeared during the first loading.

Relations (8.49)–(8.53) form a new boundary problem for asterisked values, which accept either the *generalized Masing-Moskvitin principle* (8.50) or the assumption (8.51) to describe the function.

When comparing relations (8.43)–(8.45) for a body loaded from the initial state to relations for asterisked values (8.49)–(8.53), it becomes evident that they agree till the designations. Therefore, the solution for the asterisked values may be obtained from a known solution corresponding to loading from the natural state by means of some replacements. For example, if we know the relation

$$u'_i = u'_i(x, \varepsilon'_u, \varepsilon'_y, a'_k) \ ,$$

a corresponding asterisked displacement will be as follows:

$$u^*_i = u'_i(x, \varepsilon^*_u, \varepsilon^*_y, a^*_k), \ \ \varepsilon^*_y = \varepsilon'_y - \varepsilon''_y \ .$$

The sought displacement at a repeated alternate displacement is found from the relation similar to (8.48)

$$u''_i = u'_i - u^*_i \ . \tag{8.54}$$

Stresses and strains are calculated also by the formulas similar to (8.54).

The result obtained may be extended to the case of any $n$-th cyclic loading. Let stresses $\sigma^n_{ij}$, strains $\varepsilon^n_{ij}$, and displacements $u^n_i$ occur at the $n$-th loading by external forces $F^n_i$, $R^n_i$ at boundary displacements $u^n_{0i}$ where equilibrium equations, boundary conditions, and Cauchy relations should be obeyed:

$$\sigma^n_{ij},_j + \rho F^n_i = 0, 2\varepsilon^n_{ij} = u^n_i,_j + u^n_j,_i \ ,$$
$$\sigma^n_{ij} l_j = R^n_i \ \text{on} \ S_\sigma, \ u^n_i = u^n_{0i} \ \text{on} \ S_u. \tag{8.55}$$

Let us introduce the following differences:

$$\sigma^{*n}_{ij} = (-1)^n(\sigma^{n-1}_{ij} - \sigma^n_{ij}), \quad \varepsilon^{*n}_{ij} = (-1)^n(\varepsilon^{n-1}_{ij} - \varepsilon^n_{ij}), u^{*n}_i = (-1)^n(u^{n-1}_i - u^n_i).$$

Then the relations similar to (8.55) turn to be valid for asterisked values:

$$\sigma_{ij}^{*n},_j + \rho F_i^{*n} = 0, \quad F_i^{*n} = (-1)^n (F_i^{n-1} - F_i^n), 2\varepsilon_{ij}^{*n} = u_i^{*n},_j + u_j^{*n},_i \,,$$

$$\sigma_{ij}^{*n} l_j = R_i^{*n} \text{ on } S_\sigma, u_i^{*n} = u_{0i}^{*n} \text{ on } S_u \,,$$

$$R_i^{*n} = (-1)^n (R_i^{n-1} - R_i^n), \quad u_{0i}^{*n} = (-1)^n (u_{0i}^{n-1} - u_{0i}^n)\,. \tag{8.56}$$

We shall assume that the relationship between spherical components of the stress and strain tensors remains elastic at any $n$-th variable loading. By repeating above assumption on the possibility of describing curves $s'_{ij} \sim \acute{y}'_{ij}$ и $s_{ij}^{*n} \sim э_{ij}^{*n}$ either by means of the generalized *Masing-Moskvitin principle* or by nonlinearity functions of a similar closed form

$$s_{ij}^{*n} = 2G\, э_{ij}^{*n}\, \alpha_n f'\left(\frac{\varepsilon_u^*}{\alpha_n}\right), s_{ij}^{*n} = 2G\, э_{ij}^{*n}\, f'\left(\varepsilon_u^{*n}, a_k^{*n}\right), \tag{8.57}$$

we may conclude that the solution for asterisked values (8.56), (8.57) at any $n$-th variable loading may be obtained by solving the problem for loading from the natural state. For example, if we know the displacement

$$u'_i = u'_i(x, \varepsilon'_u, \varepsilon'_y, a'_k)\,,$$

a corresponding asterisked value is $u_i^{*n} = u'_i(x, \varepsilon_u^{*n}, \varepsilon_y^{*n}, a_k^{*n})$.

The sought displacement $u_i^n$ will be found by generalizing (8.54):

$$u_i^n = u'_i - \sum_{k=2}^{n} (-1)^k u_i^{*k}\,. \tag{8.58}$$

Stresses and strains are calculated by the formulas like (8.58).

Notice that $\alpha_n \equiv \alpha(n)$ in (8.57) are *scale factors* of cyclic hardening found from experiments. For *the cyclically hardening materials* we take that $\alpha_n > 2$, for *the softening materials*—$\alpha_n < 2$, for *the cyclically perfect materials*—$\alpha_n = 2$ (the latter corresponds to Masing's principle). Table 8.1, borrowed from (Moskvitin, 1981) gives the values of constants $\alpha_2$ and $\kappa$ at power approximation of function $\alpha(n)$:

$$\alpha_n \equiv \alpha(n) = \alpha_2(n-1)^\kappa.$$

**Table 8.1.**

| **Material** | $\alpha_2$ | $\kappa$ |
|---|---|---|
| Aluminum alloy B-96 | 2.08 | 0.047 |
| " " B-95 | 1.95 | 0 |
| " " Д16Т | 2.02 | 0.030 |
| Steel TS ($n \leq 11$) | 1.93 | 0.011 |
| Steel TS ($n > 11$) | 2.10 | –0.024 |

This implies that aluminum alloys B-96 and Д16T are cyclically hardening material, alloy B-95—cyclically perfect material, steel TC after first 11 cycles becomes a cyclically softening material.

## CYCLIC LOADING IN TEMPERATURE FIELD

Let us consider a process of nonisothermal loading of a solid deformed body from the natural state by external forces $F_i$, $R_i$ at boundary displacement $u_{0i}$ within the framework of the theory of small elastoplastic deformations. Temperature $T(x, t)$ shall be counted from some initial state $T_0$.

The dependence between the mean stress, relative volume variation, and temperature is assumed to be as follows:

$$\sigma = 3K(T)(\varepsilon - \alpha_0 T),\ \alpha_0 = \frac{1}{T}\int_{T_0}^{T_0+T} \alpha(T)\,dT, \tag{8.59}$$

where $\alpha(T)$ is the coefficient of linear thermal expansion; $K$—modulus of volume strain dependent in general on temperature.

The stress versus strain deviator relations differ depending on whether the active loading, unloading or a new (secondary) plastic deformation occurs. If in some area of the body $V_e''$ no variations take place in the plastic components of deformation, a corresponding dependence will be

$$s_{ij} - s'_{ij} = 2G(T)(э_{ij} - э'_{ij}), \tag{8.60}$$

where $s_{ij}'$, $э_{ij}'$ are stress and strain deviators before unloading, $G$—temperature-dependent shear modulus.

Hooke's law is valid in the area $V_e'$ where no plastic deformations appeared

$$s_{ij} = 2G(T)\,э_{ij}. \tag{8.61}$$

During active loading from the natural state in the area $V_p'$ of plastic deformations, the following relation is valid:

$$s_{ij} = 2G\,э_{ij}\, f(\varepsilon_u, T), \tag{8.62}$$

where the unimodular plasticity function may contain temperature $T$.

Finally, for the area $V_p''$ of the secondary plastic deformations caused by unloading, we shall use the relations of the form:

$$s_{ij} - s'_{ij} = 2Gf^*(\varepsilon_u^*, T)(э_{ij} - э'_{ij}). \tag{8.63}$$

It follows at $\varepsilon_u^* \le \varepsilon_y^*$ that $f^* = 1$, and the dependence (8.63) converts into (8.60).

Let us write the relations on the boundaries of above areas. On the boundary $\Gamma_1$ of areas $V_e'$ and $V_p'$ the following condition should be met:

$$\varepsilon_u = \varepsilon_y(T) = \frac{\sigma_y(T)}{2G(T)}.$$

Unloading starts on the boundary $\Gamma_2$ of areas $V_p$' and $V_e$ by obeying condition $\partial\varepsilon_u/\partial t<0$. On the boundary of areas $V_e$" and $V_p$" the following relation is valid:

$$\varepsilon_u^* = \varepsilon_y^*(T) = \frac{\sigma_y^*(T)}{2G(T)}.$$

It should be noted in addition, that the corresponding strain intensities must not attenuate in the areas of active loading:

$\frac{\partial\varepsilon_u}{\partial t}\geq 0$ in domain $V_p$', $\frac{\partial\varepsilon_u^*}{\partial t}\geq 0$ in domain $V_p$".

Above relations help to distinguish between the processes of loading and unloading.

Let us add apparent relations and boundary conditions to derived equations (8.59)–(8.63)

$$\sigma_{ij},_j+\rho F_i=0, 2\varepsilon_{ij}=u_i,_j+u_j,_i, \sigma_{ij}l_j=R_i \; on \; S_\sigma, u_i=u_{0i} \; on \; S_u. \tag{8.64}$$

Temperature $T(x, t)$ in the case of uncoupled thermoplasticity shall meet the heat conductivity equation

$$\rho c\frac{\partial T}{\partial t}=\mathrm{div}(\lambda_0(T)\mathrm{grad}T)+W^*, \; T(x, 0) = T_0, \tag{8.65}$$

where ρ—density, $c$—specific heat capacity, $\lambda_0$—heat conduction factor, $W^*$—intensity of heat sources. We shall add boundary conditions to equations (8.65) not specified here.

Thus, we have stated the problem of thermoplasticity with regard to unloading and secondary plastic strains within the frames of the theory of small elastoplastic deformations. Certain restrictions are imposed in this problem on variations of external loads, boundary displacements, and temperature in time requiring that the corresponding load paths be not related to the class of essentially complex loads.

Notice that we have considered a general case when the external loads are interacting with nonstationary temperature fields. The case of free bodies, in which to variable loading is performed exclusively due to variations of the temperature field gradients in time, is also of specific interest. Here appears a phenomenon of thermal fatigue when structural elements fail after a few temperature cycles.

Moskvitin has considered the case of variable heat-and-force loading reduced to the problem of isothermal loading of some fictitious body from the natural state. Let the external forces start varying from time $t = t_1$ so that all points of plastically deformed areas of the body unloading and a subsequent variable loading occurs by the volume forces $F_i$" and surface forces $R_i$" (on $S_\sigma$) at boundary displacement $u_{0i}$" (on $S_u$). We shall also assume that the temperature remains invariable throughout the body during the whole time interval of unloading and subsequent variable loading, and coincident with the temperature field $T_1(x) = T(x, t_1)$ by time $t_1$ of unloading start. This condition is true when unloading and variable loading are exercised within a limited time period, during which the temperature in each point of the body does not practically change. Let us denote as earlier the corresponding stresses, strains, and displacements in terms of $\sigma_{ij}''$, $\varepsilon_{ij}''$, $u_i''$. The values before unloading will retain designations $\sigma_{ij}'$, $\varepsilon_{ij}'$, $u_i'$.

We shall write the constitutive equation. Relations (8.60) are valid in the areas of unloading and elastic deformation $V_e'$ and $V_e''$ which in this case may be expressed as

$$s_{ij}^* = 2G(T_1)\,\text{э}_{ij}^*,\ \varepsilon_u^* \le \varepsilon_y^*(T_1),\ s_{ij}^* = s_{ij}' - s_{ij}'', \text{э}_{ij}^* = \text{э}_{ij}' - \text{э}_{ij}''. \tag{8.66}$$

Plastic deformations vary in $V_p''$ area under a variable loading, where relations (8.63) are true

$$s_{ij}^* = 2Gf^*(\varepsilon_u^*, T_1)\text{э}_{ij}^*. \tag{8.67}$$

Equation (8.59) is retained in all points of the body. Let us rewrite it for the state prior to unloading and for the current state:

$$\sigma' = 3K(T_1)(\varepsilon' - \alpha_0 T_1),\ \sigma'' = 3K(T_1)(\varepsilon'' - \alpha_0 T_1),$$

where from

$$\sigma^* = 3K(T_1)\varepsilon^*. \tag{8.68}$$

Equilibrium equations, boundary conditions, and Cauchy relations for values $\sigma_{ij}^*$, $\varepsilon_{ij}^*$, $u_i^*$ are given by relations (8.53), (8.64)

$$\sigma_{ij,j}^* + \rho F_i^* = 0,\quad F_i^* = F_i' - F_i'',\ 2\varepsilon_{ij}^* = u_{i,j}^* + u_{j,i}^*,$$

$$\sigma_{ij}^* l_j = R_i^*,\quad R_i^* = R_i' - R_i'',\ \text{on } S_\sigma,\quad u_i^* = u_{0i}^* = u_{0i}' - u_{0i}'',\ \text{on } S_u. \tag{8.69}$$

When evaluating equations (8.66)–(8.69), we come to a conclusion that values $\sigma_{ij}^*$, $\varepsilon_{ij}^*$ are the stresses and strains appearing in some inhomogeneous elastoplastic fictitious body at its isothermal loading from the natural state by external forces $F_i^*$, $R_i^*$ at boundary displacement $u_{0i}^*$. The fictitious body complies geometrically with the considered one, and its elastoplastic properties are characterized by the variant in the coordinate shear modulus $G(T_1(x))$, volume modulus $K(T_1(x))$, unimodular nonlinearity function $f^*(\varepsilon_u^*, T_1)$, and yield point $\varepsilon_y^*(T_1)$. Once we have solved the specified problem on deformation of the inhomogeneous elastoplastic body, the sought values may be found from relations:

$$\sigma_{ij}'' = \sigma_{ij}' - \sigma_{ij}^*,\ \varepsilon_{ij}'' = \varepsilon_{ij}' - \varepsilon_{ij}^*. \tag{8.70}$$

It should be noted, that solution of the boundary problem for above-considered inhomogeneous elastoplastic body may be significantly simplified if to average the temperature through the volume and express nonlinearity function $f^*$ in terms of $f'$ using equations (8.50), (8.51). In this case, we get rid of inhomogeneity and may use the conclusions of the previous paragraph.

We can now apply this result to defining residual stresses and strains remaining in the body after complete unloading. In this case, external forces $F_i''$, $R_i''$, and boundary displacement $u_{0i}$" are varying from the corresponding values $F_i(t_1)$, $R_i(t_1)$, $u_{0i}(t_1)$ till zero. Therefore, values $F_i^*$, $R_i^*$, $u_{0i}^*$ are changing from zero till $F_i(t_1)$, $R_i(t_1)$, $u_{0i}(t_1)$. So, it is necessary to distinguish between two options. In the first case no plastic strain areas appear in above-mentioned fictitious body at

$$F_i^* = F_i(t_1), \quad R_i^* = R_i(t_1), \quad u_{0i}^* = u_{0i}(t_1),$$

that is $\varepsilon_u^* \leq \varepsilon_y^*(T_1)$ in all points of the body. In the second case, the areas of the secondary plastic strain occur at above-mentioned values of external parameters, where $\varepsilon_u^* > \varepsilon_y^*(T_1)$.

In the first case the residual stresses $\sigma_{ij}^0$ and residual strains $\varepsilon_{ij}^0$ are calculated by formulas (8.70)

$$\sigma_{ij}^0 = \sigma_{ij}' - \sigma_{ij}^f, \varepsilon_{ij}^0 = \varepsilon_{ij}' - \varepsilon_{ij}^f, \tag{8.71}$$

where $\sigma_{ij}', \varepsilon_{ij}'$ are the stresses and strains before unloading; $\sigma_{ij}^f, \varepsilon_{ij}^f$—stresses and strains arising in above-specified inhomogeneous fictitious body subjected to elastic deformation $F_i(t_1)$, $R_i(t_1)$ at boundary displacement $u_{0i}(t_1)$.

In the second case, values $\sigma_{ij}^0$, $\varepsilon_{ij}^0$ can also be found by formulas (8.71). However, $\sigma_{ij}^f$ and $\varepsilon_{ij}^f$ are here the stresses and strains arising in the inhomogeneous fictitious body loaded by forces $F_i(t_1)$, $R_i(t_1)$ at boundary displacement $u_{0i}(t_1)$ if the areas of plastic deformations have already appeared.

As we see, the analog of the corresponding Moskvitin's theorems disclosed above is used to define residual stresses and strains in the considered case of thermoplasticity.

***Example 8.1.*** Let us consider deformation of a circular sandwich elastoplastic plate at alternating heat and force loading. The only difference here is that the plate under consideration is a piecewise-homogeneous body.

We assume that at the initial moment the external distributed loads $p'$ and $q'$ (Fig. 6.11) and a heat flow of $q_t$ intensity directed perpendicularly to the first bearing layer are acting on the plate in a natural state. As earlier, forces $N_{\alpha\beta 0}'^{(k)}$, $Q_{\alpha 0}'^{(k)}$, $M_{\alpha\beta 0}'^{(k)}$ may be set on the boundary. The layer number is denoted by $k = 1, 2, 3$. We shall assume that temperature $T(z, t)$ is known. Let the bearing layers of the plate be elastoplastic, and the filler—nonlinearly elastic. The physical equations of the state will be as follow:

$$
\begin{aligned}
&s_\alpha^{(k)} = 2G_k(T) f^{(k)}(\varepsilon_u^{(k)}, T)\, \text{э}_\alpha^{(k)}, \quad \alpha = r, \varphi; \\
&\sigma^{(k)} = 3K_k(T)(\varepsilon^{(k)} - \alpha_{0k}T), \quad k = 1, 2; \\
&f_1(\sigma^{(3)}, T^{(3)}) s_{\alpha\beta}^{(3)} = 2G_3(T) f^{(3)}(\varepsilon^{(3)}, T)\, \text{э}_{\alpha\beta}^{(3)}, \quad (\alpha, \beta = r, \varphi); \\
&\varphi_2(\sigma^{(3)}, T)\sigma^{(3)} = 3K_3(T)(\varepsilon^{(3)} - \alpha_{0k}T).
\end{aligned} \tag{8.72}
$$

Where $f^{(k)}(\varepsilon_u^{(k)}, T)$ are plasticity functions of the bearing layer materials and the function of physical nonlinearity of the filler. Volume deformations in the bearing layers are assumed thermoelastic. Relations (8.72) differ from earlier accepted ones for the sandwich plate (8.32) by the presence of nonlinearity functions $\varphi_n(\sigma^{(3)}, T)$ for the filler, accounting additionally for the influence of hydrostatic stress $\sigma^{(3)}$.

Let us single out the linear and nonlinear constituents in the stress tensor components:

$$\sigma_{\alpha}^{(k)} = \sigma_{\alpha e}^{(k)} - \sigma_{\alpha\omega}^{(k)}, \sigma_{\alpha e}^{(k)} = 2G_k \text{э}_{\alpha}^{(k)} + 3K_k \varepsilon^{(k)},$$

$$\sigma_{\alpha\omega}^{(k)} = 2G_k \omega^{(k)} \text{э}_{\alpha}^{(k)} + 3K_k \alpha_{0k} T \; (k = 1, 2),$$

$$\sigma_{\alpha}^{(3)} = \sigma_{\alpha e}^{(3)} - \sigma_{\alpha\omega}^{(3)}, \sigma_{\alpha e}^{(3)} = 2G_3 \text{э}_{\alpha}^{(3)} + 3K_3 \varepsilon^{(3)},$$

$$\sigma_{rz}^{(3)} = \sigma_{rze}^{(3)} - \sigma_{rz\omega}^{(3)}, \sigma_{rze}^{(3)} = 2G_3 \text{э}_{rz}^{(3)}, \sigma_{rz\omega}^{(3)} = 2G_3 \omega^{(3)} \text{э}_{rz}^{(3)} + \omega_1^{(3)} s_{rz}^{(3)},$$

$$\omega_{\gamma}^{(3)} = \varphi_{\gamma}(\sigma^{(3)}) - 1 \; (\gamma = 1, 2). \tag{8.73}$$

By repeating the procedure of derivation of the equilibrium equation for the elastoplastic plate and using the method of linear approximation (refer to Item 7.10), we come to a set of iteration differential equations of equilibrium coinciding in form with (8.38):

$$\mathrm{L}_2(a_1 u^n + a_2 \psi^n - a_3 w,_r^n) = -p + p_{\omega}^{n-1},$$

$$\mathrm{L}_2(a_2 u^n + a_4 \psi^n - a_5 w,_r^n) - 2cG_3 \psi^n = h_{\omega}^{n-1},$$

$$\mathrm{L}_3(a_3 u^n + a_5 \psi^n - a_6 w,_r^n) = -q + q_{\omega}^{n-1}. \tag{8.74}$$

Coefficients $a_i$ are calculated not by (6.59), but by the integral relations derived from (6.57) since elasticity module of materials in the layers vary through thickness. However, the integration is performed with respect to coordinate $z$, so $a_i$ remain constants. Therefore, a series of differential equations (8.74) is solved by formulas (8.41) with regard to the fact that nonlinear components of stresses entering additional components $p_{\omega}^{n-1}$, $h_{\omega}^{n-1}$, and $q_{\omega}^{n-1}$ (8.39), (8.40) at each step, are set now by relations (8.73). Hence, the solution of the problem of the heat and force loading of the plate from the natural state is considered to be known.

Let us assume further that beginning from moment $t = t_1$, instantaneous unloading and subsequent loading by alternating forces $p''$, $q''$, $N_{\alpha\beta 0}''^{(k)}$, $Q_{\alpha 0}''^{(k)}$ is performed. We shall also suppose that within unloading and subsequent loading the temperature in all points of the body remains invariable, being coincident with the temperature field by the time the unloading starts: $T_1(z) = T(z, t_1)$.

Similarly to previous solutions, let us introduce the differences for all layers (overprime—direct loading, double prime—inverse loading):

$$\sigma_{\alpha\beta}^{(k)*} = \sigma_{\alpha\beta}'^{(k)} - \sigma_{\alpha\beta}''^{(k)}, \quad \varepsilon_{\alpha\beta}^{(k)*} = \varepsilon_{\alpha\beta}'^{(k)} - \varepsilon_{\alpha\beta}''^{(k)}, q^* = q' - q'', \quad u^* = u' - u'', \tag{8.75}$$

where single-primed values are stresses, strains, and displacements in the plate prior to unloading.

For the asterisked values let us introduce the equations of the state similar to (8.72):

$$s_{\alpha\beta}^{(k)*} = 2G_1^{(k)} f^{(k)*} \text{э}_{\alpha\beta}^{(k)*}, \sigma^{(k)*} = 3K_1^{(k)} \varepsilon^{(k)*} \quad k = 1, 2,$$

$$\varphi_1^* s_{\alpha\beta}^{(3)} = 2G_1^{(3)} f^{(3)*} \text{э}_{\alpha\beta}^{(3)*}, \varphi_2^* \sigma^{(3)*} = 3K_1^{(3)} \varepsilon^{(3)*}. \tag{8.76}$$

Here $G_1^{(k)}(z) \equiv G^{(k)}(T_1(z));\ K_1^{(k)}(z) \equiv K^{(k)}(T_1(z)$ .

The corresponding unimodular nonlinearity functions for the bearing layers in (8.76) are supposed to be expressed in terms of plasticity functions at loading from the natural state:

$$f^{(k)*} \equiv f^{(k)}(\varepsilon_u^{(k)*}, \varepsilon_y^{(k)*}, T_1, a_m^*). \tag{8.77}$$

Where $f^{(k)*} = 1$ at $\varepsilon_u^{(k)*} \le \varepsilon_y^{(k)*} = \alpha_2^{(k)} \varepsilon_y^{(k)}(T_1)$, and the case $\alpha_2^{(k)} > 2$ corresponds to a cyclically hardening material, $\alpha_2^{(k)} < 2$—to softening one, $\alpha_2^{(k)} = 2$—cyclically perfect material (the subscript indicates the second half-cycle). The elastic characteristics of the polymeric filler vary at cyclic deformation.

In view of linearity of the stress–strain relationship in the plate layers (6.49), equilibrium equations (8.74), and boundary conditions (6.56), these relations will be valid for the asterisked values as well (8.75). So far, the boundary problem for the asterisked values (8.75)–(8.77) agrees with that of some fictitious sandwich elastoplastic plate subjected to isothermal loading from the natural state by external forces $p^*$, $q^*$, $N_{\alpha\beta 0}^{(k)*}, Q_{\alpha 0}^{(k)*}, M_{\alpha\beta 0}^{(k)*}$. The fictitious plate complies geometrically with the considered one. Its elastoplastic properties are characterized by shear modulus $G^{(k)}(z)$ varying in coordinates, volume modulus $K^{(k)}(z)$, and unimodular nonlinearity functions (8.77). If we solve the problem on deformation of an inhomogeneous elastoplastic plate, the sought displacements may be obtained from relations $u'' = u' - u^*$, $\psi'' = \psi' - \psi^*$, $w'' = w' - w^*$.

The recurrent solution of the problem for the asterisked values results from (8.41), in which appropriate replacements of nonlinearity functions, stresses, and coefficients $a_i$ should be made. For example, if to take the asterisked plasticity function in the form indicated in Ch. 7.6 to describe the behavior of duralumin in the plastic area, the constants of the material will be as follows: $A^* = 0.924$, $\alpha^* = 2.27$, $\varepsilon_y^* = 1.485\%$.

In numerical studies it has been assumed that the bearing layers of the plate are made of Д16T alloy (duralumin), the filler from polytetrafluoroethylene (PTFE, Teflon). The required mechanical characteristics of these materials are given in Ch. 11.

Let us presume that the heat flow of intensity $q_t$ falls perpendicularly to the external surface of the first layer of the considered plate (Fig. 6.11). The heat spent on heating of the external metal layer may be ignored (in view of its thinness and small heat capacity). The temperature of this layer is assumed equal to the temperature of the filler in the splice $T^{(1)} = T^{(3)}(c, t)$. All the heat absorbed by the plate within time $t$, is spent on heating of the filler. The temperature of the second layer is also assumed equal to the temperature of the filler in the splice $T^{(2)} = T^{(3)}(-c, t)$. In this case, the temperature field in the filler may be calculated by formula (11.37),

$$T = \frac{qH}{\lambda}\left\{\tau + \frac{1}{2}\left(s + \frac{c+h_2}{H}\right)^2 - \frac{1}{6} - \frac{2}{\pi^2}\sum_{n=1}^{\infty}\frac{(-1)^n}{n^2}\cos\left[\pi n\left(s + \frac{c+h_2}{H}\right)\right]e^{-n^2\pi^2\tau}\right\},$$

where $a_k = \lambda_k/(C_k\rho_k)$ is thermal diffusivity of the $k$-th layer; $\lambda_k$, $C_k$—heat conductivity and heat capacity factor; $\rho_k$—material density.

In the given case we are to set specific conductivity as $\lambda = \lambda_3$, and heat capacity as $c_0 = c_3$. At $q_t = 5000$ J/m²s ($t_1 = 30$ min) the temperature in the outer bearing layer reaches 510 K at the moment of unloading, in the second bearing layer the temperature remains constant and equal to 293 K. Thus, a variable heat and force loading is observed in one of the plate layers, while in the other—an isothermal one.

The value of loading ($p = 0$, $q = 3.0{\cdot}10^7$), intensity of the heat flow, time of exposure ($t_1 = 30$ min), and relative thickness of the layers ($h_1 = h_2 = 0.04$, $h_3 = 0.2$) were selected so that nonlinear and rheonomic properties of the materials are shown adequately.

Figure 8.20 shows a deflection of a sandwich beam symmetric in thickness (single prime—direct loading, double prime—inverse loading). Curves (*1*, *2*) are plotted for the case of a constant temperature (*1*—linear elasticity, *2*—elastoplasticity). Since the coefficient of cyclic hardening for duralumin is $\alpha_2 = 2.02$ (Table 8.1), and the material in the second half-cycle becomes more rigid, the maximum deflection of the plate at isothermal repeated loading somewhat decreases.

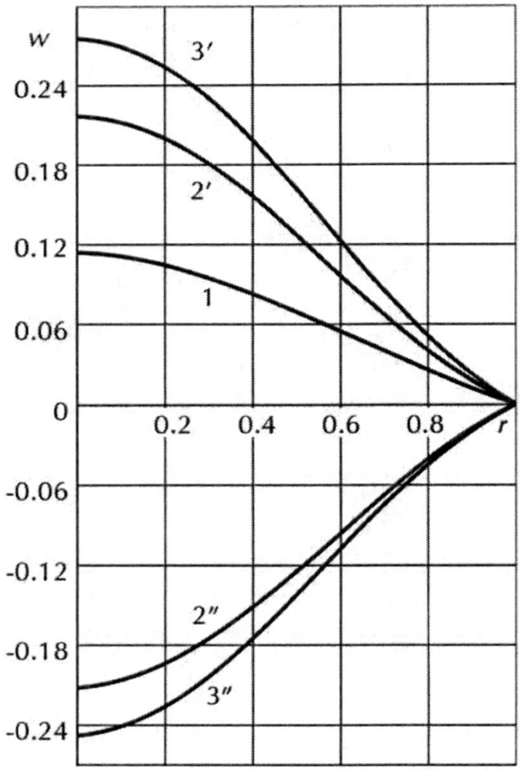


**Figure 8.20.**

Under the heat and force effect within time $t_1$, curve *3'* corresponds to deflection the plate. The temperature in this period in the outer bearing layer reaches 510 K and remains constant. This leads to decreasing of the elasticity moduli and yield point of the materials. As a result, the deflection increases at direct loading (*3'*) as compared to the isothermal case. After changing of the load sign ($t > t_1$), its maximum value decreases a little (*3"*) due to a cyclic work-hardening of the materials.

The numerical study has shown practical convergence of the method of linear approximation. The maximal difference of the results in the fifth approximation (assumed as the sought solution) from the preceding ones is less than 1%.

## CYCLIC DEFORMATION OF ELASTOPLASTIC BODIES IN NEUTRON FIELD

Exposure of solids to radiation is accompanied by numerous effects, as a result of which there appears volume deformation ($\theta_I$) while elastic, and especially plastic characteristics of the materials alter.

A neutron with sufficient kinetic energy creates on its way primary, secondary, and other recoiling atoms when passing through the lattice. Knocked-out atoms leave vacancies and are finally trapped in voids leading to creation of Frenkel's coupled defects in the lattice named «interstitial atom or vacancy». An atom may be knocked-on from the site when it is imparted a certain threshold energy $E_d$. If the atom receives an energy less than $E_d$, it is dissipated at excitation of the lattice oscillations (heating) causing no displacements.

Interactions between neutrons and nuclei may be accompanied by the neutron capture and nuclear fission apart from elastic scattering. The energy in every decay event is released and new chemical elements are created.

We are interested in the mechanical aspect of this process, that is, in the possibility of reflecting the influence of neutron irradiation on the SSS mode of the elastoplastic body in statements and solutions of the boundary problems.

Let us consider an initially homogeneous isotropic body occupying a half-space $z \geq 0$. If the neutrons with identical mean energy and intensity $\varphi_0$, neuron/(m$^2$·s), impinge the boundary ($z = 0$) parallel to axis $z$, the intensity of the neutron flux, reaching plane $z$ = const, will be (Ilyushin and Ogibalov, 1960)

$$\varphi(z) = \varphi_0 e^{-\mu z}. \tag{8.79}$$

The value of $\mu$ is known as a *macroscopic cross-section*. For any chemical element

$$\mu = \sigma n_0 = \sigma \frac{A_0 \rho}{A}$$

being of the order of 1/cm, where $\sigma$—cross-section related to a nucleus, $n_0$—number of nuclei in 1 cm$^3$, $A_0$—Avogadro number, $\rho$—density, $A$—atomic weight.

If $\varphi_0$ is independent of time according to (8.79), the integrated flux given below will pass through section $z$ in time $t$:

$$I(z) = \varphi_0 t e^{-\mu z} \cdot \tag{8.80}$$

It may be assumed in linear approximations that the change in the material volume is directly proportional to flux $I(z)$ (Ilyushin and Ogibalov, 1960) and, therefore,

$$\theta_I = BI(z),$$

where $B$—constant obtained from the experiment.

Value $I_0 = \varphi_0 t$ gives a compound neutron flux per unit area of the body surface. In the reactors, $\varphi_0$ is of the order of $10^{17}$–$10^{18}$ neutron/(m$^2$·s), and $I_0$ reaches $10^{23}$–$10^{27}$ neutron/m$^2$, where $\theta_I$ may reach the order of 0.1. Consequently, depending on the energy of neutrons and the exposed material value $B$ may be about $10^{-28}$–$10^{-24}$ m$^2$/neutron.

The dependence of the elasticity modulus, yield point, strength, and the total tensile stress–strain diagram upon $I_0$ of various energies has been investigated experimentally on the specimens pre-exposed in a nuclear reactor. The experiments have proved that the elasticity modulus varies slightly, as a rule (increase by 1.5–5%). As for the ultimate strength and especially yield strength, they are highly sensitive to irradiation.

For the bulky bodies with a plane boundary, the number of neutrons passing through depth $z$ beneath this boundary in time $t$ is calculated by formula (8.80), so the yield point will be variable through $z$ thickness. On the body surface ($z = 0$) the influence of radiation on the plastic limit $\sigma_y$ is described satisfactorily by the *irradiation hardening formula* (Starovoitov, 2004)

$$\sigma_y = \sigma_{y0}\left[1 + A\left(1 - \exp\left(-\xi I_0\right)\right)^{1/2}\right],$$

where $\sigma_{y0}$—plastic limit of unexposed material; $A$, $\xi$—constants of material, obtained from the experiments. At depth $z$ this formula is in the form

$$\sigma_y = \sigma_{y0}\left[1 + A\left(1 - \exp\left(-\xi I\right)\right)^{1/2}\right],$$

where the flux value $I(z)$ is described by the formula (8.80). Let us denote the corresponding values of deformation by $\varepsilon_{y0}$, $\varepsilon_y$.

It is of interest to consider the process of combined effect of the external force and radiation field on the deformed body.

We shall assume that in the initial moment external forces $F_i'$, $R_i'$ at boundary displacement $u_{0i}'$ and, simultaneously, the neutron flux of $I_0 = \varphi t$ value start acting on the body in the natural state. It is anticipated that the areas of plastic deformations appear in the body under this combined effect. Variations of the elasticity modules of the material due to neutral irradiation are ignored. The stresses, strains, and displacements arising in the body are single primed.

Hooke's law is valid in the elastic domains of the solid and the following relations are met:

$$s'_{ij} = 2G\,э'_{ij}, \sigma' = K(3\varepsilon' - BI),$$

For the domains of the solid body where plastic deformations are present, the relationship between deviators at simple stresses may be presented as follows:

$$s'_{ij} = 2G\, э'_{ij}\, f'\left(\varepsilon'_u, I, a'_k\right), \sigma' = K(3\varepsilon' - BI).$$

Where $f'(\varepsilon'_u, I, a'_k)$—unimodular function of plasticity depending on the magnitude of strain intensity $\varepsilon_u{}'$ of neutron flux $I$ and approximation parameters $a_k{}'$.

Thus, the stress–strain relation in a deformed body at active loading from the natural state with simultaneous influence of the neutron flux may be expressed as

$$s'_{ij} = 2G\, э'_{ij}\, f'\left(\varepsilon'_u, I, a'_k\right), \sigma' = K(3\varepsilon' - BI), \tag{8.81}$$

where the plasticity function shall be set as $f'(\varepsilon'_u, I, a'_k) = 1$ in those areas where $\varepsilon_u{}' \leq \varepsilon_y{}'$, $\varepsilon_y{}'$—yield point at the moment of deforming.

Under a sufficiently fast "instantaneous" application of the load, hardening effects of irradiation have no time to manifest themselves, and the arisen domains of plastic deformations will be the same as the ones without the neutron flux. However, if the active loading proceeds rather slowly, the outer layers of the body will be hardened in due course, and the domains of plastic deformation may turn to be smaller or absent at all as compared to the unexposed body. This effect may occur when the first plastic deformations appear not on the external hardened surface, but beneath it, where the strain intensity is great, and the yield point has no time to augment.

Hence, irradiation is opposite to the heating effect on elastoplastic bodies, so it decreases the yield point of the materials and leads to expansion of the areas of plastic deformation at statically equivalent loads.

Let us add differential equilibrium equations and boundary conditions, as well as Cauchy relations to equations (8.81) in the assumption of smallness of deformations.

$$\sigma'_{ij},_j + \rho F'_i = 0, \sigma'_{ij} l_j = R'_i \text{ on } S_\sigma, u'_i = u'_{0i} \text{ on } S_u,$$

$$2\varepsilon'_{ij} = u'_i,_j + u'_j,_i. \tag{8.82}$$

We shall further accept that the behavior of external loads and boundary displacements in time is such that the corresponding load paths cannot be related to the class of essentially complex loading, and the radiation-induced growth of the plastic threshold does not exceed the growth of the strain intensity in the exposed points of the solid body, which may prevent origination of plastic deformations.

Let as assume that beginning from time $t = t_1$ the effect of neutron flux ceases ($\varphi = 0$) and the external forces vary so that unloading and subsequent variable loading by volume $F_i{}''$ and surface $R_i{}''$ forces (on $S_\sigma$) at boundary displacement $u_{0i}{}''$ (on $S_u$) take place in all points of plastically deformed areas $V_p{}'$ of the body. The level of the body exposure remains constant and equal to its value prior to unloading $I_1 = \varphi t_1$. The plastic limit in the body points depends on coordinate $z$ and becomes equal to $\sigma_y{}'(I_1(z))$.

We assume that the boundary problem (8.79)–(8.82) has been solved and denote corresponding stresses, strains, and displacements, like previously, by $\sigma_{ij}{}''$, $\varepsilon_{ij}{}''$, $u_i{}''$. For the values before unloading we retain designations $\sigma_{ij}{}'$, $\varepsilon_{ij}{}'$, $u_i{}'$.

Let us write the constitutive equations. In zones $V_e$' and $V_e$" of unloading and elastic deformation the following relations are valid:

$$s_{ij}^* = 2G\,\text{э}_{ij}^*,\ \varepsilon_u^* \pounds\ \varepsilon_y^*(I_1),\ s_{ij}^* = s_{ij}' - s_{ij}'', \text{э}_{ij}^* = \text{э}_{ij}' - \text{э}_{ij}'' . \tag{8.83}$$

In domain $V_p$", where plastic deformations vary in the process of variable loading, the following relations are valid:

$$s_{ij}^* = 2Gf^*\left(\varepsilon_u^*, I_1, a_k^*\right)\text{э}_{ij}^* . \tag{8.84}$$

The volume deformation in all points of the body remains elastic. Therefore, prior to unloading and for the current state the next relations should be obeyed

$$\sigma' = K(3\varepsilon' - BI_1),\ \sigma'' = K(3\varepsilon'' - BI_1),$$

This implies that

$$\sigma^* = 3K\varepsilon^* . \tag{8.85}$$

The equilibrium equations, boundary conditions, and Cauchy relations for values $\sigma_{ij}^*$, $\varepsilon_{ij}^*$, $u_i^*$ will be

$$\sigma_{ij},_j^* + \rho F_i^* = 0,\quad F_i^* = F_i' - F_i'',$$

$$\sigma_{ij}^* l_j = R_i^*,\quad R_i^* = R_i' - R_i'',\ \text{on } S_\sigma,\quad u_i^* = u_{0i}^* = u_{0i}' - u_{0i}'',\ \text{on } S_u,$$

$$2\varepsilon_{ij}^* = u_i^*,_j + u_j^*,_i . \tag{8.86}$$

Relations (8.83)–(8.86) create a new boundary problem for the asterisked values. By assuming that function $f^*$ may be approximated by function $f'$ in any point of the deformation curve means that it is described by the same expression of a closed-form but with other parameters $a_k^* : f^* = f'\left(\varepsilon_u^*, a_k^*, I_1\right)$.

By comparing relations (8.81)–(8.82) for the body at loading from the natural state to the relations for the asterisked values (8.83)–(8.86), we may see that they coincide till designations. That is why the solution of the problem for the asterisked values may be obtained from a known solution of the problem corresponding to the loading from the natural state by means of some substitutions. For example, if we know displacement $u_i' = u_i'(x, \varepsilon_u', \varepsilon_y', I, a_k')$, a corresponding displacement is $u_i^* = u_i'(x, \varepsilon_u^*, \varepsilon_y^*, I_1, a_k^*)$, and the sought displacement at repeated alternating loading is found on the base of relation (8.83)

$$u_i'' = u_i' - u_i^* .$$

Stresses and strains are calculated by the formulas of the same type.

The obtained results may be extended to the case of any *n*-th cyclic loading (*theorem of cyclic loading of elastoplastic bodies in a neutron flux*). The level of irradiation after initial loading is fixed, and the formal proof of the theorem shows no difference from that specified in 8.12.

It follows from above that application of the linear approximation methods for homogeneous and piecewise inhomogeneous bodies makes it possible to obtain closed-form solutions for the problems of the theory of small elastoplastic deformations.

## NOTES

1. See Bauschinger (1881).
2. *Viedemann, G.H.* (1826–1899), German physicist.
3. *Thurston, R.H.* (1839–1903), American mechanical engineer.
4. *Tresca, Henri Edouard*, French scientist in the field of mechanics. The main works in the plasticity theory. The main series of works on the solid body flow were published in the 1960s.
5. Notice that a world-famous scientist in the field of plasticity Saint-Venant, born in 1797, was 71, when he was elected a member of French Academy. It is explained by his antiwar activities when a student expressed in opposition to Napoleon in 1814.
6. *Hyber, M.T.* (1872–1950), published an article in 1904 in Poland [24], in which he relates the transition into the plastic state to the forming energy. This work was published a few decades later.
7. *Mises*, Richard in 1913 published an article, in which he proposed to assume the relation (8.3) as a simplification of (8.2), considering his condition to be approximate.
8. *Hencky, H* (1885–1951) published a work in 1924, in which he related independently of Hyber the transition into the plastic state to the energy of form-changing.
9. *Coulomb, Charles Augustin* (1736–1806), French engineer and physicist, founder of electrostatics; worked in mechanics of friction and fiber twisting.
10. *Lenskiy, Victor S.* (1913–1998), Russian mechanical engineer, Professor of Moscow State University; is known for his theoretical and especially experimental works in the field of plasticity.
11. *Moskvitin, Viktor Vasil'evich* (1922–1983), Russian mechanical engineer, Professor of Moscow State University; works in the field of plasticity.

## KEYWORDS

- **Ilyushin theory of plasticity**
- **Material embrittlement**
- **Mises condition of plasticity**
- **Strain hardening**
- **Translational hardening**
- **Yield surface**

# Chapter 9

## Linear Viscoelastic Continua

### CREEP AND RELAXATION

We assumed in the previous chapters that stress–strain state (SSS) of the body considered remains constant under constant in time effects. However, many materials, for example, polymers, concrete, composites, and so forth, possess the ability to slowly undergo deformation with time even at room temperatures under constant stresses. This property of materials is called *creep*.

Starting from the middle of twentieth century there arose a necessity to take into account creep in design models in response to engineering demands, first of all turbine engineering. Later on, these models have found application in nuclear power and chemical engineering, aircraft construction, jet machines, civil construction. The demands of mechanical engineering have led to a sharp growth of experimental and theoretical research in the field of creep.

Creep may eventually result in significant changes in SSS of buildings and constructions. For example, as consequence of nonuniform settlement of the basement soil with time the strains redistribute between separate elements of constructions. Sometimes, this results in cracks and breakage under most unfavorable conditions. One more example is massive concrete dams of modern hydroelectric power stations in which exothermal processes during concrete hardening play a significant role. Creep is playing a positive role in this case as it reduces the resultant stresses.

Notice, that metals also show *rheonomic* properties when heated during operation. These are by their essence the mechanical properties of materials depending much on time.

Creep can be investigated experimentally at stretching cylindrical samples. The diagrams of strain growth as a function of time under constant stresses are called *creep curves*.

Test conditions and a creep curve are schematically shown in Fig. 9.1. The upper end of the sample is fixed, and the bottom is loaded. Variations in the rated part length are recorded and the curve of strain variations $\varepsilon$ in time $t$ is constructed. The strain increases from its initial value $\varepsilon_0$, thus reflecting elastic properties of the material and corresponds to an instantly applied load $P$.

Further, three characteristic sections of the creep curve can be selected. In the first one ($AB$), the creep rate is high, but decreases monotonously starting from some high value. This is *a section of nonstationary creep*. The second section of the curve ($BC$) is almost rectilinear and characterizes *a stationary creep (secondary creep)*. Here, the rate of creep is practically constant. Finally, a section is observed ($CD$) that precedes failure and *the creep rate increases monotonously*.

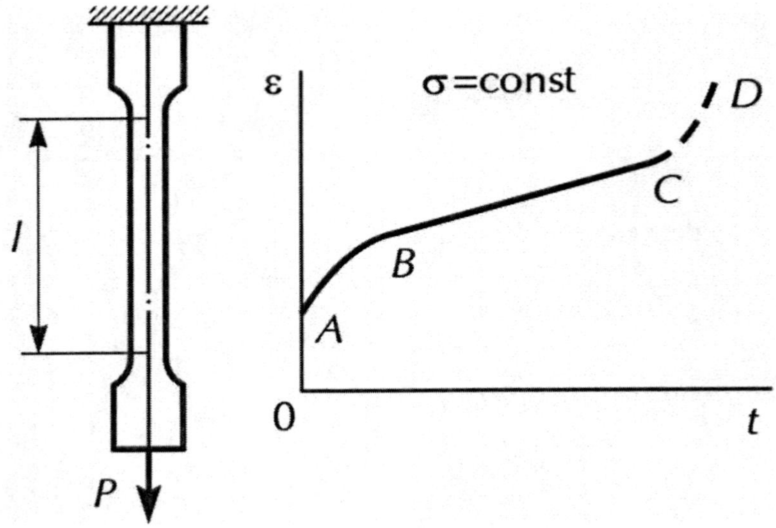


Figure 9.1.

We should underline that creep occurs at any kind of stress, therefore, it is impossible to specify a threshold for the creep.

The phenomenon of stress drop in a body at a constant strain is called *relaxation*. We shall assume that the sample was instantly stretched by way of applying force $P_1$ so that the design length $l_0$ became equal to $l$ (Fig. 9.2).

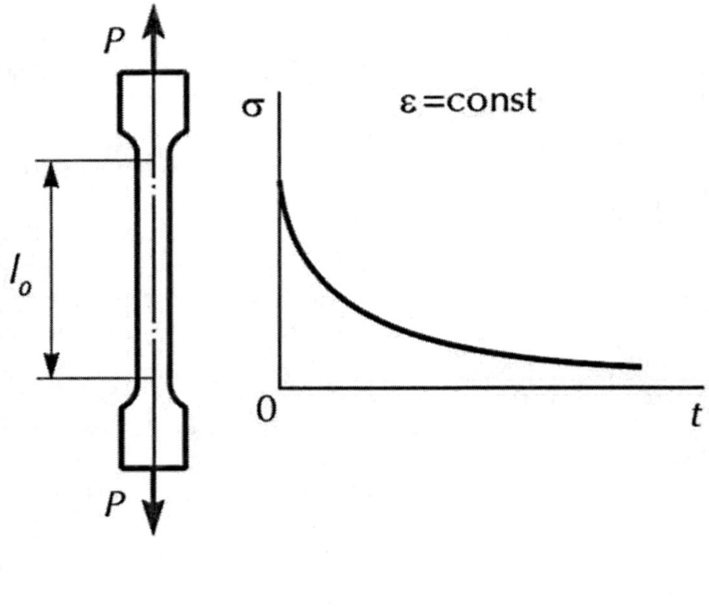


Figure 9.2.

The sample was fixed in the stretched state for a while, then released and a load was applied again to stretch the rated part of the sample until value $l$. It turned out that the required load $P_2$ was smaller than the initial one $P_1$. Hence, the stress needed for maintaining a constant strain drops, decreases and relaxes.

Creep and relaxation as rheonomic properties of materials were display in the real elements of constructions simultaneously, and are interrelated. They can describ analytically by introduction of time into the stress–strain relation of a body. The description of this interrelation suggested by Boltzmann[1] is based on the assumption of the influence of the whole previous period of stress action on strain at a given moment. Such continua are called *linear viscoelastic continua of the hereditary type.*

The onset of the viscoelasticity theory was connected with the names of Boltzmann, Maxwell, Kelvin[2] and Voigt.[3] We should recollect here Maxwell's (Fig. 9.3) and Voigt's elementary models (Fig. 9.4) for the viscoelastic continuum representing a viscoelastic body as a combination of elastic and viscous elements. At a constant force a piston may move with a constant velocity or, otherwise, $\sigma = \lambda\dot{\varepsilon}$ ($\lambda$—proportionality factor). The elastic element looks like a spring with a linear characteristic, that is, $\sigma = E\varepsilon$. The viscous element is a cylinder with a viscous liquid in which the piston with an aperture or a clearance along the cylinder wall moves thus making liquid leak from one part of the cylinder into the other.

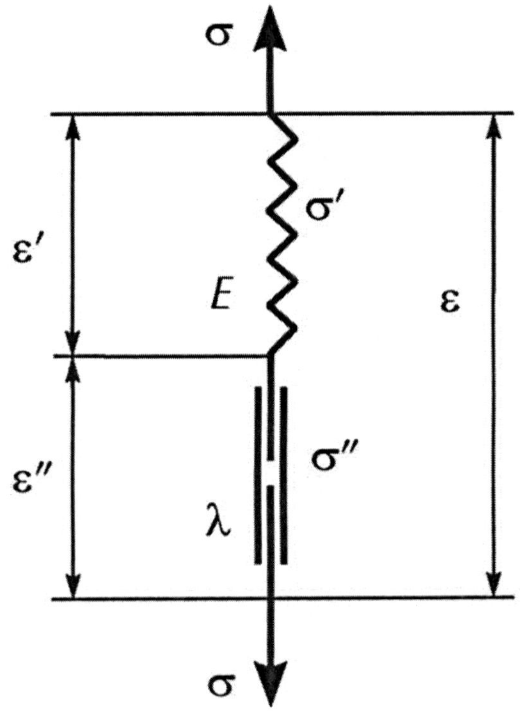


**Figure 9.3.**

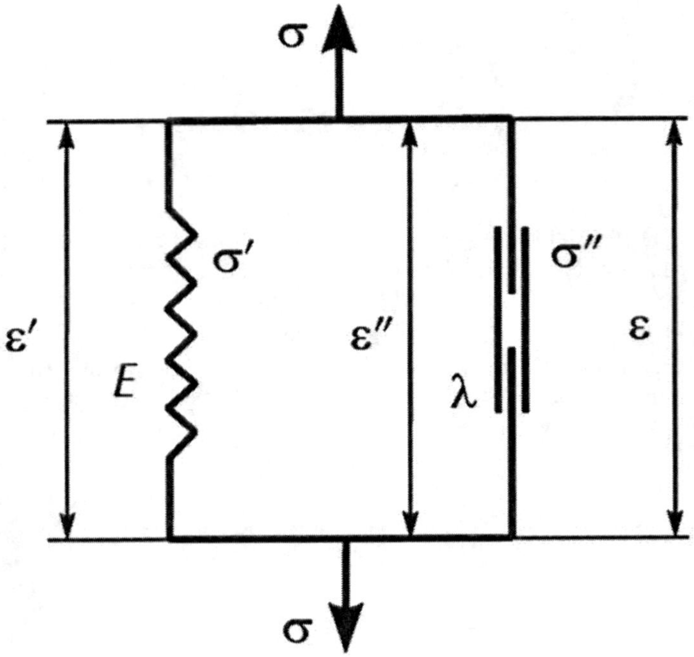


**Figure 9.4.**

In a unidimensional Maxwell's model the connection of elastic and viscous elements is sequential (Fig. 9.3). Stresses in the elements are identical, while strains are summarized. We may obtain the stress–strain relation by differentiating the equation of Hooke's law in time and adding the viscous component to the received strain rate:

$$\dot{\varepsilon} = \frac{\dot{\sigma}}{E} + \frac{\sigma}{\lambda}.$$

Stresses in the elements are summarized Voigt's model, and their strains are taken identical. One can obtain such a picture if to connect the elements in parallel (Fig. 9.4). Then,

$$\sigma = E\varepsilon + \lambda\dot{\varepsilon}.$$

*Volterra's principle*[4] is strongly important for recalculating the results of static problems of the elasticity theory in the hereditary viscoelastic states. Numerous achievements in the modern theory of viscoelasticity belong to *Ilyushin*, *Ishlinskii*,[5] *Koltunov*,[6] *Moskvitin*, *Rabotnov*, *Slonimskii*, *Rzhanitsyn*,[7] *Pobedrya*, and other home scientists. In particular, Ilyushin has presented a comprehensive development of the general theory of thermoviscoelasticity and offered an effective method of solving particular problems, namely, the method of approximations (Ilyushin and Pobedrya, 1970).

## STATEMENT OF LINEAR VISCOELASTICITY PROBLEMS

Physical relations between stress and strain deviators ($s_{ij} = \sigma_{ij} - \sigma\delta_{ij}$, $э_{ij} = \varepsilon_{ij} - \varepsilon\delta_{ij}$) in linear viscoelasticity are recorded as follows:

$$2G\, э_{ij}(t) = s_{ij}(t) + \int_0^t \Gamma(t-\tau) s_{ij}(\tau)\,\mathrm{d}\tau,$$

$$K\theta(t) = \sigma(t). \tag{9.1}$$

We have denoted $\theta = \varepsilon_{kk}$ as a relative volume variation, $\sigma = \sigma_{kk}/3$—mean (hydrostatic) stress, $G$—instant elastic shear modulus, $K$—instant modulus of volume strain, function $\Gamma(t)$ characterizes rheological properties of the material and is called *creep kernel.*

Physical equations (9.1) express the following: the field of strain $э_{ij}$ is determined presently not only by the instant stress $s_{ij}$ (related to strains by a generalized Hooke's law), but also by the previous stress values via some hereditary function. Volume strain $\theta$ is accepted as elastic since volume creep is negligible as compared to the shear-induced one. Notice that the hereditary function has a difference $(t - \tau)$ as its argument, that is, equation (9.1) are invariant relative to time reference.

Kernel $\Gamma(t)$ is a positive monotonously decreasing function of its arguments. It asymptotically tends to zero at infinity, and at negative values of the argument, it is identically equal to zero.

Let us assume that equation (9.1) can be solved relative to $s_{ij}$ and $\sigma$:

$$s_{ij}(t) = 2G\left( э_{ij}(t) - \int_0^t R(t-\tau)\, э_{ij}(\tau)\,\mathrm{d}\tau \right),$$

$$\sigma(t) = K\theta(t). \tag{9.2}$$

Function $R(t)$ is called a *relaxation kernel* of the material. It is *a resolvent of* kernel $\Gamma(t)$. Function $\Gamma(t)$, in its turn, is *a resolvent* of kernel $R(t)$.

Let us substitute $э_{ij}(t)$ from (9.1) in the first equation (9.2) and demand its identical obeyance at any $s_{ij}(t)$. We may receive the following integrated formula interrelating the kernel:

$$R(t) - \Gamma(t) = \int_0^t \Gamma(t-\tau) R(\tau)\,\mathrm{d}\tau. \tag{9.3}$$

Hence, it is quite enough to determine experimentally one of the kernels to obtain another analytically. If all kernels have been determined in experiments, then equation (9.3) serves to check if the strain description by the accepted model is sufficient.

We may write the deviator relations (9.1), (9.2) in some other way:

$$э_{ij}(t) = \int_0^t J(t-\tau) ds_{ij}(\tau), \quad s_{ij}(t) = \int_0^t \Pi(t-\tau) d\, э_{ij}(\tau)\cdot \tag{9.4}$$

In above-mentioned equations $J(t)$ is a *function of creep* (pliability), $\Pi(t)$ is a *function of relaxation.* A relation can be written between these functions and the kernel introduced earlier by integrating, for example, equation (9.4) in parts. As a result, we have

$$\text{э}_{ij}(t) = J(t-\tau)s_{ij}(\tau)\Big|_0^t + \int_0^t \frac{dJ}{d\tau} s_{ij}(\tau)d\tau .$$

*Taking* $s_{ij}(0) = 0$ and obeying $J(0) = 1/2G$, $\Gamma(t) = 2G\mathrm{d}J(t)/\mathrm{d}t$, we come to equation (9.1.). Similarly, the identity of equation (9.2) and the second from (9.4) can established.

## TYPES OF CREEP AND RELAXATION KERNELS

In order to determine experimental values of functions $\Gamma(t)$ and $R(t)$, we shall consider the stress state at a pure shear. For this aim, creep curves $\varepsilon_{12}(t)/\sigma_{12}^{\ 0} \sim t$ are constructed from trials. In this case, all other components of stress and strain tensors are equated to zero. Equation (9.1) will be reduced to a single one:

$$2G\,\text{э}_{12}(t) = s_{12}(t) + \int_0^t \Gamma(t-\tau)s_{12}(\tau)d\tau ,$$

if creep ($\sigma_{12} = \sigma_{12}^{\ 0} = \text{const}$), it follows

$$2G\frac{\text{э}_{12}(t)}{\sigma_{12}^0} = 1 + \int_0^t \Gamma(\xi)d\xi .$$

This relation allows us to determine the sought function $\Gamma(t)$:

$$\Gamma(t) = 2G\frac{\mathrm{d}J(t)}{\mathrm{d}t} ,$$

where

$$J(t) = \frac{\text{э}_{12}(t)}{\sigma_{12}^0} .$$

Value $J(t)$ is called *pliability* (*creep function*) at pure shear and is calculated on the base of experimental curves of creep.

The values of relaxation kernel $R(t)$ can be similarly determined from the tests on pure shear.

In practice, the following analytical expressions for the kernels of creep and relaxation are basically used for approximation of experimental data.

*1. Exponential creep kernel:*

$$\Gamma(t) = Ae^{-pt} \quad (A > 0, p > 0).$$

The presence of only two constants of the material $A$ and $p$ complicates application of this representation for describing experimental data within a wide time range, especially at the initial stage of loading.

*The kernel resolvent* (kernel of relaxation) can be also presented as

$$R(t) = Be^{-qt}.$$

Substitution of exponential kernel of creep and its resolvent in relation (1.44), and demanding its identity for $t$, will give $B = A$, $q = A + p$.

Representation that is more exact gives a kernel and its resolvent accepted as a sum of exponential functions

$$\Gamma(t) = \sum_{i=1}^{m} A_i e^{-p_i t}, \quad R(t) = \sum_{i=1}^{m} B_i e^{-q_i t}.$$

Substitution of these equations in the integrated relation between the kernels of creep and relaxation (1.44) and demanding its identity, will give *2m* of the equations relating the constants of resolvents $B_i$ and $q_i$ and constants of kernels $A_i$ and $p_i$:

$$\sum_{i=1}^{m} \frac{A_i}{q_j - p_i} = 1, \quad \sum_{i=1}^{m} \frac{B_i}{q_i - p_j} = 1 \quad (j = 1, 2, ..., m).$$

These equations can be useful in finding one of the kernels when the other is known. From the first system of equations we extract roots $q_i$. They should be material, unequal and distinct from $p_i$. Then, from the second system of linear equations, factors $B_i$ are calculated.

2. *Duffing's*[8] *exponential kernel:*

$$\Gamma(t) = \frac{C}{t^{1-\beta}} \quad (0 < \text{b} < 1).$$

Constant β is restricted because at β = 0 the strain rate and strain itself at the moment of loading becomes infinitely large. The kernel impairs at quite longer time of loading. Its resolvent is a function

$$R(t) = \sum (-1)^{k+1} \frac{C^k \Gamma_*^k(\beta) t^{k\beta-1}}{\Gamma_*(k\beta)},$$

where $\Gamma_*(\beta)$ is a gamma-function of argument β. Under specified restrictions for β this series converges everywhere. The kernels of a similar exponential type are called kernels with a weak singularity or weak feature, meaning that at $t \to 0$, $\Gamma(t) \to \infty$, which corresponds to one infinitely high strain rate at the moment of loading. A weak tendency to infinity is due to a restriction $0 < \beta < 1$. Duffing's kernel cannot describe satisfactorily the experimental data within a wide time interval because of a limited number of constants.

3. *Rshanitsyn's kernel.* A typical relaxation kernel uniting the exponential and weak singular properties was suggested by Rzhanicyn (1968):

$$R(t) = \frac{Ae^{-pt}}{t^{1-\beta}} \quad (p > 0,\ 0 < \text{b} < 1). \tag{9.5}$$

The resolvent of kernel (9.5) is the following function:

$$\Gamma(t)=\frac{e^{-pt}}{t}\sum_{k=1}^{\infty}(-1)^{k+1}\frac{A^k\Gamma_*^k(\beta)t^{k\beta}}{\Gamma_*(k\beta)}.$$

It differs from the resolvent of Duffing's kernel only by a multiplier $e^{-pt}$. This kernel is often applied at present for specific numerical calculations. It has been perfectly tabulated by Koltunov, whose diagrams, tables and variants of calculations are given in monograph (Koltunov, 1976).

A more general kernel was considered by Koltunov (1976):

$$R(t)=\frac{Ae^{-pt^{\alpha}}}{t^{1-\beta}} \quad (p>0,\ 0<a<1,\ 0<b<1). \tag{9.6}$$

The resolvent of this kernel is a function

$$\Gamma(t)=\frac{e^{-pt^{\alpha}}}{t^{\alpha}}\sum_{k=1}^{\infty}(-1)^{k+1}\frac{\left[A\Gamma_*\left(\dfrac{\alpha+\beta-1}{\alpha}\right)\right]^k t^{k(\alpha+\beta-1)}}{\Gamma_*\left(\dfrac{\alpha+\beta-1}{\alpha}k\right)}.$$

A kernel of (9.6) type was suggested much earlier by Bronskii (1941) and Slonimskii. This kernel is derived from (9.6) for $\alpha = \beta$.

$$R(t)=\frac{Ae^{-pt^{\beta}}}{t^{1-\beta}}.$$

The resolvent of this kernel we may find from the resolvent of kernel (9.6) by accepting that $\alpha = \beta$.

*4. Abel's kernel of the type:*

$$J_{\alpha}(t)=\frac{t^{\alpha}}{\Gamma_*(1+\alpha)} \quad (-1<a<0).$$

The resolvent this kernel is a function obtained by Rabotnov at b = –1:

$$э_{\alpha}(\beta,t)=t^{\alpha}\sum_{n=0}^{\infty}\frac{\beta^n t^{n(1+\alpha)}}{\Gamma_*[(n+1)(1+\alpha)]}.$$

A series of this resolvent converges for all $t$ and b values. Apparently, $э_{\alpha}(0, t) = J_{\alpha}(t)$.

Rabotnov has called function $э_{\alpha}(b, t)$ *a fractional-exponential kernel*. This kernel is extensively used in literature due to wide opportunities in describing rheological properties of materials and due to the presence of developed enough special algebra of corresponding operators (Rabotnov, 1977, 1979).

## VOLTERRA'S PRINCIPLE

We shall present further an effective method for solving boundary problems of the linear theory of viscoelasticity analogous to (9.2), (9.5). For this purpose, we shall write equation (9.2) in a symbolical form by introducing a designation for the integrated operator $G^*$:

$$s_{ij} = 2G^* \ni_{ij}, \sigma = K\theta, \tag{9.7}$$

where

$$G^* = G(1 - R^*), \; R^* f = \int_0^t R(t-\tau) f(\tau) \mathrm{d}\tau .$$

Based on the comparing of (9.7) and (2.9), we may conclude that the physical relations of the linear viscoelasticity are of the same kind as that of the generalized Hooke's law but having shear modulus $G$ substituted for operator $G^*$.

Equation (9.1) can also be presented in the operation form:

$$\ni_{ij} = \frac{1}{2G^*} s_{ij}, \, K\theta = \sigma, \; K\theta = \sigma,$$

where operator $1/G^*$ is equal to

$$\frac{1}{G^*} = \frac{1}{G}(1 + \Gamma^*), \; \Gamma f = \int_0^t \Gamma(t-\tau) f(\tau) \mathrm{d}\tau .$$

Substitution of $G^* = G(1 - R^*)$ from (9.8) gives

$$\frac{1}{1 - R^*} = 1 + \Gamma^* .$$

Thus, the operation of dividing by some operator becomes definite.

Hence, we come to an important *Volterra principle*: the solution of a linear problem of viscoelasticity can received from the corresponding solution of a linear elasticity problem by substitution of elasticity constants in the latter with some operators. In our case, it is necessary to substitute shear modulus $G$ by operator $G^*$.

However, solution of the elasticity theory problem can contain some other constants, for example, Young's modulus $E$ and Poisson's ratio ν. Then, we should first substitute $E$ and ν for constants $G$ and $K$ in the elastic solution, and then substitute constant $G$ by operator $G^*$.

By having used Volterra's principle, we shall come to a solution that includes algebraic or transcendental functions of operators in time and this solution is to be interpreted. In general, the interpretation connected with certain difficulties. In some cases, these difficulties are surmountable by using LaplaceCarson's integrated transformation, Ilyushin's (Ilyushin and Pobedrya, 1970) method of approximations and Rabotnov's operators (1977, 1979).

In particular, we shall call function $f^*(p)$ of a real parameter $p$ by the Laplace-Carson transform of some function $f(t)$:

$$f^*(p) = p\int_0^\infty f(t)e^{-pt}dt.$$

If $f(t) = f_0$, then $f^*(p) = f_0$. Function $f(t)$ presents convolution of two functions $F$ and $\psi$, then

$$f(t) = \int_0^t F(t-\tau)d\psi(\tau), \tag{9.8}$$

its transform is equal to a product of the transforms of subintegral functions:

$$f^*(p) = F^*(p)\psi^*(p). \tag{9.9}$$

Further, if the transform of function $f(t)$ is found from Eq. (9.9), then the function itself is represented by formula (9.8).

Let us write out the full system of equations and add equilibrium equations, Cauchy relations and boundary conditions to the linear viscoelasticity equation (9.4) under the elastic volume strain:

$$э_{ij}(t) = \int_0^t J(t-\tau)ds_{ij}(\tau),\ K\theta = \sigma,$$

$$\sigma_{ij},_j + \rho F_i = 0,\ \varepsilon_{ij} = (u_i,_j + u_j,_i)/2,$$

$$\sigma_{ij}l_j = R_i \text{ on } S_\sigma,\ u_i = u_{0i} \text{ on } S_u. \tag{9.10}$$

If to apply LaplaceCarson's transform to equation (9.10) and take into account (9.9), we shall have

$$э^*_{ij} = J^*s^*_{ij}, K\theta^* = \sigma^*,$$

$$\sigma^*_{ij},_j + \rho F^*_i = 0, 2\varepsilon^*_{ij} = u^*_i,_j + u^*_j,_i,$$

$$\sigma^*_{ij}l_j = R^*_i \text{ on } S_\sigma,\ u^*_i = u^*_{0i} \text{ on } S_u. \tag{9.11}$$

Hence, the problem of the linear viscoelasticity in transformations coincides with a corresponding problem for the elastic body. This means that solutions of these problems coincide too with a difference that constant (1/2) $G$ is substituted by transform $J^*$. The modulus of volume compression retained in the solution. So, if we know the solution of the linear elasticity problem, transforms $\sigma_{ij}^*$, $\varepsilon_{ij}^*$ will be known too. Once we have found the inverse transforms, we may come to the sought solutions of $\sigma_{ij}(t, x)$, $\varepsilon_{ij}(t, x)$.

Notice that the general problem of recovering inverse transforms has certain difficulties. Some elementary functions have the formulas for finding inverse transforms usually given in manuals for operational calculations. In more intricate cases, one should use specially developed approximation methods. Further, an effective method of approximations described and an example of its application for a circular three-layer plate is given. We should underline that in recovering the transforms of boundary conditions (9.11) we implicitly assumed the body boundary to be invariable during loading. Therefore, this method is inapplicable, for example, for the contact problems on the action of a punch and the problems with the boundary burn out.

## ILYUSHIN'S EXPERIMENTAL AND THEORETICAL METHOD

Let us consider a method for the problems of linear viscoelasticity based on solution of the corresponding elasticity theory problem and application of some functions $g_\beta(t)$ additionally determined in experiments. When analyzing the analytical forms of solving the problems on ideal elasticity, Ilyushin has formulated a theorem stating that the solution of the problem of ideal elasticity can be presented in the following symbolical form (Ilyushin and Pobedrya, 1970):

$$u = f + \omega\varphi + \frac{1}{\omega}\psi + \frac{1}{1+\beta_k\omega}\chi_k .$$

Where $f$, $\varphi$, $\psi$, $\chi_k$ are the functions of coordinates, external forces, specified boundary displacements and temperatures; $\omega = 2G/(3K)$. Parameters $\beta_k$ are constant. Summation over index $k$ is made in the last summand.

Then, solution of the corresponding problem of linear viscoelasticity in transforms will be ($\omega^* = 2G^*/(3K)$):

$$u^* = f^* + \omega^*\varphi^* + \frac{1}{\omega^*}\psi^* + \frac{1}{1+\beta_k\omega^*}\chi_k^* . \tag{9.12}$$

Our task is to represent the solution of (9.12) in inverse transforms.

Let us denote the operator (Ilyushin's)

$$g_{\beta k}^* \equiv \frac{1}{1+\beta_k\omega^*} .$$

We take the transform $g_{\beta k}(t)$ corresponding to some function $g_{\beta k}^*$. In this case, there is no problem in converting the solution into inverse transforms:

$$u(t) = f(t) + \frac{2G}{3K}\left(\varphi(t) - \int_0^t R(t-\tau)\varphi(\tau)\,\mathrm{d}\tau\right)$$

$$+\frac{3K}{2G}\left(\psi(t) + \int_0^t \Gamma(t-\tau)\psi(\tau)\,\mathrm{d}\tau\right) + \int_0^t g_{\beta k}(t-\tau)\,\mathrm{d}\chi_k(\tau). \tag{9.13}$$

Ilyushin has disclosed an experimental procedure in (Ilyushin and Pobedrya, 1970) for finding these functions; so we shall consider them known in our further discourse.

## TIME-TEMPERATURE ANALOGY

Creep characteristics of filled polymers are usually strongly dependent on temperature. The effect of creep grows with temperature rise, thus enabling experimentation at high temperatures over limited time intervals to forecast rheological properties for prolonged periods. The experiments at a constant temperature show that curves $\sigma \sim \varepsilon$ are the higher the greater is the loading rate. If the tests are carried out at a constant loading rate, but at different temperatures, diagrams $\sigma \sim \varepsilon$ are the higher the lower the temperature. This fact has allowed us to determine a time-temperature relation called the *time-temperature analogy*. We can comprehend its essence when consider a family of creep curves at different temperatures and similar stresses (Fig. 9.5). These curves in coordinates $E\varepsilon/\sigma \sim \lg t$ are shown in Fig. 9.6 (Koltunov, 1976).

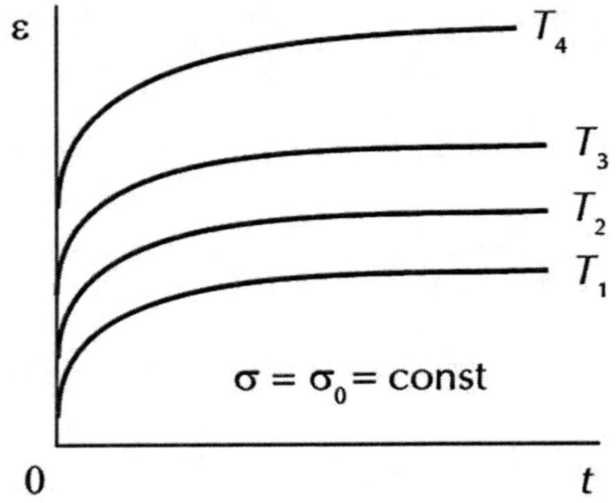


**Figure 9.5.**

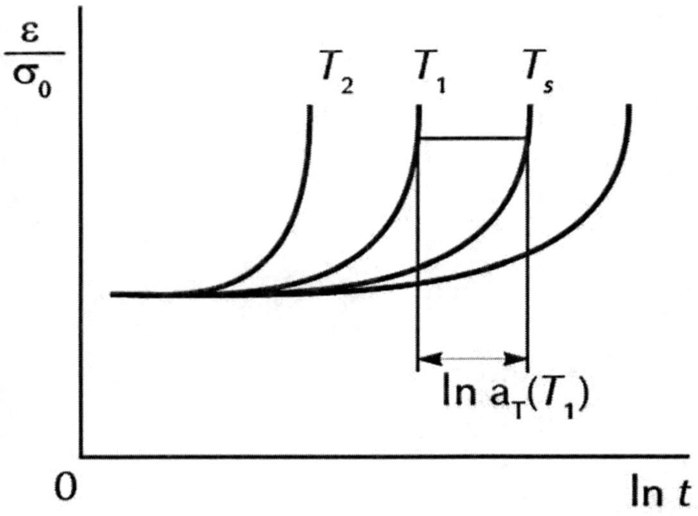


**Figure 9.6.**

According to the principle of time-temperature analogy (or time-temperature superposition) as applied to the phenomenon of creep, the creep curves at temperatures $T_1 > T_s$, $T_2 > T_1$, $T_3 > T_2$, can be superimposed with the creep curve at $T_s$ by shifting along the time logarithm axis (Fig. 9.6) to certain segments $\psi(T_1)$, $\psi(T_2)$, $\psi(T_3)$, accordingly. Obviously, $\psi(T_s) = 0$ and function $\psi(T)$ are found for a given material experimentally according to its definition.

Aside of forecasting, the principle of time-temperature analogy enables to consider the influence of temperature on physico-mechanical properties of the material by introduction *of a modified time t'*. Indeed, let $J_1 = J_1(t) = J^*\ln(t)$ be the equation for the creep curve $\varepsilon/\sigma_0 \sim t$ at a given temperature $T_s$. According to the principle of time-temperature analogy, we may write

$$J_1(t,T)=J^*(\ln t+\psi(T)\ ,\ J_1(t,T_s)=J^*(\ln t)\,. \tag{9.14}$$

Function $\psi(T)$ shows the following properties: $\psi(T_s) = 0$, $\psi(T) > 0$ at $T > T_s$, $\psi(T) < 0$ at $T < T_s$, $d\psi/dT > 0$. Therefore, if to accept $\psi(T) = -\ln a_T(T)$, so $a_T(T_s) = 1$, $0 < a_T(T) < 1$ at $T > T_s$, $a_T(T) > 1$ at $T < T_s$. These properties are peculiar to the function

$$\ln a_T = C_1 \frac{T - T_s}{C_2 + T - T_s}, \tag{9.15}$$

where $C_1$, $C_2$ are positive constants determined for each material experimentally. Above relation is known as a *Williams-Landel-Ferry equation* [80]. Value $a_T$ is sometimes called the *function of time-temperature shift* or a *factor of displacement.*

A graphical dependence $\lg a_T \sim T$ for a typical polymer is presented in Fig. 9.7. If to use analytical dependence (9.15) and accept the temperature of reduction $T_s = 300$ K, we have for this curve

$$\ln a_T = -8{,}86 \frac{T - T_s}{101{,}6 + T - T_s}\,.$$

Now, taking into account equation.

$$t' = \frac{t}{a_T(T)}\,. \tag{9.16}$$

(9.14) we shall introduce a modified time

$$\ln t' = \ln t - \ln a_T(T)\,,$$

or It follows from (9.14)

$$J_1(t,T)=J^*(\ln t')=J_1(t')=J_1\left(\frac{t}{a_T}\right). \tag{9.17}$$

It is important that $J_1 = J_1(t)$ is pliability at temperature $T_s$.

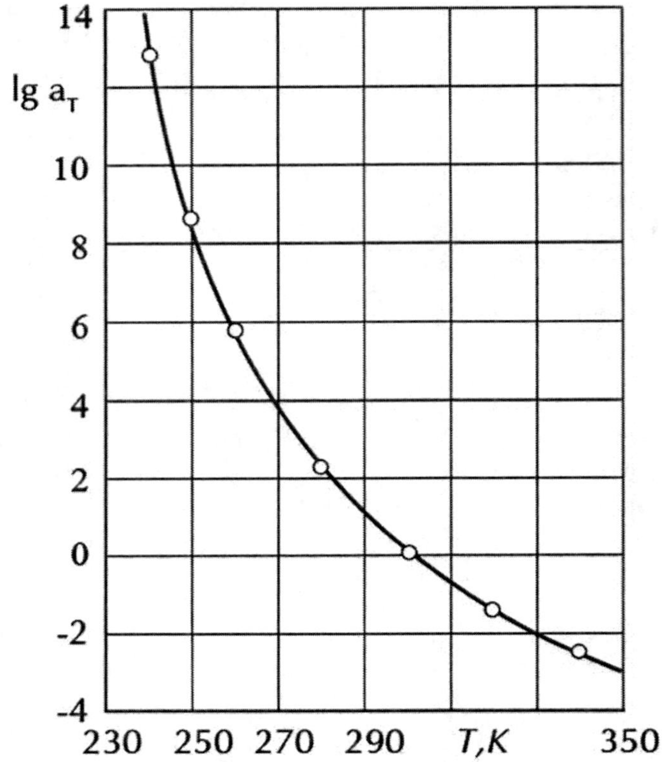


**Figure 9.7.**

Hence, introduction of the modified time helps us to avoid formulation of the analytical expression for $J_1(t, T)$ at temperature $T$. Furthermore, this method enables to expand the efficient methods of solving the linear viscoelasticity problems for the case of thermoviscoelastic models.

Relation (9.16) for the modified time is true when the temperature is invariable during the whole experiment. If the temperature is a function of time, this method can be used for a small time interval $dt$, similarly to (9.16)

$$dt' = \frac{dt}{a_T(T)}, \quad t' = \int_0^t \frac{dt}{a_T(T)}. \tag{9.18}$$

Introduction of the modified time is based on the principle of time-temperature analogy. This principle was verified experimentally more than once for a number of polymeric materials at physical linearity and corroborated well with other results. The principle of time-temperature analogy (9.16)–(9.18) works well for solving certain problems at physical nonlinearity of materials as well.

## THE THEORY OF AGEING

*The technical theories of creep* have gained further development in engineering practice. They reflect rheonomic properties of materials with account of time in stress–strain relations. *The theory of ageing* is used for calculations of concrete. According to this theory, unidimensional determinants of equations are the following:

$$\varepsilon=\frac{\sigma}{E}+f(\sigma,t)\,,\;\; f(\sigma,t)=f_1(\sigma)f_2(t)=B\sigma^n f_2(t)\,. \tag{9.19}$$

Constants $B$, $n$ and function $f_2(t)$ are found from the curves of creep of the material. Relations (9.19) show that the material is likely to vary its properties with time, that is, "become older," which explains the name of the theory. The theory of ageing takes into account the age of concrete, but it does not account for the loading duration.

As additional material on the linear viscoelasticity and creep problems, we recommend (Blend, 1960, Christensen, 1971, Filonenko-Borodich, 1947).

## A CIRCULAR LINEARLY VISCOELASTIC THREE-LAYER PLATE

We may solve the linear viscoelasticity problem from solution of the elastic plate (6.67) by using Ilyushin's theoretical method of approximations. Additionally, the condition $\beta < 1$ should be obeyed, which takes place, for example, when the elastic constants of the filler, $G_3$, $K_3$ cede to the ones in the bearing layers. This makes possible to describe the behavior of the modified Bessel's functions (6.64) over segment $0 < x < 1$ with sufficient accuracy by using the following formulas:

$$I_n(x)=\frac{1}{\Gamma(n+1)}\left(\frac{x}{2}\right)^n-\frac{1}{\Gamma(n+2)}\left(\frac{x}{2}\right)^{n+2}.$$

Let us check the accuracy of this representation for the functions $I_0(x)$, $I_1(x)$. In this case, the remainder (6.64) will be accepted in the Lagrangian form

$$R_0(x)=\frac{I_0^{\mathrm{IV}}(\vartheta x)}{4!}x^4,\; R_1(x)=\frac{I_1^{\mathrm{V}}(\vartheta x)}{5!}x^5\;(0<\vartheta<1),$$

since

$$I_0^{\mathrm{IV}}(x)=\tfrac{1}{8}(3I_0(x)+4I_2(x)+I_4(x)),$$

$$I_1^{\mathrm{V}}(x)=\tfrac{1}{32}(10I_0(x)+15I_2(x)+6I_4(x)+I_6(x)),$$

where both functions $I_0(x)$, $I_1(x)$ grow monotonously with their argument, the following estimates are true:

$$R_0(x)=\frac{I_0^{\mathrm{IV}}(1)}{4!}x^4,\; R_1(x)=\frac{I_1^{\mathrm{V}}(1)}{5!}x^5.$$

The tables of Besselian functions (Yanke et al., 1979) help to estimate in the numerical form: $R_0(x) = 0{,}0240x^4$, $R_1(x) = 0{,}00432x^5$.

The sought solution will be constructed for a circular linear-viscoelastic plate pinched over the contour. For this aim, the solution of the corresponding elasticity theory problem (6.69) and required coefficients (6.14) are given in the form:

$$\psi = -\frac{b_2 q}{4cb_3G_3}\left(\frac{I_1(\beta r)}{I_1(\beta)} - r\right),$$

$$u = \frac{a_3}{a_1a_6 - a_3^2}\left[\left(a_5 - \frac{a_2a_6}{a_3}\right)\psi + \frac{qr}{16}(r^2-1)\right],$$

$$w = \frac{b_2^2 q}{4cb_3^2G_3}\left(\frac{I_0(\beta r) - I_0(\beta)}{\beta I_1(\beta)} - \frac{1}{2}(r^2-1)\right) + \frac{q}{64b_3}(r^2-1)^2,$$

where

$$a_1 = \sum_{k=1}^{3} h_k K_k^+,$$

$$a_2 = c(h_1K_1^+ - h_2K_2^+),$$

$$a_3 = h_1(c+\frac{h_1}{2})K_1^+ - h_2(c+\frac{h_2}{2})K_2^+,$$

$$a_4 = c^2\left[h_1K_1^+ + h_2K_2^+ + \frac{2}{3}cK_3^+\right],$$

$$a_5 = c\left[h_1(c+\frac{h_1}{2})K_1^+ + h_2(c+\frac{h_2}{2})K_2^+ + \frac{2}{3}c^2K_3^+\right],$$

$$a_6 = h_1(c^2 + ch_1 + \frac{h_1^2}{3})K_1^+ + h_2(c^2 + ch_2 + \frac{h_2^2}{3})K_2^+ + \frac{2}{3}c^3K_3^+,$$

$$b_1 = \frac{a_1a_4 - a_2^2}{a_1}, \quad b_2 = \frac{a_1a_5 - a_2a_3}{a_1}, \quad b_3 = \frac{a_1a_6 - a_3^2}{a_1},$$

$$\beta^2 = \frac{2cb_3G_3}{b_1b_3 - b_2^2}, \quad K_k + \tfrac{4}{3}G_k \equiv K_k^+. \tag{9.20}$$

Let us introduce the following hypothesis on the *similarity of viscoelastic properties of the materials of the plate*: the relaxation kernels of the bearing layers are similar to those of the filler and differ by a constant multiplier $c_k$, that is

$$R_k(t) = c_k R(t), \quad c_k \le 1, \quad k = 1,2,3, \quad R_3(t) \equiv R(t). \tag{9.21}$$

Let us transform elasticity constants $a_i$ (9.20) into the viscoelasticity operators $a_i^*$ by substitution of shear modulus $G_k$ for operator $G_k^*$ (9.7), where operators $\overset{*}{g}_{\beta k}$ introduced by Ilyushin will be isolated:

$$g^*_{\beta k} = \frac{1}{1+\beta_k \omega^*},\ \omega^* = \frac{2G^*}{3K},\ \beta_k = \text{const}, \tag{9.22}$$

using relations (9.13) and the result of the similarity hypothesis of the relaxation kernel from (9.21)

$$G_k^* = G_k(1-c_k) + \frac{c_k G_k}{G} G^*,\quad G \equiv G_3,\quad G^* \equiv G_3^*,\quad K \equiv K_3.$$

Where from,

$$K_k^{+*} = K_{k1} + K_{k2}\omega^*,\quad K_{k1} = K_k + \tfrac{4}{3}G_k(1-c_k),\quad K_{k2} = 2c_k G_k K / G,$$

and operators $a_i^*$ ($i = 1,\ldots,6$) will be transformed into

$$a_i^* = a_{i1} + a_{i2}\omega^*,$$

where ($\alpha = 1, 2$)

$$a_{1\alpha} = \sum_{k=1}^{3} h_k K_{k\alpha}^+,$$

$$a_{2\alpha} = c(h_1 K_{1\alpha} - h_2 K_{2\alpha}),$$

$$a_{3\alpha} = h_1(c + \frac{h_1}{2})K_{1\alpha} - h_2(c + \frac{h_2}{2})K_{2\alpha},$$

$$a_{4\alpha} = c^2\left[h_1 K_{1\alpha} + h_2 K_{2\alpha} + \frac{2}{3}cK_{3\alpha}\right],$$

$$a_{5\alpha} = c\left[h_1(c + \frac{h_1}{2})K_{1\alpha} + h_2(c + \frac{h_2}{2})K_{2\alpha} + \frac{2}{3}c^2 K_{3\alpha}\right],$$

$$a_{6\alpha} = h_1(c^2 + ch_1 + \frac{h_1^2}{3})K_{1\alpha} + h_2(c^2 + ch_2 + \frac{h_2^2}{3})K_{2\alpha} + \frac{2}{3}c^3 K_{3\alpha}.$$

Other operators are expressed through $a_i^*$:

$$b_1^* = \frac{a_1^* a_4^* - a_2^{*2}}{a_1^*},\quad b_2^* = \frac{a_1^* a_5^* - a_2^* a_3^*}{a_1^*},\quad b_3^* = \frac{a_1^* a_6^* - a_3^{*2}}{a_1^*},$$

$$\beta^{*2} = \frac{2cb_3^* G_3^*}{b_1^* b_3^* - b_2^{*2}}.$$

Then, from the elastic solution (9.20) we obtain the one for the linear viscoelasticity problem in LaplaceCarson's transforms:

$$\psi^* = -\frac{b_2^* q^*}{4cb_3^* G_3^*}\left(\frac{I_1(\beta^* r)}{I_1(\beta^*)} - r\right),$$

$$u^* = \frac{a_3^*}{a_1^* a_6^* - a_3^{*2}} \left[ \left( a_5^* - \frac{a_2^* a_6^*}{a_3^*} \right) \psi^* + \frac{q^* r}{16}(r^2 - 1) \right], \tag{9.23}$$

$$w^* = \frac{b_2^{*2} q^*}{4c b_3^{*2} G_3^*} \left( \frac{I_0(\beta^* r) - I_0(\beta^*)}{\beta^* I_1(\beta^*)} - \frac{1}{2}(r^2 - 1) \right) + \frac{q^*}{64 b_3^*}(r^2 - 1)^2 .$$

Let us use Bassel's approximations and present the operators from (9.23) in a canonical form for the experimental and theoretical method:

$$\frac{1}{b_3^*} = \alpha_1 g_{\beta 1}^* + \alpha_2 g_{\beta 2}^*, \frac{b_2^*}{b_3^* G_3^*} = \alpha_3 g_{\beta 3}^* + \alpha_4 g_{\beta 4}^* + \alpha_0 \frac{1}{\omega^*} .$$

With account of approximation

$$\frac{b_2^*}{b_3^* G_3^*} \frac{I_1(\beta^* r)}{I_1(\beta^*)} = \frac{b_2^*}{b_3^* G_3^*} r \left( r^2 + \frac{8(1 - r^2)}{8 + \beta^{*2}} \right),$$

the sought operator will be

$$\frac{b_2^*}{b_3^* G_3^* (8 + \beta^{*2})} = \alpha_5 g_{\beta 3}^* + \alpha_6 g_{\beta 4}^* + \alpha_7 g_{\beta 5}^* + \alpha_8 g_{\beta 6}^* + \alpha_9 g_{\beta 7}^* + \alpha_{10} g_{\beta 8}^* + \alpha_{11} \frac{1}{\omega^*},$$

$$\frac{b_2^{*2}}{b_3^* G_3^*} = \alpha_{12} \frac{1}{\omega^*} + \alpha_{13} g_{\beta 3}^* + \alpha_{14} g_{\beta 4}^* + \alpha_{15} g_{\beta 3}^{*2} + \alpha_{16} g_{\beta 4}^{*4} .$$

The operator with Basselian functions included in the deflection equations will be present as follows

$$\frac{b_2^{*2}}{b_3^{*2} G_3^*} \frac{I_0(\beta^* r) - I_0(\beta^*)}{\beta^* I_1(\beta^*)} = \frac{4 b_2^{*2} (r^2 - 1)}{b_3^* G_3^* (8 + \beta^{*2})} .$$

Where from,

$$\frac{b_2^{*2}}{b_3^{*2} G_3^* (8 + \beta^{*2})} = \alpha_{a7} \frac{1}{\omega^*} + \alpha_{18} g_{\beta 3}^{*2} + \alpha_{19} g_{\beta 4}^{*2} + \sum_{n=3}^{8} \alpha_{17+n} g^*_{n} .$$

Due to their bulkiness (Ch. 13.4) coefficients $\alpha_m$ ($m = 1, 2, \ldots, 41$) and $\beta_k$ ($k = 1, 2, \ldots, 8$) are presented in the appendix.

The operators included in the equations for the radial displacement will be:

$$\frac{a_3^*}{a_1^* a_6^* - a_3^{*2}} = \alpha_{26} g_{\beta 1}^* + \alpha_{27} g_{\beta 2}^*,$$

$$\frac{a_3^* a_5^* - a_2^* a_6^*}{a_1^* a_6^* - a_3^{*2}} \frac{b_2^*}{b_3^* G_3^*} = \alpha_{28} g_{\beta 1}^* + \alpha_{29} g_{\beta 2}^* + \alpha_{30} g_{\beta 3}^* + \alpha_{31} g_{\beta 4}^* + \alpha_{32} \frac{1}{\omega^*},$$

$$\frac{a_3^* a_5^* - a_2^* a_6^*}{a_1^* a_6^* - a_3^{*2}} \frac{b_2^* I_1(\beta^* r)}{b_3^* G_3^* I_1(\beta^*)} = \frac{a_3^* a_5^* - a_2^* a_6^*}{a_1^* a_6^* - a_3^{*2}} \frac{b_2^*}{b_3^* G_3^*} r \left( r^2 + \frac{8(1 - r^2)}{8 + \beta^{*2}} \right),$$

$$\frac{a_3^* a_5^* - a_2^* a_6^*}{a_1^* a_6^* - a_3^{*2}} \frac{b_2^*}{b_3^* G_3^* (8+\beta^{*2})} = \alpha_{33} \frac{1}{\omega^*} + \sum_{n=1}^{8} \alpha_{33+n} g_{\beta n}^* .$$

According to above-stated, the solution of the linear viscoelasticity problem for a circular three-layer pinched plate can be written in transforms:

$$\psi^* = \frac{r(r^2-1)q^*}{4c}\left((\alpha_0 - 8\alpha_1)\frac{1}{\omega^*} + \sum_{n=3}^{4} \alpha_n g_{\beta n}^* - 8\sum_{n=3}^{8} \alpha_{n+2} g_{\beta n}^*\right),$$

$$u^* = \frac{r(r^2-1)q^*}{4c}\left(\left(\frac{c}{4}\alpha_{26} + \alpha_{28}\right)g_{\beta 1}^* + \left(\frac{c}{4}\alpha_{27} + \alpha_{29}\right)g_{\beta 2}^* + \alpha_{30} g_{\beta 3}^*\right.$$

$$\left. + \alpha_{31} g_{\beta 4}^* + (\alpha_{32} - 8\alpha_{33})\frac{1}{\omega^*} - \sum_{n=1}^{8} \alpha_{n+33} g_{\beta n}^*\right),$$

$$w^* = \frac{(r^2-1)^2 q^*}{64}(\alpha_1 g_{\beta 1}^* + \alpha_2 g_{\beta 2}^*) + \frac{(r^2-1)q^*}{8c}\left((8\alpha_{17} - \alpha_{12})\frac{1}{\omega^*}\right.$$

$$\left. + 8\sum_{n=3}^{8} \alpha_{n+17} g_{\beta n}^* + \sum_{n=3}^{4}((8\alpha_{n+15} - \alpha_{n+12}) g_{\beta n}^{*2} - \alpha_{n+10} g_{\beta n}^*)\right). \tag{9.24}$$

Transform $1/\omega^*$ responds to the creep kernel $\Gamma(t)$. The transforms $g_{\beta n}(t)$ (9.22) correspond to the inverse transforms $g_{\beta n}(t)$ determined experimentally. The sought solution of the problem of the linear viscoelasticity in inverse transforms follows from (9.24) after decoding the operators by presenting (9.13) as:

$$\psi(t) = \frac{r(r^2-1)}{4c}\left[(\alpha_0 - 8\alpha_1)\left(q(t) + \int_0^t R(t-\tau)q(\tau)\,\mathrm{d}\tau\right)\right.$$

$$\left. + \sum_{n=3}^{4} \alpha_n \int_0^t g_{\hat{a}n}(t-\tau)\,\mathrm{d}q(\tau) - 8\sum_{n=3}^{8} \alpha_{n+2} \int_0^t g_{\hat{a}n}(t-\tau)\,\mathrm{d}q(\tau)\right),$$

$$u(t) = \frac{r(r^2-1)}{4c}\left[(\alpha_{32} - 8\alpha_{33})\left(q(t) + \int_0^t R(t-\tau)q(\tau)\,\mathrm{d}\tau\right)\right.$$

$$+ \sum_{n=1}^{2}\left(\left(\frac{c}{4}\alpha_{n+25} + \alpha_{n+27}\right)\int_0^t g_{\beta n}(t-\tau)dq(\tau) + \alpha_{n+29}\int_0^t g_{\beta(n+2)}(t-\tau)dq(\tau)\right)$$

$$\left. - \sum_{n=1}^{8} \alpha_{n+33} \int_0^t g_{\beta n}(t-\tau)dq(\tau)\right],$$

$$w(t) = \frac{(r^2-1)^2}{64}\sum_{n=1}^{2} \alpha_n \int_0^t g_{\hat{a}n}(t-\tau)\,\mathrm{d}q(\tau) + \frac{r^2-1}{8c}\left[(8\alpha_{17} - \alpha_{12})\left(q(t)\right.\right.$$

$$\left. + \int_0^t R(t-\tau)q(\tau)d\tau\right) + 8\sum_{n=3}^{8} \alpha_{n+17} \int_0^t g_{\beta n}(t-\tau)dq(\tau) + \tag{9.25}$$

$$+\sum_{n=3}^{4}\left(\left(8\alpha_{n+15}-\alpha_{n+12}\right)\int_0^t g_{\beta n}(t-\tau)d\int_0^t g_{\beta n}(t-\tau)dq(\tau)-\alpha_{n+10}\int_0^t g_{\beta n}(t-\tau)dq(\tau)\right)\Bigg]\Bigg].$$

The solution of (9.25) is true for the case of a pinched over the edge plate. In the case of a hinge a support there appears an additional summand in the deflection according to (6.70) without any new functions $g_{\beta n}(t)$.

Proceeding from the above, to describe deformation of a circular three-layer linear-viscoelastic plate it is enough to have a creep kernel and eight experimental functions $g_{\beta n}(t)$. If they are multiple among roots $\omega_{n,}$, the quantity of these functions may diminish.

## NOTES

1. *Boltzmann, L.* (1844–1906), Austrian physicist, one of the founders of statistical physics and mechanics.
2. *Thomson, W.*, given the title of Baron *Kelvin* in 1892 for scientific contributions—(Kelvin, 1824–1907), English physicist famous for works in thermodynamics, electromagnetics, suggested an absolute temperature scale, invented numerous scientific instruments.
3. *Voigt, W.* (1850–1919), German physicist.
4. *Volterra, V.* (1860–1940), Italian mathematician, is famous for works in mathematical physics, integral equations, functional analysis.
5. *Ishlinskii, A.Yu.* (1913–2003), Russian mechanical engineer; worked in the theory of elasticity and plasticity.
6. *Koltunov, M.A.* (1918–1980), Russian mechanical engineer; worked in the theory of viscoelasticity.
7. *Rzhanicyn, A.R.* (1911–1987), Russian mechanical engineer; worked in the theory of viscoelasticity.
8. *Duffing, G.* (1861–1944), German engineer-physicist.

## KEYWORDS

- **Rheonomic**
- **thermoviscoelasticity**
- **Unidimensional**
- **Viscoelasticity**
- ***Volterra principle***

# Chapter 10

## Thermoviscoelastoplasticity

### NONLINEAR VISCOELASTIC CONTINUA

A great number of polymeric materials do not obey the linear model of viscoelastic continuum under elevated stresses (8.1), (8.2) but display physically nonlinear properties. Various analytical models used to describe them have been studied at length by Moskvitin (1972). We are going to touch upon only some of them here.

Let us take a simplest opportunity of accounting for the nonlinear viscoelastic properties of materials. To this end, we shall accept the stress–strain equation in the form

$$s_{ij} = 2G(f(\varepsilon_u)\,э_{ij} - \int_0^t R(t-\tau) f(\varepsilon_u)\,э_{ij}(\tau)\,d\tau,$$

$$\sigma = 3K\varepsilon\,. \qquad (10.1)$$

where $s_{ij} = \sigma_{ij} - \sigma\delta_{ij}$ is a stress deviator, $э_{ij} = \varepsilon_{ij} - \varepsilon\delta_{ij}$—strain deviator, $\sigma = \sigma_{kk}/3$—mean (hydrostatic) stress, $\varepsilon = \varepsilon_{kk}/3$—mean strain; $G$—instantaneous elastic shear modulus, $K$—instantaneous modulus of volume strain, function $R(t)$—relaxation kernel. Physical nonlinearity of the material is described by a universal function $f(\varepsilon_u)$, where $f(\varepsilon_u) = 1$ if the strain rate $\varepsilon_u$ does not exceed some threshold value $\varepsilon_s$: $\varepsilon_u \leq \varepsilon_s$.

Physical equations (10.1) imply that as soon as the strain rate in a considered point reaches some threshold value $\varepsilon_s$ the strain field $э_{ij}$ at this moment is determined not only by the instantaneous and previous stress $s_{ij}$ values but by the kind of the strain state itself. Variations in the volume are accepted to be elastic.

Having solved (10.1) relative to $s_{ij}$, we shall obtain ($\Gamma(t)$—creep kernel)

$$2Gf(\varepsilon_u)\,э_{ij} = s_{ij} + \int_0^t \Gamma(t-\tau) s_{ij}(\tau)\,d\tau,$$

$$3K\varepsilon = \sigma\,. \qquad (10.2)$$

Let us consider a procedure of finding a function of nonlinearity $f(\varepsilon_u)$ on the experimental creep curves. May, for example, the results of experiments on creep at pure shear be known (Fig. 10.1). Six equations (10.2) will be reduced to one, since $э_{ij} = 0$, except for $э_{12} = \varepsilon_{12}$, and $s_{ij} = 0$, except for $s_{12} = \sigma_{12}$. In this case, $2\varepsilon_{12}$—relative shear; $\sigma_{12}$—corresponding tangential stress. In our case, $\varepsilon_u = \frac{2\sqrt{3}}{3}\varepsilon_{12}$. The lower curve in Fig. 10.1 corresponds to the linear behavior of the material ($\varepsilon_u \leq \varepsilon_s$) at a constant stress $\sigma_{12}^{(0)}$. The next relation is true along this curve

$$2G\varepsilon_{12}^{(0)}(t)=\sigma_{12}^{(0)}\left(1+\int_0^t\Gamma(t-\tau)\,\mathrm{d}\,\tau\right). \tag{10.3}$$

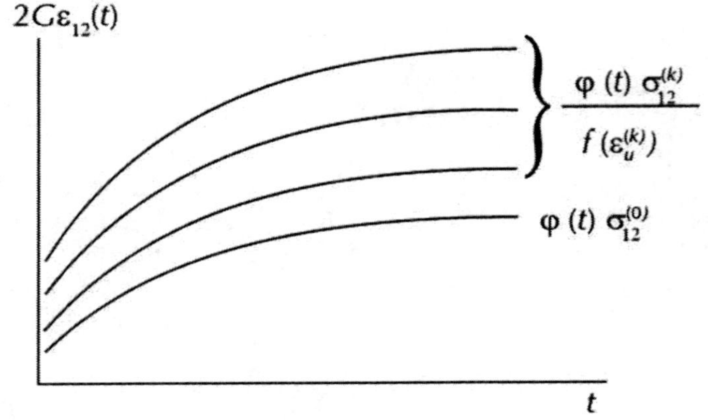


**Figure 10.1.**

This curve can be also used for determining the creep kernel:

$$\int_0^t\Gamma(t-\tau)d\tau=\frac{2G\varepsilon_{12}^{(0)}(t)}{\sigma_{12}^{(0)}}-1.$$

For other curves we have from (10.2)

$$2Gf(\varepsilon_u^{(k)})\varepsilon_{12}^{(k)}(t)=\sigma_{12}^{(k)}\left(1+\int_0^t\Gamma(t-\tau)d\tau\right). \tag{10.4}$$

where $\sigma_{12}^{(k)}$ is a constant stress along $k$ curve ($k = 1, 2, 3$).

Let us divide the left and right-hand sides of equation (10.4) into corresponding parts of equation (10.3). After elementary transformations we obtain the experimental values of nonlinearity function at various moments

$$f(t)=\frac{\sigma_{12}^{(k)}\varepsilon_{12}^{(0)}(t)}{\sigma_{12}^{(0)}\varepsilon_{12}^{(k)}(t)}.$$

Since the strain rate at each moment of time is $\varepsilon_u = \frac{2\sqrt{3}}{3}\varepsilon_{12}^{(k)}(t)$ and the function of nonlinearity *f(t)* is known, the experimental curve $f \sim \varepsilon_u$ can be drawn by comparing their values for one and the same *t*. Then, the constants in the accepted approximating formula are determined for the function of nonlinearity.

Nonlinear equations for the stress–strain ratio can be taken in some other form:

$$2G\,э_{ij}=g(\sigma_u)s_{ij}+\int_0^t\Gamma(t-\tau)g(\sigma_u)s_{ij}(\tau)\,\mathrm{d}\,\tau,$$

$$3K\varepsilon = \sigma , \tag{10.5}$$

where $g(\sigma_u)$ is a universal function of the stress rate determined experimentally.

Equations (10.1) and (10.5) are not equivalent to each other. We choose some way of expressing physical nonlinearity of the material depending on its properties and on available experimental data.

## NONLINEAR VISCOELASTICITY EQUATIONS ACCOUNTING FOR THE TYPE OF STRESS–STRAIN STATE (SSS) EFFECT

High-filled polymers display peculiar physicomechanical properties, for example, the dependence of deformation on the value and sign of hydrostatic pressure $\sigma$ (enlargement of microdefects at the all-sided stretching and their healing at compression). These properties are not taken into account by the models (10.1) and (10.5) in which the relations between the deviators and spherical tensor parts of stresses and strains are divided. The elementary equations taking into account the effect of the volume stress and temperature $T = T(x, t)$ counted from some initial value $T_0$ can be introduced by natural generalization of the previous relations:

$$2Gf(\varepsilon_u, T)\,\text{э}_{ij} = \varphi_1(\sigma, T)s_{ij} + \int_0^t \Gamma(t-\tau)\varphi_1(\sigma, T)s_{ij}(\tau)\,\mathrm{d}\tau ,$$

$$3K(\varepsilon - \alpha T) = \varphi_2(\sigma, T)\sigma . \tag{10.6}$$

The equations for the deviator values and their second invariants (strain rates $\varepsilon_u$) are now containing the first invariants of stress tensors $\sigma$ as well. This allows us to consider different types of polymer behavior at tension and compression. After solving equation (10.6) for strains, we receive

$$s_{ij} = 2G(T)\left( \varphi(\varepsilon_u, T)\,\text{э}_{ij} - \int_0^t R(t-\tau, T) f(\varepsilon_u, T)\,\text{э}_{ij}(\tau)\,\mathrm{d}\tau \right),$$

$$\sigma = 3K(T)(\varepsilon - \alpha T) . \tag{10.7}$$

It is supposed that functions f, $\varphi_1$ and $\varphi_2$ in (10.6), (10.7) are universal and independent of the kind of SSS. They are determined experimentally. The influence of temperature on physicomechanical properties is known to be taken into account in some cases by the time-temperature analogy or, as it will be shown further, by a direct processing of experimental data.

## VISCOELASTOPLASTIC CONTINUA

Along with plasticity, a number of metals and alloys display pronounced *rheonomic properties at high temperatures*. Such continua we shall call *viscoelastoplastic*. We shall consider a model, which enables to describe a similar behavior of materials. We shall take the following physical equations of state at the presence of temperature field $T(x, t)$:

$$s_{ij} = 2G(T)\left( \varphi(\varepsilon_u, T)\,Э_{ij} - \int_0^t R(t-\tau, T) f(\varepsilon_u, T)\,Э_{ij}(\tau)\,d\tau \right),$$

$$\sigma = 3K(T)(\varepsilon - \alpha T)\,. \quad (10.8)$$

The expressions for strains through stresses will be an alternative to them

$$2G(T)\,Э_{ij} = \varphi_1(\sigma_u, T)s_{ij} + \int_0^t \Gamma(t-\tau, T) g(\sigma_u, T) s_{ij}(\tau)\,d\tau\,,$$

$$3K(T)(\varepsilon - \alpha T) = \sigma\,. \quad (10.9)$$

Notice that the instantaneous elasticity modules $G(T)$ and $K(T)$, creep $\Gamma(t, T)$ and relaxation $R(t, T)$ kernels as well as plasticity function $\varphi(\varepsilon_u, T)$, $\varphi_1(\sigma_u, T)$, and the universal functions of physical nonlinearity $f(\varepsilon_u, T)$, $g(\sigma_u, T)$ may depend on temperature $T$, whereas $\varphi(\varepsilon_u, T) = 1$, if $\varepsilon_u \le \varepsilon_y(T)$; $\varphi_1(\sigma_u, T) = 1$, if $\sigma_u \le \sigma_y(T)$; $f(\varepsilon_u, T) = 1$, if $\varepsilon_u \le \varepsilon_s(T)$; $g(\sigma_u, T) = 1$, if $\sigma_u \le \sigma_s(T)$.

For a general statement of the problem on loading from the initial state, the equilibrium equations, Cauchy relations, and boundary conditions should be added to equation (10.8) or (10.9), being true only at active loading.

## A METHOD OF SUCCESSIVE APPROXIMATION IN VISCOELASTOPLASTICITY PROBLEMS

Let us add the equilibrium equations, Cauchy relations, and boundary conditions to equation (10.8):

$$s_{ij} = 2G(T)\left( \varphi(\varepsilon_u, T)\,Э_{ij} - \int_0^t R(t-\tau) f(\varepsilon_u, T)\,Э_{ij}(\tau)\,d\tau \right),$$

$$\sigma = 3K(T)(\varepsilon - \alpha T)\,,$$

$$\sigma_{ij,j} + \rho F_i = 0,\ \varepsilon_{ij} = \tfrac{1}{2}(u_{i,j} + u_{j,i})\,,$$

$$u_i = u_{i0}(x) \text{ on } S_u\,,\ \sigma_{ij} l_j = R_i \text{ on } S_\sigma\,. \quad (10.10)$$

Solution of the problems on viscoelastoplasticity is interrelated with solution of a system of nonlinear integral differential equations in partial derivatives (10.10). This presents an intricate mathematical problem and its analytical decision is out of question. Therefore, we shall use here *the method of successive approximations,* which is based on Ilyushin's method of elastic solutions.

Let us present for this purpose the function of plasticity as

$$\varphi(\varepsilon_u, T) = 1 - \omega(\varepsilon_u, T)\,,\ f(\varepsilon_u, T) = 1 - \omega_1(\varepsilon_u, T)\,, \quad (10.11)$$

where $0 \le \omega < 1$; $0 \le \omega_1 < 1$; and $\omega(\varepsilon_u) = 0$, if $\varepsilon_u \le \varepsilon_y$; $\omega_1(\varepsilon_u) = 0$, if $\varepsilon_u \le \varepsilon_y$.

Let us substitute (10.11) into (10.10). Then, the physical relations become the following:

$$s_{ij} = 2G(T)\left( (1-\omega(\varepsilon_u, T))\,\text{э}_{ij} - \int_0^t R(t-\tau)(1-\omega_1(\varepsilon_u, T))\,\text{э}_{ij}(\tau)\,\mathrm{d}\tau \right),$$

$$\sigma = 3K(T)(\varepsilon - \alpha T). \qquad (10.12)$$

Thus, at $\omega = \omega_1 = 0$, equation (10.12) describe the linear thermoviscoelastic properties of the materials.

Similarly to the plasticity theory, differential equilibrium equations, and boundary conditions will be written by expanding the stress tensor into the deviator and spherical parts:

$$s_{ij},_j + \sigma,_i + \rho F_i = 0, \qquad (10.13)$$

$$u_i = u_{i0}(x) \text{ on } S_u, \quad s_{ij} l_j + \sigma l_i = R_i \text{ on } S_\sigma. \qquad (10.14)$$

We shall solve the problem of thermoviscoelastoplasticity in displacements. For this purpose, we shall substitute the components of the deviator and spherical part of the stress tensor (10.12) in equilibrium equation (10.13) and boundary conditions of force (10.14), and take into account Cauchy relations. As a result, we shall come to the following generalized Lame equations and boundary conditions:

$$(\lambda + \mu)\theta,_i + \mu \Delta u_i + \rho F_i - F_{\omega i} - 3K\alpha T,_i = 0,$$

$$\lambda \theta l_i + \mu(u_i,_j + u_j,_i) l_j = R_i + R_{\omega i} + 3K\alpha T l_i.$$

where

$$F_{\omega i} = 2G\left( \omega\,\text{э}_{ij} + \int_0^t R(t-\tau, T) f(\varepsilon_u, T)\,\text{э}_{ij}(\tau)\,\mathrm{d}\tau \right),_j;$$

$$R_{\omega i} = 2G\left( \omega\,\text{э}_{ij} + \int_0^t R(t-\tau, T) f(\varepsilon_u, T)\,\text{э}_{ij}(\tau)\,\mathrm{d}\tau \right) l_j.$$

We shall take $\omega^{(0)} = f^{(0)} = 0$ as a zero approximation. Then, $F_{\omega i} = R_{\omega i} = 0$, so to find the first approximation $u_i^{(1)}$, we use a common problem of the linear thermoelasticity. Based on the determined displacements $u_i^{(1)}$, we find the values

$$\text{э}_{ij}^{(1)},\ \varepsilon_u^{(1)},\ \omega^{(1)} \equiv \omega(\varepsilon_u^{(1)}, T),\ f^{(1)} \equiv f(\varepsilon_u^{(1)}, T),\ F_{\omega i}^{(1)},\ R_{\omega i}^{(1)}.$$

For any *k-th* approximation, there exist the equations and boundary conditions

$$(\lambda + \mu)\theta,_i^{(k)} + \mu \Delta u_i^{(k)} + \rho F_i - F_{\omega i}^{(k-1)} - 3K\alpha T,_i = 0, \qquad (10.15)$$

$$\lambda \theta^{(k)} l_i + \mu(u_i^{(k)},_j + u_j^{(k)},_i) l_j = R_i + R_{\omega i}^{(k-1)} + 3K\alpha T l_i, \qquad (10.16)$$

where $F_{\omega i}^{(k-1)}$, $R_{\omega i}^{(k-1)}$ are known from the previous $(k-1)$-th approximation.

Equation (10.15) and boundary conditions (10.16) are linear relative to unknown displacements $u_i^{(k)}$. They differ from the corresponding equations of thermoelasticity by fictitious forces $F_{\omega i}^{(k-1)}$, $R_{\omega i}^{(k-1)}$ added to the external forces $F_i$, $R_i$. This enables to solve the elastoplastic problem in a recurrent variant based on the corresponding known solution of the elasticity theory.

To provide the convergence of the considered method of successive approximations, parameter $\omega$ is related to function $\varphi(\varepsilon_u)$ via equation (10.11) and the function of nonlinearity $f(\varepsilon_u)$ should be small as compared to a unity. If we additionally take $\varphi(\varepsilon_u) \equiv f(\varepsilon_u)$, then the described method of approximation in transforms will not differ from the approximation by Ilyushin's method of elastic solutions considered earlier in the theory of small elastoplastic deformations (7.7). In the latter case, the arguments of convergence of the method (e.g., Ilyushin, 1963) are known. Hence, we can also speak about the convergence in a given case.

The proof of convergence of the method of elastic solutions for an incompressible nonlinear viscoelastic continuum is presented in (Ilyushin and Pobedrya, 1970). The convergence is shown by way of example on bending of a three-layer plate described below.

When solving a problem in transforms for any approximation, it is convenient to expand the volume $F_i + F_{\omega i}^{(k-1)}$ and surface $R_i + R_{\omega i}^{(k-1)}$ forces in a series of known functions. In this case, the solution in any approximation will differ only by a constant.

We can obtain one more method of successive approximations if we take $\omega^{(0)} = 0$ and $\omega_1^{(0)} = 1 - f^{(0)} = 0$ as a zero approximation. In this case, to determine the first approximation $u_i^{(1)}$ a usual problem of the linear thermoviscoelasticity is taken. Further, for the *k-th* approximation we take a problem of the linear thermoviscoelasticity with some other additional external loads $F_{\omega i}^{(k-1)}$, $R_{\omega i}^{(k-1)}$ calculated from the results of previous approximations, being a correction for the plasticity and physical nonlinearity of the material.

## A CIRCULAR VISCOELASTOPLASTIC THREE-LAYER PLATE

Let us continue the study on the influence of material properties on SSS of a circular three-layer plate (6.14, 7.10, 8.4). We shall assume that the materials of the bearing layers can show viscoelastoplastic properties during deformation. For their description, we use the hereditary relations between stresses and strains like (10.8):

$$s_{\alpha}^{(k)} = 2G_k(T)\left( f_1^{(k)}(\varepsilon_u^{(k)}, T)\,\text{э}_{\alpha}^{(k)} - \int_0^t R_k(t-\tau, T) f_2^{(k)}(\varepsilon_u^{(k)}, T)\,\text{э}_{\alpha}^{(k)}(\tau)\,d\tau \right),$$

$$\sigma^{(k)} = 3K_k(T)(\varepsilon^{(k)} - \alpha_{0k}T). \quad (10.17)$$

We take into account the influence of SSS type in a physically nonlinear filler:

$$\varphi_1(\sigma^{(3)}, T) s_{\alpha\beta}^{(3)} = 2G_3(T)\left( f^{(3)}(\varepsilon_u^{(3)}, T)\,\text{э}_{\alpha\beta}^{(3)} - \int_0^t R_3(t-\tau, T) f^{(3)}(\varepsilon_u^{(3)})\,\text{э}_{\alpha\beta}^{(3)}(\tau)\,d\tau \right),$$

$$\varphi_2(\sigma^{(3)}, T)\sigma^{(3)} = 3K_3(T)(\varepsilon^{(3)} - \alpha_{03}T), (\alpha, \beta = r, \varphi)\cdot \tag{10.18}$$

In this case, the equilibrium equations in iterations (7.26) retain their type:

$$\mathrm{L}_2(a_1u^n + a_2\psi^n - a_3w,_r^n) = -p + p_\omega^{n-1},$$
$$\mathrm{L}_2(a_2u^n + a_4\psi^n - a_5w,_r^n) - 2cG_3\psi^n = h_\omega^{n-1},$$
$$\mathrm{L}_3(a_3u^n + a_5\psi^n - a_6w,_r^n) = -q + q_\omega^{n-1}. \tag{10.19}$$

Where the factors $a_i$ and differential operators $\mathrm{L}_2$, $\mathrm{L}_3$ are determined by relations (6.59), $n$ is the approximation number. The additional external loads $p_\omega^{n-1}$, $h_\omega^{n-1}$, $q_\omega^{n-1}$ are calculated by the formulas

$$p_\omega^{n-1} = T_{\text{rù}}^{n-1},_r + \frac{1}{r}(T_{\text{rù}}^{n-1} - T_{\varphi\text{ù}}^{n-1});$$

$$h_{\text{ù}}^{n-1} = H_{\text{rù}}^{n-1},_r + \frac{1}{r}(H_{\text{rù}}^{n-1} - H_{\varphi\text{ù}}^{n-1}) - Q_{\text{ù}}^{n-1};$$

$$q_{\text{ù}}^{n-1} = M_{\text{rù}}^{n-1},_{rr} + \frac{1}{r}(2M_{\text{rù}}^{n-1},_r - M_{\text{rù}}^{n-1},_r),$$

where

$$T_{\alpha\omega}^{n-1} \equiv \sum_{k=1}^{3} T_{\alpha\omega}^{(k)n-1} = \sum_{k=1}^{3}\int_{h_k} \sigma_\alpha^{(k)\omega n-1}\,\mathrm{d}z;$$

$$M_{\alpha\omega}^{n-1} \equiv \sum_{k=1}^{3} M_{\alpha\omega}^{(k)n-1} = \sum_{k=1}^{3}\int_{h_k} \sigma_\alpha^{(k)\omega n-1} z\,\mathrm{d}z;$$

$$H_{\alpha\omega}^{n-1} = M_{\alpha\omega}^{(3)n-1} + c\left(T_{\alpha\omega}^{(1)n-1} - T_{\alpha\omega}^{(2)n-1}\right);$$

$$Q_\omega^{n-1} = \int_{-c}^{c} \sigma_{rz}^{(3)n-1}\,\mathrm{d}z, (\alpha = r, \varphi). \tag{10.20}$$

According to (10.17) and (10.18), the nonlinear components of stresses for the bearing layers look like ($k = 1, 2$)

$$\sigma_\alpha^{(k)\omega n\text{-}1} = 2G_k\left(\omega_1^{(k)n\text{-}1}\,\text{э}_\alpha^{(k)n\text{-}1} + \int_0^t R_k(t\text{-}\tau)(1\text{-}\omega_2^{(k)n\text{-}1})\,\text{э}_\alpha^{(k)n\text{-}1}(\tau)d\tau\right)$$

$$+3K_k\alpha_{0k}T.$$

So, we have for the filler

$$\sigma_\alpha^{(3)\omega n-1} = 2G_3\left(\omega^{(3)n-1}\,\text{э}_\alpha^{(3)n-1} + \int_0^t R_3(t-\tau)(1-\omega^{(3)n-1})\,\text{э}_\alpha^{(3)n-1}(\tau)\,\mathrm{d}\tau\right)$$

$$+3K_3\alpha_{03}T+3K_3\alpha_{03}T+\omega_1^{(3)n-1}s_\alpha^{(3)n-1}+\omega_2^{(3)n-1}\sigma^{(3)n-1},$$

$$\sigma_{rz}^{(3)\omega n-1}=2G_3\left(\omega^{(3)n-1}\,\ni_{rz}^{(3)n-1}+\int_0^t R_3(t-\tau)(1-\omega^{(3)n-1})\,\ni_{rz}^{(3)n-1}(\tau)\,\mathrm{d}\,\tau\right)$$

$$\omega_1^{(3)n-1}s_{rz}^{(3)n-1},\ \omega_\gamma^{(3)n-1}=\phi_\gamma(\sigma^{(3)n-1})-1,\quad (\gamma=1,2).$$

Thus, we have as earlier a linear problem of thermoelasticity at each step of approximation with the known additional external loads (10.20). The type of the iterative system solution (10.19) with account of its smoothness in the center of the plate does not differ from that of the corresponding elastoplastic plate (7.29):

$$\psi^n=C_2^n I_1(\beta r)+\psi_r^n\,;$$

$$u^n=\frac{a_3}{a_1a_6-a_3^2}\left[\mathrm{L}_3^{-1}(q-q_\omega^n)-\frac{a_6}{a_3}\mathrm{L}_2^{-1}(p-p_\omega^n)+\left(a_5-\frac{a_2a_6}{a_3}\right)\psi^n+C_7^n r\right];$$

$$w^n=\frac{1}{b_3}\left[b_2\left(\frac{C_2^n}{\beta}I_0(\beta r)+\int\psi_r^n dr\right)-\int\left(\frac{a_3}{a_1}\mathrm{L}_2^{-1}(p-p_\omega^n)-\mathrm{L}_3^{-1}(q-q_\omega^n)\right)\mathrm{d}\,r\right.$$

$$\left.+\frac{C_5^n r^2}{4}+C_4^n\right]. \qquad (10.21)$$

The partial solution of the modified inhomogeneous Bassel equation looks like

$$\psi_r^n=-K_1(\beta r)\int I_1(\beta r)f^n r\,\mathrm{d}r+I_1(\beta r)\int K_1(\beta r)f^n r\,\mathrm{d}r,$$

where

$$f^n(r)=\frac{b_3}{b_1b_3-b_2^2}\left[h_\omega^{n-1}+\frac{a_2b_3-a_3b_2}{a_1b_3}(p-p_\omega^{n-1})+\frac{b_2}{b_3r}\int(q-q_\omega^{n-1})r\,\mathrm{d}\,r\right].$$

However, the additional loads in above-given solution also serve us corrections for the rheonomic properties of the materials of the layers.

The integration constants for the case of boundary conditions of closing the plate over its contour are found from relations (7.30).

*At numerical analysis* of solution (10.21) we assume that the heat flow of intensity $q_t$ falls perpendicularly onto the external surface of the three-layer plate. We shall consider two types of plates with different materials of the layers of the pack: *metal-polymer-metal* and *ceramics-polymer-metal* ones.

We disregard the heat spent on heating of the external metal layer for the plates *of the first type* (because of its thinness and low heat capacity). We accept the temperature of this layer equal to that of the filler in the splice: $T^{(1)}=T^{(3)}(c,t)$. The total heat perceived by the plate in time $t$ is spent on heating of the polymeric filler. The temperature of the second bearing layer is also accepted equal to the temperature of the filler in their splice $T^{(2)} = T^{(3)}(-c, t)$. By taking the plate heatproof over its contour, the temperature field in the filler can be calculated by equation (10.37), providing that the specific heat conductivity is $\lambda = \lambda_3$ and heat capacity $c_0 = c_3$.

The rated temperature variation of the pack through the plate thickness at $q_t = 5000$ J/(m$^2$ c) is shown in Fig. 10.2. The curve number corresponds to various moments of time at a 10 min interval after the beginning of the heat flow effect. At the initial moment $t_1 = 0$, and at the final one $t_7 = 60$ min. At $t_4 = 30$ min, the temperature in the external bearing layer reaches 510 K.

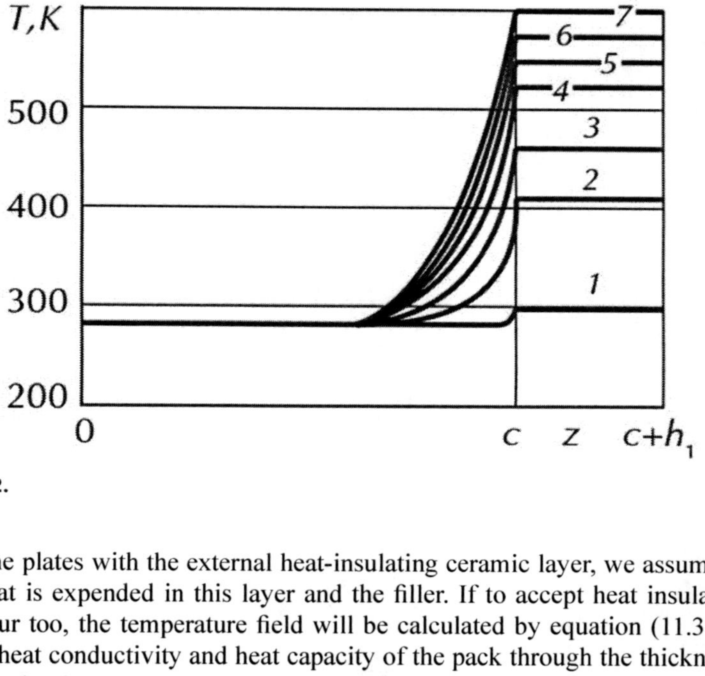


**Figure 10.2.**

For the plates with the external heat-insulating ceramic layer, we assume that the whole heat is expended in this layer and the filler. If to accept heat insulation over the contour too, the temperature field will be calculated by equation (11.37), which averages heat conductivity and heat capacity of the pack through the thickness of the heat-insulating layer and the filler of the plate:

$$T(s,\tau) = \frac{qH}{\lambda}\left\{\tau + \frac{1}{2}\left(s + \frac{c+h_2}{H}\right)^2 - \frac{1}{6} - \right.$$

$$\left. \frac{2}{\pi^2}\sum_{n=1}^{\infty}\frac{(-1)^n}{n^2}\cos\left[\pi n\left(s + \frac{c+h_2}{H}\right)\right]e^{-n^2\delta^2\hat{o}}\right\},$$

where $\tau = at/h^2$; $s = z/h$; $h = h_1 + h_3$; $a = \lambda/\rho c_0$; $\lambda = (\lambda_1 h_1 + \lambda_3 h_3)/h$; $\rho c_0 = (\rho_1 c_1 h_1 + \rho_3 c_3 h_3)/h$.

The value of loading ($q_0 = 3.0\times10^7$), heat flow intensity, time of their influence ($t_0$ = 30 min), and relative thicknesses of the layers [$h_1 = 0.04 \rightarrow 0.02$ (at ablation, see 11.5); $h_2 = 0.04$; $h_3 = 0.2$] are selected so as nonlinear, thermophysical, and rheonomic properties of the materials to make evident.

The numerical results have shown actual convergence of the modified method of the elastic solutions applied here for the inhomogeneous laminated continuum. The maximal difference of the results in the 6-th approximation (for the *metal-polymer-metal* plate) accepted as the sought solutions from the previous one is about 3%, and from the 7-th is less than 0.5% (Fig. 10.3). The number of the curve corresponds here to the iteration step. The shear in the filler is shown in Fig. 10.3a, and deflection of the plate in moment $t = 30$ min in Fig. 10.3b.

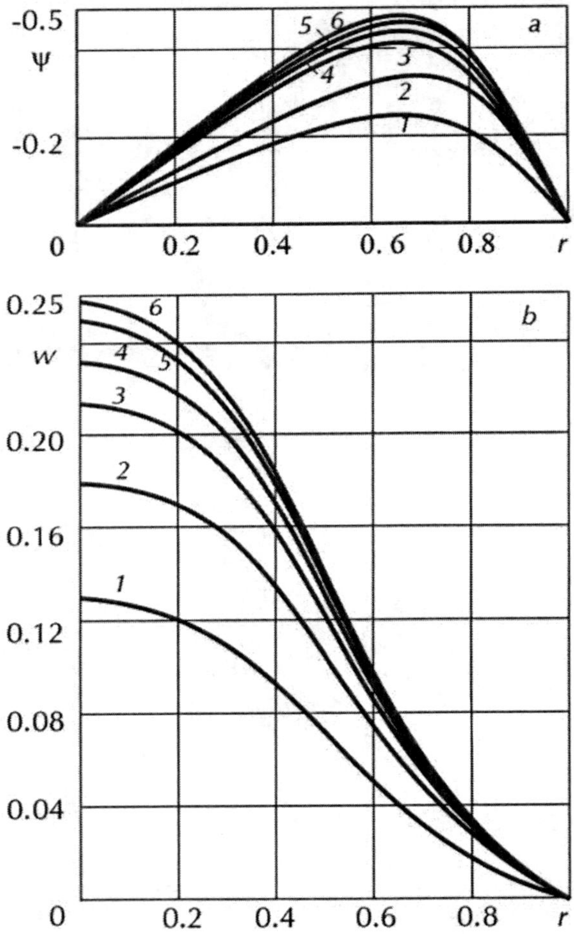


**Figure 10.3.**

If the external heatproof layer is ceramic, the convergence is accelerated and very often the 4-th approximation becomes suitable enough. This effect is explained by the physical linearity of the most rigid in the pack ceramic layer.

The stress and strain deviators were calculated numerically at various moments of time, and their coaxiality was confirmed. This proves the fulfillment of the simple loading conditions.

Figure 10.4 illustrates enlargement of the deflection due to the heat flow influence on the plate. We can analyze the results of calculations by various physical equations of the state. Curve *1* corresponds to isothermal loading of the elastic plate; *2*, *3* are the thermal force loading on the linearly viscoelastic plate in moments $t = 0$ and $t = t_0 = 30$ min. Other curves are calculated for a general case of physical equations of the state (10.17), (10.18). Curves *4*, *5* correspond to isothermal loading at $t = 0$ and $t = t_0$; 6, *7* correspond to a thermal force influence in similar moments.

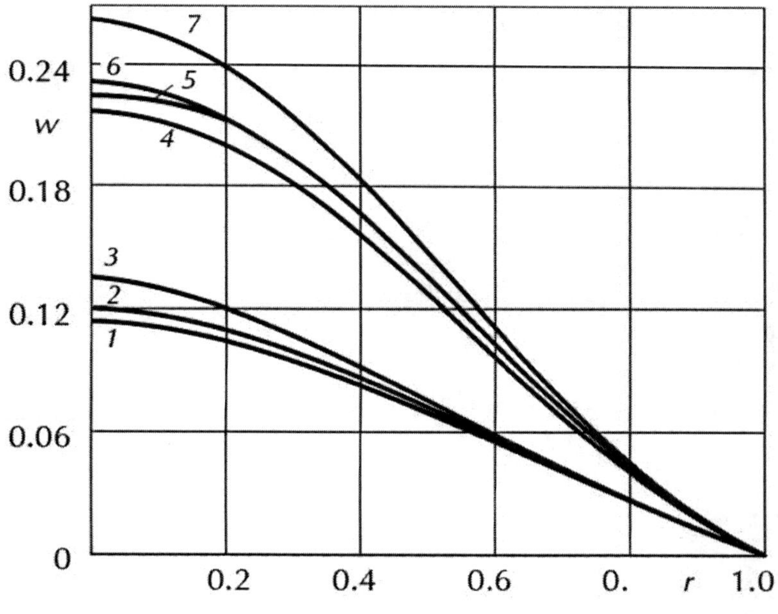


**Figure 10.4.**

The linear thermal viscoelasticity adds 10.3% to deflection of the elastic plate. The deflection of the plate with physically nonlinear and rheonomic properties under the heat effect surpasses by 16.2% the corresponding isothermal deflection. In contrast to the elastic deflection of the viscoelastoplastic plate, the one induced by a thermal force increases with time $t_0$ by 123%.

The influence of the external loading $q$ value on the maximal deflection of the plate is shown in Fig. 10.5. With its increase, the nonlinear character of curves *2*, *3* intensifies (a general case of physical equations of state at thermal loading in moments $t = 0$ and $t = t_0 = 30$ min, accordingly). The elastic deflection *1* is linearly proportional to the loading.

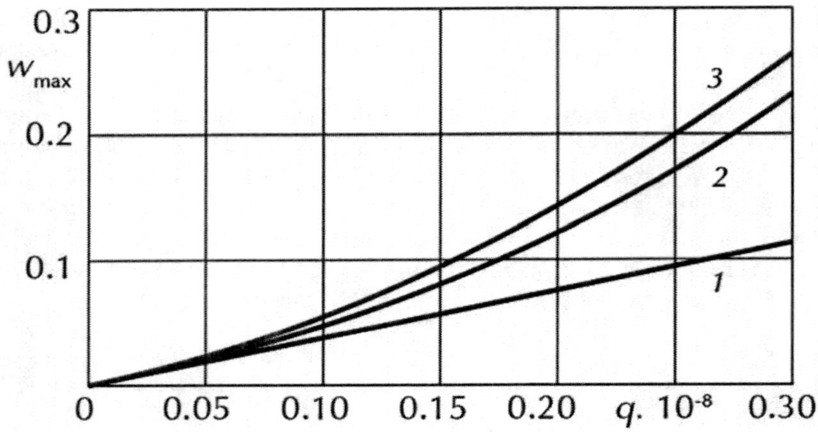


**Figure 10.5.**

The influence of various geometric parameters of the plate on the deflection is shown in Fig. 10.6. When thickness of the filler is constant as well as the total thickness of the bearing layers ($h_3 = 0.2$; $h_1 + h_2 = 0.08$), the two-layer plate *3* shows the greatest bending strength, while the three-layer plate *1* symmetric by thickness has the least one. Curve *2* corresponds to a deflection with parameters $h_1 = 0.02$; $h_2 = 0.06$. The values of deflections for other possible thickness ratios within specified limits are found between curves *1* and *3*.

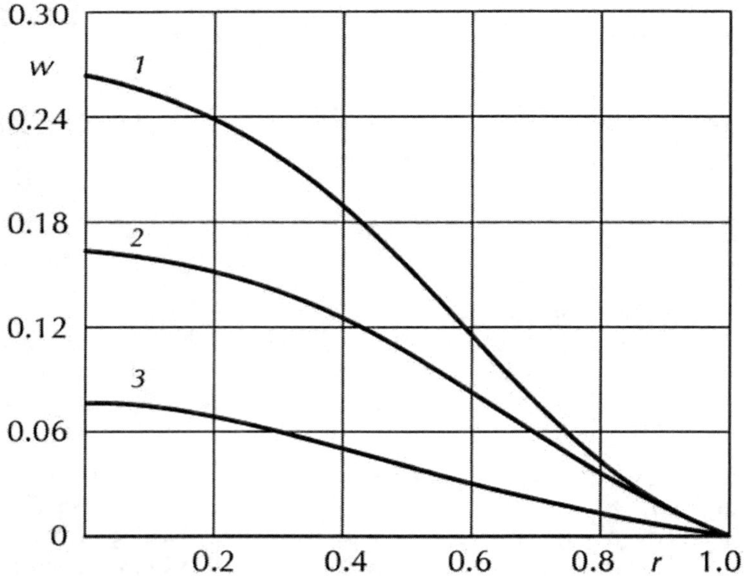


**Figure 10.6.**

Distribution of the areas of the plastic and nonlinear deformations of the materials of the layers over the plate cross-section depending on the external force load is shown in Fig. 10.7a where $q_t = 0$. The numerical analysis has shown that the first plastic deformations appear on the external surfaces of the plate contour at closure. With increasing load and time, they start to distribute inside thickness and along the boundary planes of the layers.

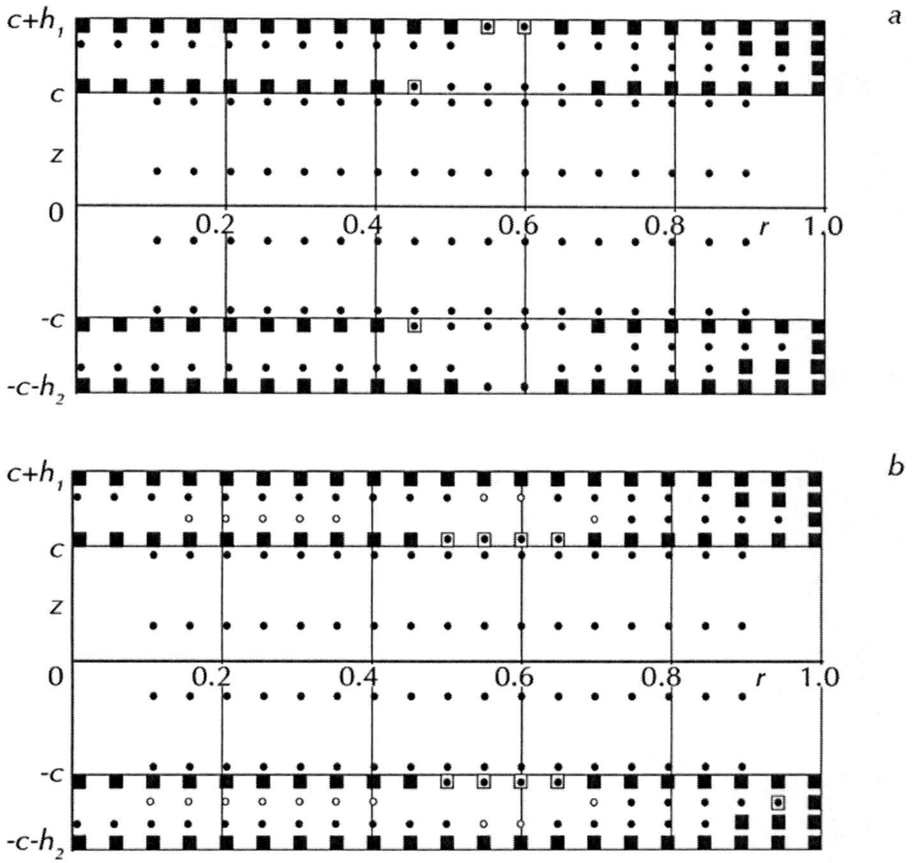


**Figure 10.7.**

As a result, one more area of plastic deformations is formed in the center of the plate ($r = 0$), which propagates towards the first one.

The effect of temperature field $q_t$ results in increasing of the plastic deformation areas of the materials (Fig. 10.7b). Dark points and rectangles in the figures below mean nonlinear creep (viscoelasticity) and plastic deformations, accordingly, during the initial moment, the light points are added within 30 min.

The cinematic model of the considered three-layer plate presumes splicing of the layers. This means that the displacements on the boundaries of the layers will be continuous while stresses break for being dependent on physical characteristics of the materials of each layer. Stress variations in the cross-section of the plate are shown in Figs. 10.8–10.10.

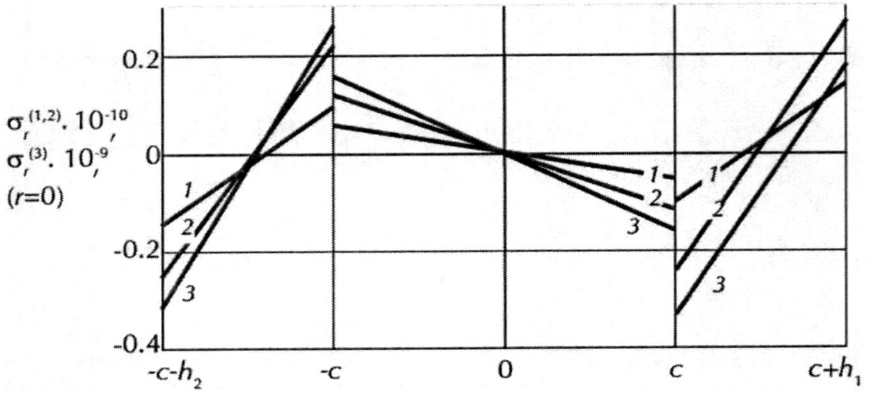


**Figure 10.8.**

The radial stresses (Fig. 10.8) $\sigma_r^{(k)}$ in the center of the plate are equal to tangential $\sigma_\varphi^{(k)}$. The radial stresses over the contour are larger by a modulus (Fig. 10.10). They are varying in the thin bearing layers over the cross-section linearly. The temperature field in the external bearing layer causes expansion of the material and stress shift into the negative area.

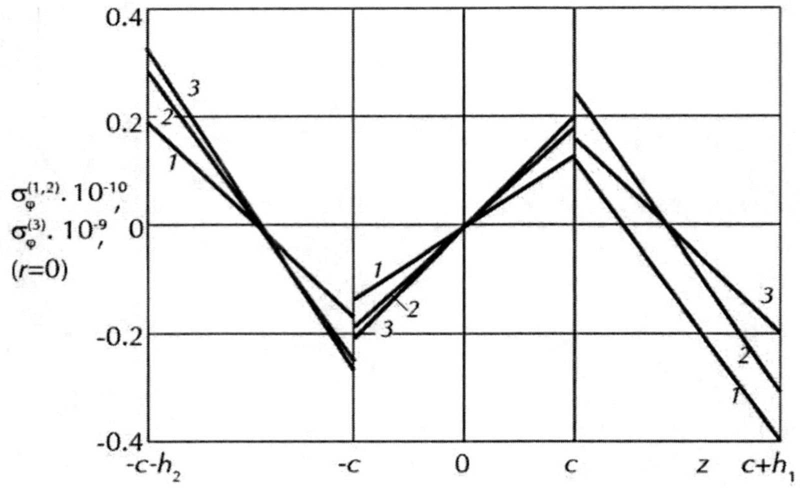


**Figure 10.9.**

The account of physical nonlinearity and rheonomic parameters of the materials results in increased maximal stresses in the second bearing layer by 55.4% as compared to the elastic stresses (isothermal loading), and by 70.1% at thermal loading. The external layer has similar values 55.4 and 79.7%. Curves *1* in these figures correspond to the elastic plate; curves *2*, *3* are a general case of physical equations of state at $t$ = 0 and $t = t_0$.

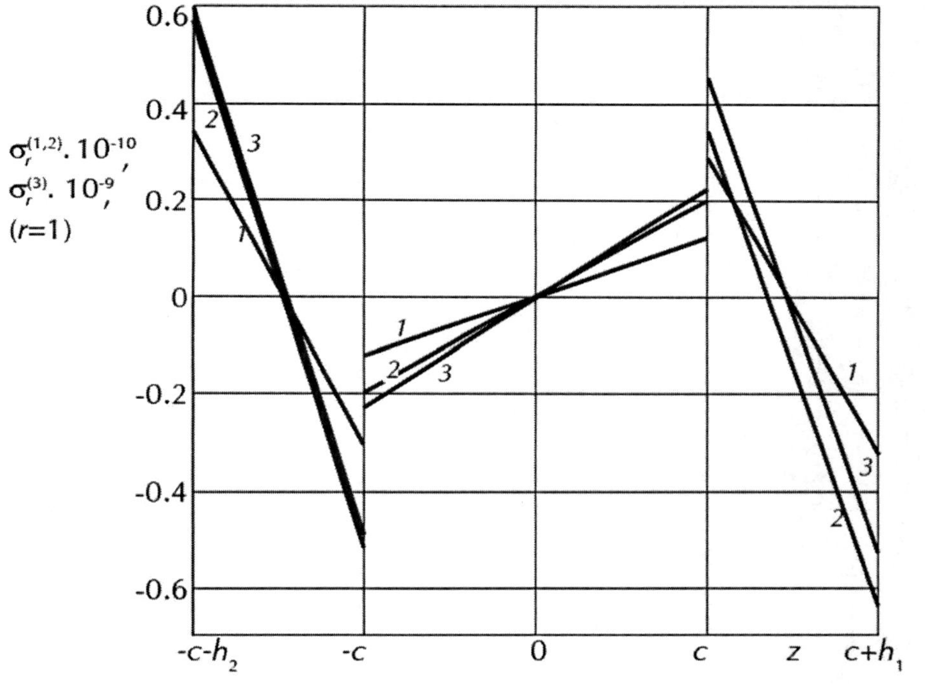


**Figure 10.10.**

Figure 10.11 illustrates the areas of the filler in which the functions of nonlinearity operate for fluoroplastics with account of hydrostatic pressure $p = -\sigma$ (curves *1*, *2* are a general case of physical equations of state at $t$ = 0 and $t = t_0$). This occurs in the whole interval $0 \leq r \leq 1$ under considered loads, because $p < p_0$ = 800 MPa, and not all defects of the material have been closed as yet.

The influence of ablation on SSS of the circular three-layer plate (*ceramics-polymer-metal*) is described for the following parameters: loads $q = 0.3\ 10^8$; $p = 0$; $q_t$ = $2.5 \cdot 10^4$ J/(m² s); time of their influence $t_0$ = 60 min; relative thickness of the layers—$h_1$ = 0.04 $\rightarrow$ 0.02; $h_2$ = 0.04; $h_3$ = 0.2.

Removal of the substance from the ceramic layer surface was supposed to be *uniform*, its beginning at the moment $t_1 = t_0/2$ corresponds to a required heating of the surface. The rate of ablation was supposed sufficient for decreasing thickness of the bearing layer twice as much. The temperature field was determined by (11.37) with account of the averaged ceramic layer thickness

$$h_1 = (h_1(t_n) + h_1(t_{n\,1}))\ /2\ ,$$

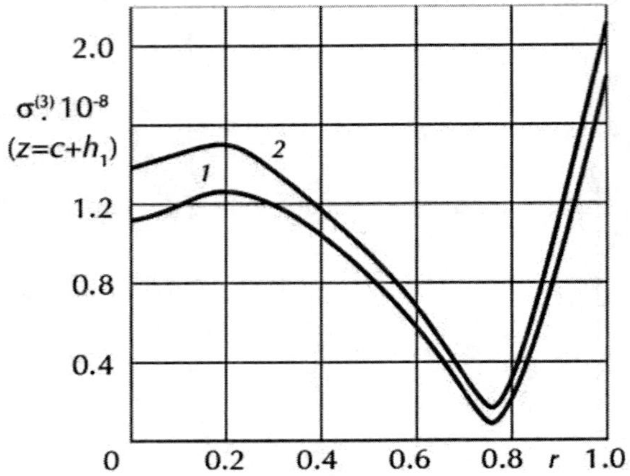


**Figure 10.11.**

where $t_i$ are the nodal moments of time, $n$—the number of the considered point.

Figure 10.12 shows that ablation is accepted beginning from moment $t_1$ when the temperature on the ceramic layer surface has reached 1200 K (curve *1*). The rated temperature on the surface at ablation (curve *2*) is lower than on the plates of a constant thickness $h_1$ = {0.04; 0.02} (curves. *3*, *4*).

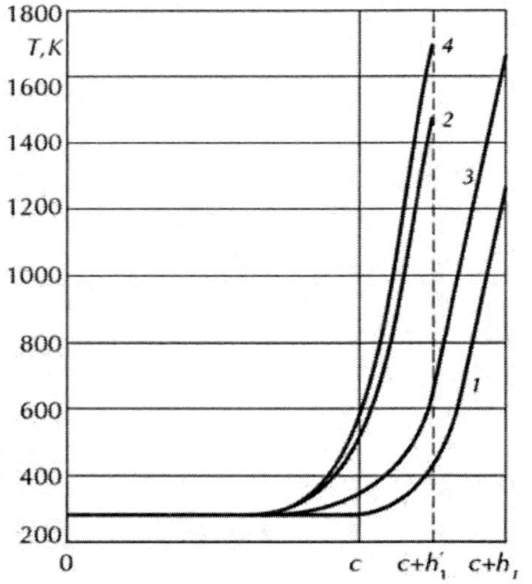


**Figure 10.12.**

The effect of ablation on the maximal deflection of the plate $w_{max}$ and the maximal relative shear $\psi_{max}$ in the filler are shown, accordingly, in Fig. 10.13a, b. These parameters in moment $t_0 = 60$ min in the presence of ablation (curve *3*) are approximately 3.5 times larger than for the plates with a constant thickness of the ceramic layer $h_1 = 0.04$ (curve *1*), and by 30% less than for the plate with $h_1 = 0.02$ (curve *2*).

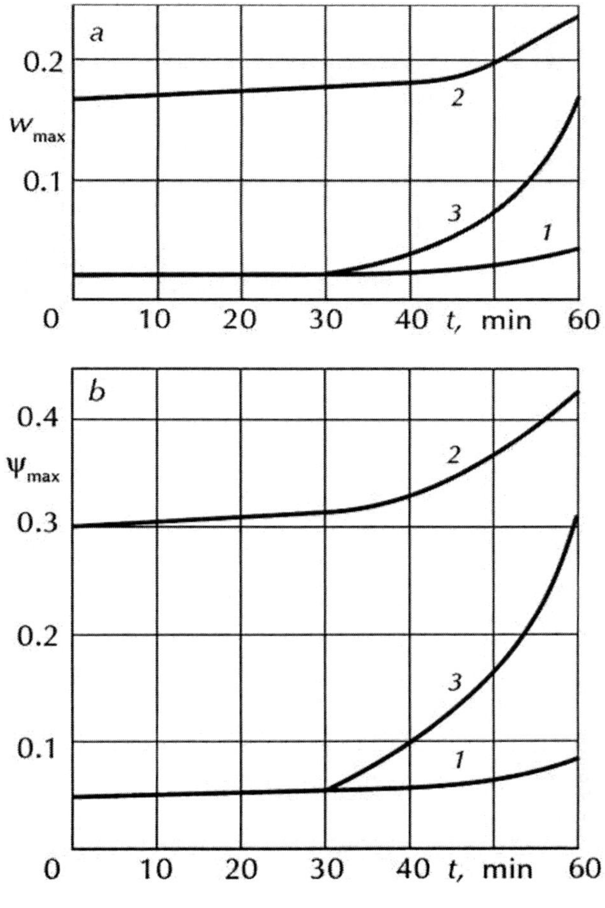


**Figure 10.13.**

The rated deflection of the plate under ablation conditions calculated by different physical equations of state is shown in Fig. 10.14. The difference in solution of the problems of thermoelasticity (curve *1*), linear viscoelasticity *2* and a general case of the considered viscoelastoplasticity *3* is by far larger than in previous cases.

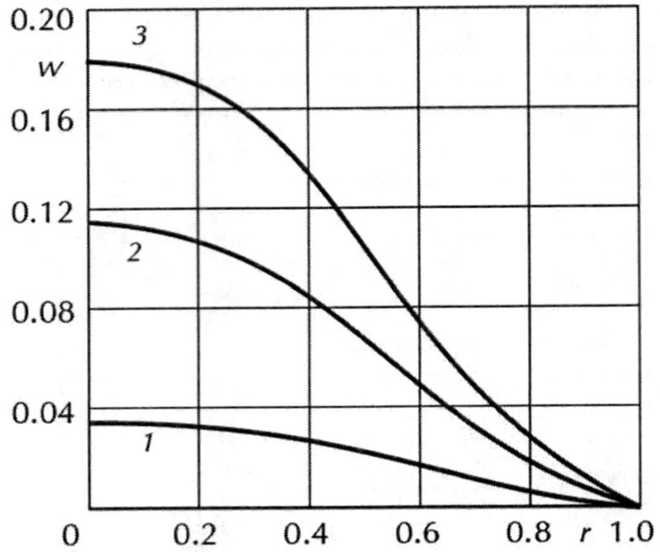


**Figure 10.14.**

Similar differences caused by ablation for the corresponding radial stresses in the closure (Fig. 10.15) are not so essential. Only a noticeable stress growth in the bearing metal layer is observed due to the loading redistribution in the layers.

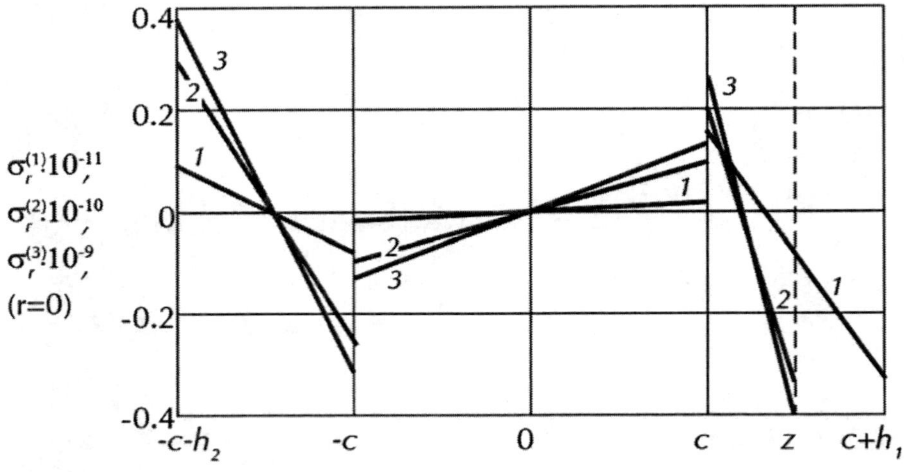


**Figure 10.15.**

Above considered phenomena are due to the fact that under identical thermal conditions a part of energy in the presence of ablation is expended on heating of the removed substance. This results in a comparative temperature drop in the plate and relative increase of rigidity of the materials of the layers.

The effect of ablation on SSS variations in the plate is observed also at numerical simulation of the heat loading of a three-layer metal-polymer-metal pack. Notice that the effect of ablation is used in the cooling equipment for of combustion chambers of rocket engines.

Proceeding from the above, the complex heat force influence, including ablation, results in greater variations of SSS of viscoelastoplastic elements of constructions in contrast to isothermal loads. This has been proved by investigations for the layered rods, plates, and shells (Gorshkov et al., 2005, Moskvitin and Starovojtov, 1986, Pleskchevsky et al., 2004, Pobedrya and Georgievskii, 1999).

## KEYWORDS

- **Ceramics-polymer-metal**
- **Nonlinearity**
- **Physicomechanical**
- **Stress–strain ratio**
- **Viscoelastic continuum**

# Chapter 11

## Thermoviscoelastoplastic Characteristics of Materials

To solve the boundary problems numerically, it is recommended to employ the mechanical or close to them characteristics of real materials. For this purpose, necessary parameters for aluminum alloy D-16T (duralumin), cordierite, silicon nitride ceramics, and polytetrafluoroethylene (PTFE, fluoroplast-4) have been obtained using known experimental data (Starovoitov, 2002, 2004) given below. Besides, an analytical variant of plasticity and physical nonlinearity functions is presented, and a numerical processing technique of experimental data is described.

### ALUMINUM ALLOY D-16T

The physical equations of state for thermoviscoelastoplastic material and are accepted in the form

$$s_{ij} = 2G(T)\left( f_1(\varepsilon_u, T)\, \text{э}_{ij} - \int_0^t R(t-\tau) f_2(\varepsilon_u, T)\, \text{э}_{ij}(\tau)\, \mathrm{d}\tau \right),$$

$$\sigma = 3K(T)(\varepsilon - \alpha T)\,. \tag{11.1}$$

Here, like previously, $s_{ij}$, $\text{э}_{ij}$, $\sigma$, $\varepsilon$ are the deviator and spherical parts of stress and strain tensors; $f_1(\varepsilon_u, T) = 1 - \omega_1(\varepsilon_u, T)$— Ilyushin's plasticity function; $\varepsilon_u$—strain rate; $f_2(\varepsilon_u, T)$—universal function of nonlinear creep; $R(t)$—relaxation kernel; $\alpha$—averaged factor of linear temperature expansion; $T$ is inhomogeneous and non-stationary temperature field counted from some initial temperature $T_0$; $G(T)$, $K(T)$—shear and volume strain moduli.

To describe the dependence of elasticity moduli on temperature we use the known Bell's (1984) formula deduced from experimental data for above five hundred pure metals and alloys:

$$\{G(T),\ K(T),\ E(T)\} = \{G(0),\ K(0),\ E(0)\}\ \varphi(T)\,,$$

$$\varphi(T) = \begin{cases} 1, & 0 < T/T_m \le 0{,}06, \\ 1{,}03(1 - T/(2T_m)), & 0{,}06 < T/T_m \le 0{,}57, \end{cases} \tag{11.2}$$

where $T_m$ is melting point of the material; $G(0)$, $K(0)$, $E(0)$—moduli at so-called zeroth stress. They can be determined experimentally based on known $G_0$ at a certain temperature $T_0$ (e.g., at room temperature 20°C) in the form

$$G(0) = \frac{G_0}{\varphi(T_0)}\,.$$

Poisson's ratio is anticipated to be independent of temperature. At higher (homologous) temperatures $T/T_m > 0.57$ a slight deviation of the material behavior from the linear law (11.2) is possible.

The analytical variant of plasticity function at some constant temperature $T_0$ is accepted:

$$\omega_1(\varepsilon_u, T_0) = \begin{cases} 0, & \varepsilon_u \le \varepsilon_{y0}, \\ A_1(1 - \frac{\varepsilon_{y0}}{\varepsilon_u})^{\alpha_1}, & \varepsilon_u \ge \varepsilon_{y0}. \end{cases} \tag{11.3}$$

Here, $\varepsilon_{y0} = \varepsilon_y(T_0)$—strain rate corresponding to yield point at $T_0$.

To determine experimental values of the plasticity function we use the experimental data on variable torsion of a circular in section rod beyond the elasticity limits under the conditions of room temperature. They were derived by A.P. Gusenkov and G.V. Moskvitin and are presented in Fig. 11.1a. The dependence of relative strain ε on torque $M$ at instant deformation (curve *1*—direct loading, *2*—back loading) is:

$$M = \frac{M_{12}}{M_y} = \frac{\sigma_{12}}{\sigma_y}, \quad \varepsilon = \frac{\varepsilon_{12}}{\varepsilon_y}, \sigma_y = 2G_0\varepsilon_y. \tag{11.4}$$

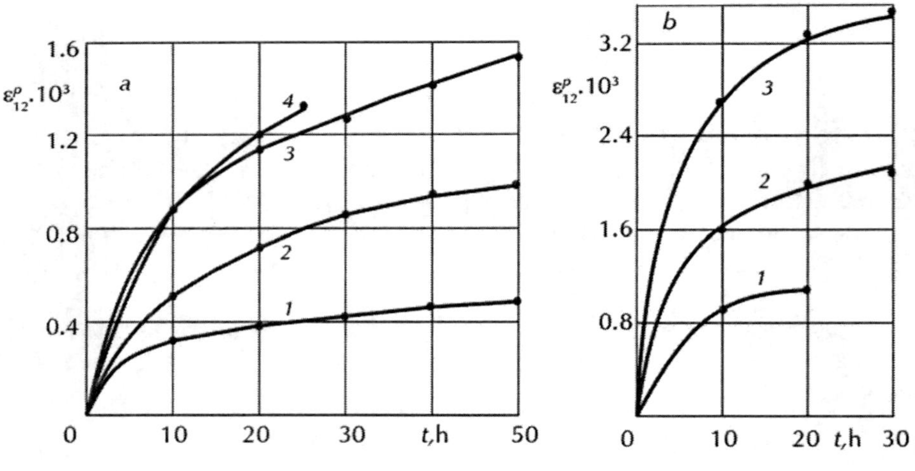


**Figure 11.1.**

At an instant torsion the integral summand should be equated to zero in relations (11.1). Out of all stress and strain tensor components only $s_{12} = \sigma_{12}$ and $э_{12} = \varepsilon_{12}$ will be nonzero. As a result, the following dependence will be true for the considered experimental curves

$$\sigma_{12} = 2G_0\varepsilon_{12}(1 - \omega_1). \tag{11.5}$$

Let us divide the left and right-hand sides of equation (11.5) by $\sigma_y$. Then, using the values introduced in (11.4), we get

$$M = \varepsilon(1-\omega_1), \quad \omega_1 = 1 - \frac{M}{\varepsilon}. \tag{11.6}$$

Based on (11.6), we find the values of plasticity function in experimental points $\omega_{1n}$ (Fig. 11.1b). In these points, the condition of approximation should be met

$$\omega_{1n} = A_1\left(1-\frac{\varepsilon_{y0}}{\varepsilon_{un}}\right)^{\alpha_1}.$$

By dividing similar relations for experimental points $n$ and $k$ we obtain

$$\frac{\omega_{1n}}{\omega_{1k}} = \frac{\left(1-\frac{\varepsilon_{y0}}{\varepsilon_{un}}\right)^{\alpha_1}}{\left(1-\frac{\varepsilon_{y0}}{\varepsilon_{uk}}\right)^{\alpha_1}},$$

where from,

$$\alpha_1 = \ln\frac{\omega_{1n}}{\omega_{1k}} \Bigg/ \ln\frac{1-\frac{\varepsilon_{y0}}{\varepsilon_{un}}}{1-\frac{\varepsilon_{y0}}{\varepsilon_{uk}}}, \quad A_1 = \frac{\omega_{1n}}{\left(1-\frac{\varepsilon_{y0}}{\varepsilon_{un}}\right)^{\alpha_1}}. \tag{11.7}$$

After averaging we accept that $A_1 = 0.96$; $\alpha_1 = 2.34$. The approximation accuracy is seen from curves *1* in Fig. 11.1a, b. Dark points are the experiment; solid line is calculations.

The plastic properties of metals increase with temperature growth, therefore the diagram of function $\omega_1(\varepsilon_u, T)$ is shifted along the abscissa axis relative to graph $\omega_1(\varepsilon_u, T_0)$ by value $\varepsilon_{y0} - \varepsilon_y$ (strain rate $\varepsilon_y$ corresponds to elasticity limit at temperature $T$). Values $\varepsilon_y$ are calculated by the relation of elasticity limit $\sigma_y$ versus temperature suggested which was by N.A. Makhutov:

$$\sigma_y = \sigma_{y0}\exp\left\{\kappa\left(\frac{1}{T}-\frac{1}{T_0}\right)\right\}, \ \varepsilon_y(T) = \frac{\sigma_y(T)}{E(T)}. \tag{11.8}$$

To determine constants $\kappa$ of the material by (11.8) we use

$$\kappa = \ln\frac{\sigma_y}{\sigma_{y0}} \Bigg/ \left(\frac{1}{T}-\frac{1}{T_0}\right).$$

The values are base on Yu. N. Rabotnov's (1977) experimental data are given in Table 11.1.

Thus, by generalizing (11.3), the plasticity function at thermal force loading from the natural state is accepted as

$$\omega_1(\varepsilon_u, T) = \begin{cases} 0, & \varepsilon_u \le \varepsilon_y, \\ A_1\left(1 - \dfrac{\varepsilon_{y0}}{\varepsilon_u + \varepsilon_{y0} - \varepsilon_y}\right)^{\alpha_1}, & \varepsilon_u > \varepsilon_y. \end{cases} \quad (11.9)$$

The analytical variant of the function of nonlinearity at a sign-variable loading $\omega_1^*(\varepsilon_u^*, T)$ is (11.9):

$$\omega_1^*(\varepsilon_u^*, T) = \begin{cases} 0, & \varepsilon_u^* \le \varepsilon_y^*, \\ A_1^*\left(1 - \dfrac{\varepsilon_{y0}^*}{\varepsilon_u^* + \varepsilon_{y0}^* - \varepsilon_y^*}\right)^{\alpha_1^*}, & \varepsilon_u^* > \varepsilon_y^*, \end{cases}$$

the corresponding constants $A_1^*$, $\alpha_1^*$, $\varepsilon_{y0}^*$ are calculated from experimental data according to above-stated procedure described in Table 11.1. Value $\varepsilon_y^*$ is calculated by (11.8). The approximation accuracy is shown by curves *2* in Fig. 11.1a, b. Dark points are the experiment; solid line is calculation.

**Table 11.1.**

| Parameter | Value | Parameter | Value |
|---|---|---|---|
| $E(0)$, MPa | $0.829 \cdot 10^5$ | $\alpha_1^*$ | 2.27 |
| $G(0)$, MPa | $0.3075 \cdot 10^5$ | $\varepsilon_{y0}^*$, % | 1.485 |
| $K(0)$, MPa | $0.9214 \cdot 10^5$ | $A_2$ | 1 |
| $E_0$, MPa | $0.72 \cdot 10^5$ | $\alpha_2$ | 0.7 |
| $G_0$, MPa | $0.267 \cdot 10^5$ | $\varepsilon_{u0}$, % | 0.27 |
| $K_0$, MPa | $0.8 \cdot 10^5$ | $A_2^*$ | 1 |
| $\alpha_0$, 1/K | $24.3 \cdot 10^{-6}$ | $\alpha_2^*$ | 0.6 |
| $T_0$, K | 293 | $\varepsilon_{u0}^*$, % | 0.535 |
| $T_m$, K | 933 | $A$, $s^{-\alpha}$ | $2.92 \cdot 10^{-4}$ |
| $\nu$ | 0.35 | $\beta$, $s^{-1}$ | $1.39 \cdot 10^{-7}$ |
| $A_1$ | 0.96 | $\alpha$ | 0.25 |
| $\alpha_1$ | 2.34 | $\sigma_{y0}$, MPa | 530 |
| $\varepsilon_{y0}$, % | 0.735 | $\rho_0$, kg/m$^3$ | 2700 |
| $\sigma_{y0}$, MPa | 340 | $C_0$, J/kg · K | 880 |
| $\kappa$, 1/K | 301 | $k_0$, J/m · s · K | 177 |
| $A_1^*$ | 0.924 | $\delta$ | 5.6 |

The universal function of a nonlinear creep is assumed as follows:

$$f_2(\varepsilon_u, T) = (1-\omega_2(\varepsilon_u)\left(\frac{T_0}{T}\right)^{\delta},$$

$$\omega_2(\varepsilon_u) = \begin{cases} 0, & \varepsilon_u \le \varepsilon_{u0}, \\ A_2\left(1-\dfrac{\varepsilon_{u0}}{\varepsilon_u}\right)^{\alpha_2}, & \varepsilon_u > \varepsilon_{u0}. \end{cases} \qquad (11.10)$$

The creep curves of D-16T alloy at direct (Fig. 11.2a) and inverse torsion (Fig. 11.2b) of thin-walled tubes are taken from Namestnikov's works without elastic components. Curve *1* in Fig. 11.2a corresponds to stress $\sigma_{12} = 113$, *2* – $\sigma_{12} = 127$, *3* – $\sigma_{12} = 144$ MPa. Respective curves in Fig. 11.2b express double stresses $\sigma_{12}^* = 2\sigma_{12}$, $T = 423$ K.

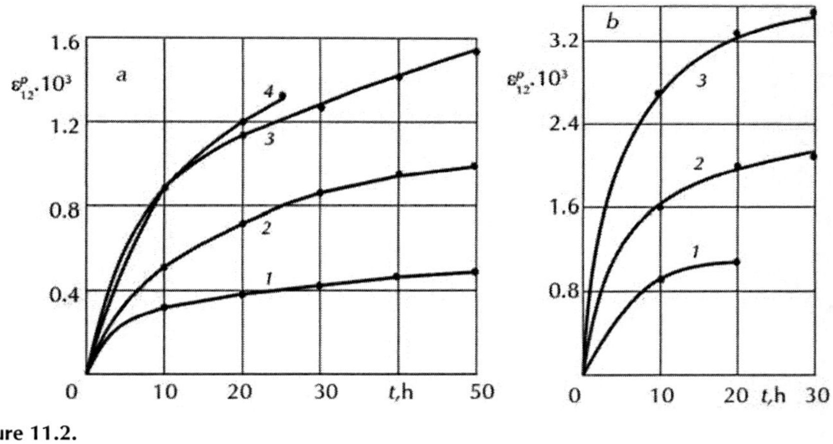


**Figure 11.2.**

To calculate temperature constant δ, we used a creep curve for the tension: *4* – $\sigma_1 = 156.8$ MPa, $T = 473$ K (Rabotnov, 1977). At reverse loading the temperature dependence of the process was taken as previously and function $\omega_2^*(\varepsilon_u^*)$ was accepted similar to (11.10):

$$\omega_2^*(\varepsilon_u^*) = \begin{cases} 0, & \varepsilon_u^* \le \varepsilon_{u0}^*, \\ A_2^*\left(1-\dfrac{\varepsilon_{u0}^*}{\varepsilon_u^*}\right)^{\alpha_2^*}, & \varepsilon_u^* > \varepsilon_{u0}^*. \end{cases}$$

It follows from the experimental data presented by A.P. Gusenkov and G.V. Moskvitin that deformation in the plastic area at loading from the natural state hardens the material since $\beta_2 = 2.02$ ($\varepsilon_y^* = \beta_2\varepsilon_y$, see Table 8.1). However, according to V.S. Namestnikov it follows that D-16T alloy weakens in response to creep. As a result of the loading sign change the creep runs faster: $\beta_2^* = 1.98$ ($\varepsilon_{u0}^* = \beta_2^*\varepsilon_{u0}$).

After processing of the experimental data using above-described method, we have derived constants $A_2$, $\alpha_2$, $\varepsilon_{u0}$, $A_2^*$, $\alpha_2^*$, $\varepsilon_{u0}^*$ included in the approximation formulas for the nonlinear creep functions at direct and back loading (Table 11.1). Figures 11.2, 11.3 (curve *1*—direct loading, *2*—back loading) reflect the conformity of theoretical curves to experimental points.

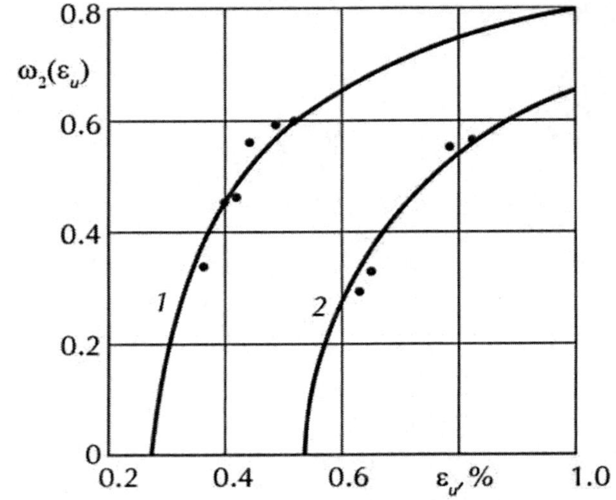


**Figure 11.3.**

The rheonomic characteristics of D-16T alloy are described by A.R. Rzhanitsyn's (1968) relaxation kernel:

$$R(t) = Ae^{-\hat{a}t}t^{\alpha-1} \ (\beta > 0, 0 < \alpha < 1). \tag{11.11}$$

This choice is conditioned by simplicity of the kernel, which nevertheless fully accounts for the weakly singular properties of the materials. The technique of determining the nucleus parameters, corresponding diagrams and tables are given in M.A. Koltunov's (1976) monograph. Based on them results according to (11.11) are shown Fig. 11.4, with a design curve of the pliability function is

$$J(t) = 1 + \int_0^t \Gamma(t-\tau)d\tau,$$

where from we can consider the approximation accuracy.

The kernel constants with density $\rho$, specific heat capacity $C$, specific heat conductivity $k_0$ and ultimate strength $\sigma_{u0}$ at a normal temperature are given in Table 11.1. We should note that all thermomechanical characteristics of D-16T alloy were received in the field of moderate temperatures ($T \leq 543$ K).

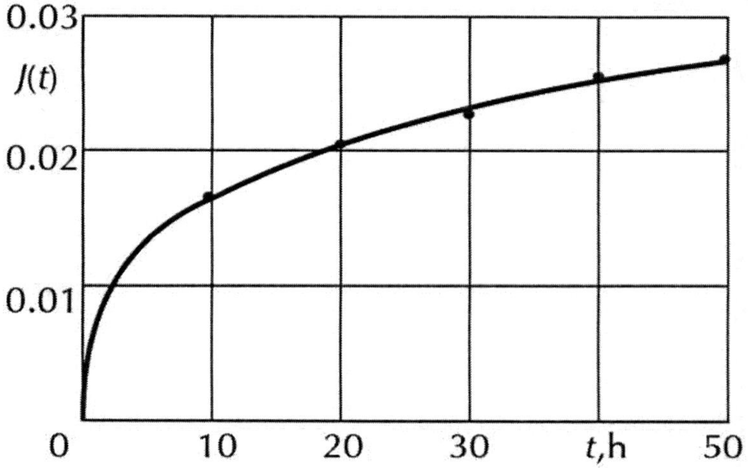


**Figure 11.4.**

## CERAMIC MATERIALS

The experimental data on the behavior of silicon nitride structural ceramics (SNSC) and cordierite during thermal loading (Fig. 11.5) have been taken from the works by Gogotsy (1982).

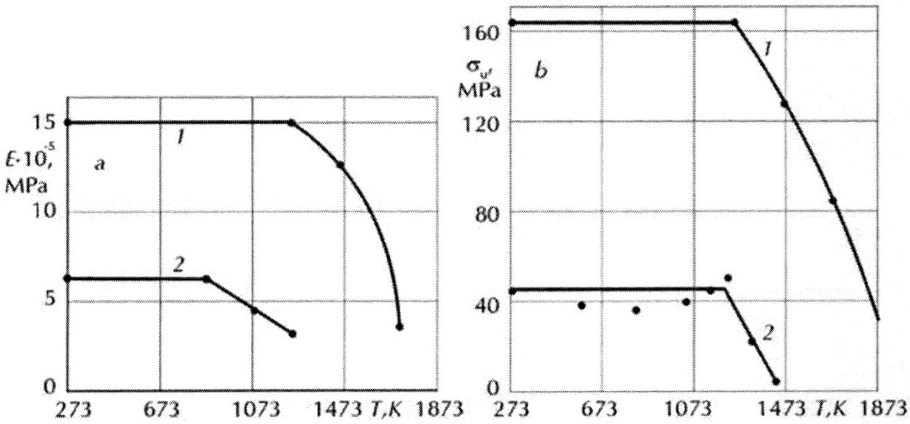


**Figure 11.5.**

We shall use the relations of the linear thermoelasticity for analytical description of these materials

$$s_{ij} = 2G(T)\,\text{э}_{ij}\,,$$

$$\sigma = 3K(T)(\varepsilon - \alpha T)\,, \qquad (11.12)$$

the temperature is counted from a certain initial value $T_0$.

Poisson's ratio ν for ceramic materials is usually accepted equal to 0.3 and it is independent of temperature, therefore

$$\{E(T), G(T), K(T)\} = \{E_0, G_0, K_0\}\phi_1(T),$$

$$\phi_1(T) = \begin{cases} 1, & T \le T_0, \\ 1 - A_1((T - T_0)/T_0)^{\alpha_1}, & T > T_0. \end{cases} \tag{11.13}$$

For the ultimate strength, we shall accept an approximation:

$$\sigma_b = \sigma_{b0}\varphi_2(T),$$

$$\varphi_2(T) = \begin{cases} 1, & T \le T_{0b}, \\ 1 - A_2(T - T_{0b})/T_{0b})^{\alpha_2}, & T > T_{0b}. \end{cases} \tag{11.14}$$

The numerical values of the constants included in (11.12)–(11.14) are obtained by processing experimental data (Table 11.2). Their agreement with the experiment is illustrated by Fig. 11.5, where the points are experimental values, solid line is calculations. Curve *1* corresponds to SNSC, *2*—cordierite.

Due to the extensive use of cermets in mechanical engineering, aerospace equipment and construction there appeared numerous publications devoted to the research of their thermomechanical properties.

**Table 11.2.**

| Parameter | Material | |
|---|---|---|
| | SNSC | Cordierite |
| $E_0$, MPa | $15.2 \cdot 10^5$ | $6.7 \cdot 10^5$ |
| $G_0$, MPa | $5.85 \cdot 10^5$ | $2.58 \cdot 10^5$ |
| $K_0$, MPa | $12.7 \cdot 10^5$ | $5.58 \cdot 10^5$ |
| $A_1$ | 9.249 | 1.073 |
| $\alpha_1$ | 2.17 | 0.843 |
| $T_0$, K | 1273 | 873 |
| $T_{0b}$, K | 1273 | 1207 |
| $\sigma_0$, MPa | 164 | 42 |
| $A_2$ | 2.33 | 6.472 |
| $\alpha_2$ | 1.34 | 1.159 |
| ρ, kg/m³ | 2510 | 2100 |
| $\alpha_0$, $K^{-1}$ | $3 \cdot 10^{-6}$ | $6 \cdot 10^{-6}$ |

## POLYTETRAFLUOROETHYLENE

The physical equations of state for polytetrafluoroethylene (PTFE, fluoroplastic-4) have a form

$$\phi_1(\sigma,T)s_{ij} = 2G_0\left( f(\varepsilon_u)\,\text{э}_{ij} - \int_0^t R(t-\tau) f(\varepsilon_u)\,\text{э}_{ij}(\tau)\,\mathrm{d}\tau \right),$$

$$\phi_2(\sigma,T)\sigma = 3K_0(\varepsilon - \alpha_0 \Delta T). \qquad (11.15)$$

Where $f(\varepsilon_u)$, $\varphi_n(\sigma, T)$ are universal functions of physical nonlinearity of the polymeric filler ($n = 1, 2$), and $f = 1$ at $\varepsilon_u < \varepsilon_s$.

If we solve equation (11.15) relative to $\varphi_1(\sigma, T)s_{ij}$ we shall have

$$2G_0 f(\varepsilon_u)\,\text{э}_{ij} = \phi_1(\sigma,T)s_{ij} + \int_0^t \Gamma(t-\tau)\phi_1(\sigma,T)s_{ij}(\tau)\mathrm{d}\tau,$$

$$3K_0(\varepsilon - \alpha_0\Delta T) = \phi_2(\sigma,T)\sigma.$$

From the first equation (11.15) at some temperature of reduction $T = T_0 = \text{const}$ and condition of pure shear $\sigma = 0$, $s_{12} = s_{12}^{\ 0} = \text{const}$, we may write

$$2G_0 f(\varepsilon_u)\,\text{э}_{12}(t) = s_{12}^0 J(t), \qquad (11.16)$$

where $\varphi_1(0, T_0) = 1$, $J(t)$ is pliability at shear

$$J(t) = 1 + \int_0^t \Gamma(t-\tau)\mathrm{d}\tau.$$

All experimental data on PTFE deformation are taken from Goldman's monographs (1979). The corresponding curves of creep ($T = T_0 = 313$ K) at shear are shown in Fig. 11.6. The first of them corresponds to the linear viscoelastic deformation, which follows from (11.16)

$$2G_0\,\text{э}_{12}^{(1)}(t) = s_{12}^{0(1)} J(t),\ \phi_1(0,T_0) \equiv 1,\ f \equiv 1. \qquad (11.17)$$

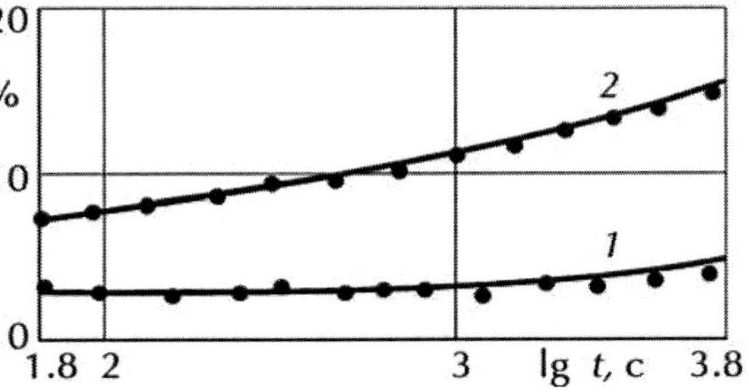


**Figure 11.6.**

Curve 2 is expressly nonlinear and we have along it

$$2G_0\,\text{э}_{12}^{(2)}(t)(1 - \omega(\varepsilon_u^{(2)}))$$

$$= s_{12}^{0(2)} J(t). \tag{11.18}$$

Here the superscript in brackets is the curve number.

Upon dividing the left and right-hand sides of equation (11.18), accordingly, by (11.17), we obtain the dependence for the nonlinearity function on time

$$\omega(t) = 1 - \frac{э_{12}^{(1)}(t) s_{12}^{0(2)}}{э_{12}^{(2)}(t) s_{12}^{0(1)}}. \tag{11.19}$$

Since we know now the strain rate at each moment of time

$$\varepsilon_u^{(2)} = \tfrac{2\sqrt{3}}{3} э_{12}^{(2)}(t)$$

and nonlinearity function $\omega(t)$ (11.19), we can draw the experimental curve $\omega \sim \varepsilon_u$ (Fig. 11.7). The analytical formula will be

$$\omega(\varepsilon_u) = \begin{cases} 0, & \varepsilon_u \le \varepsilon_s, \\ A_1(1 - \varepsilon_s / \varepsilon_u)^{\alpha_1}, & \varepsilon_u > \varepsilon_s. \end{cases} \tag{11.20}$$

The approximation accuracy is evident from Figs. 11.6 and 11.7a.

Using (11.17), we obtain the experimental values for pliability at shear

$$J(t) = \frac{2G_0 э_{12}^{(1)}(t)}{s_{12}^{0(1)}}.$$

The kind of the relaxation kernel is accepted in the form suggested by Rzhanitsyn:

$$R(t) = Ae^{-\hat{a}t} t^{\alpha-1} \quad (\beta > 0, 0 < \alpha < 1).$$

The parameters of the kernel are calculated using Koltunov's method. The agreement between the experimental (dots) and design (solid line) values is illustrated by Fig. 11.7b.

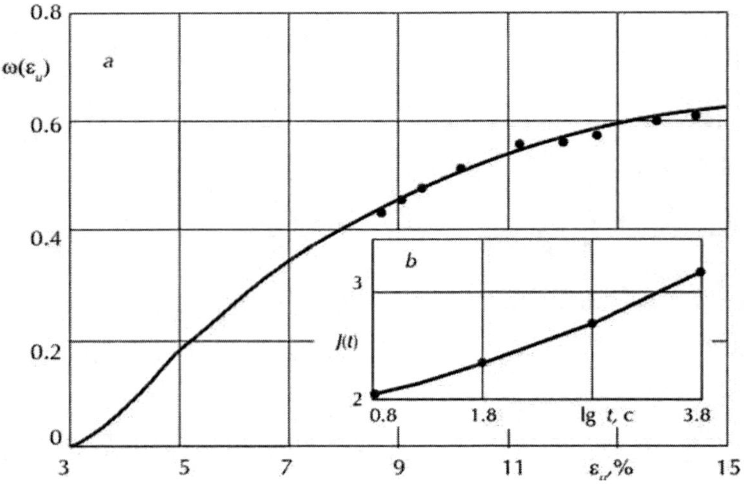


**Figure 11.7.**

The results of experimental research of PTFE creep at shear under various temperatures $T_k$ (curve *1*—293 K, *2*—303 K, *3*—313 K, *4*—323 K, *5*—333 K) are shown in Fig. 11.8, where $s_{12}^{0} = 3$ MPa is tangential stress, $p = 0$—hydrostatic stress.

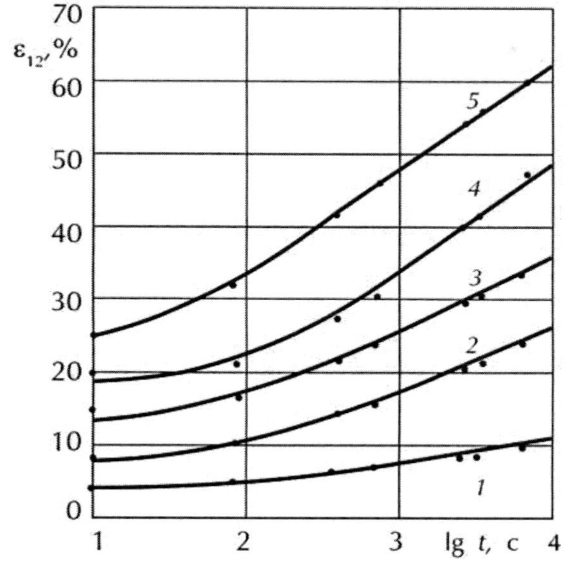


Figure 11.8

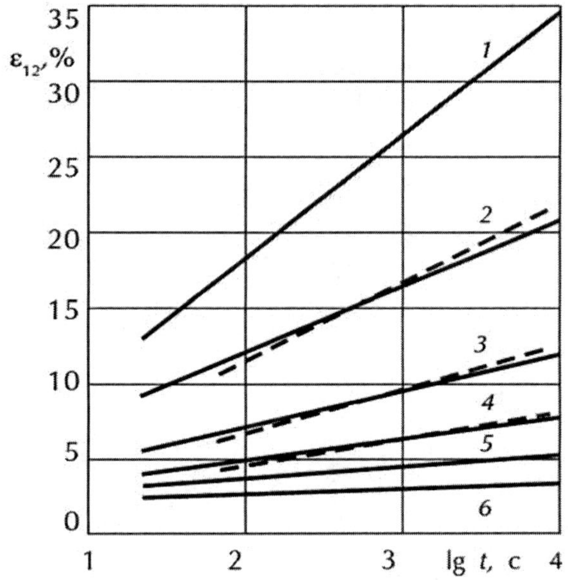


Figure 11.9.

The temperature is equal to $T_0$ along the third curve, hence, the function of nonlinearity is $\varphi(0, T_0) = 1$, therefore

$$2G_0\, э_{12}^{(3)}(t)(1-\omega(\varepsilon_u^{(3)})) = s_{12}^0 J(t)\,. \tag{11.21}$$

By assuming for other curves that

$$\varphi_1(\sigma,T) \equiv \varphi_{11}(\sigma)\varphi_{12}(T),$$

we obtain

$$2G_0(1-\omega(\varepsilon_u^k))\, э_{12}^k = \varphi_{12}(T_k)s_{12}^0 J(t). \tag{11.22}$$

Upon dividing the left and right-hand sides of equation (11.22) by, accordingly, (11.21), we shall derive the experimental values for a part of the function of physical nonlinearity at various temperatures $T_k$:

$$\varphi_{12}(T_k) = \frac{(1-\omega(\varepsilon_u^k))\, э_{12}^k}{(1-\omega(\varepsilon_u^{(3)}))\, э_{12}^{(3)}}\,. \tag{11.23}$$

As a result of calculations of function $\varphi_{12}(T_k)$ using (11.23) for each temperature $T_k$ at different moments of time and averaging, we may obtain the experimental points (Fig. 11.10a).

We should meet the relation presented below along the creep curves (Fig. 11.9) received at shear by superposition of various hydrostatic pressure values $p$ under the condition of constant temperature ($T = 313$ K, $\varphi_{12}(T_0) = 1$) and tangential stress $s_{12}{}^0 =$ 4 MPa (curve *1*—$p = 0$, *2*—20 MPa, *3*—50 MPa, *4*—100 MPa, *5*—150 MPa, *6*—200 MPa)

$$2G_0 f(\varepsilon_u^k)\, э_{12}^k = \varphi_{11}(\sigma_k)s_{12}^0 J(t),$$

where $\sigma_k = -p$—mean stress corresponding to the $k$-th curve. This makes possible to calculate the values of nonlinearity function $\varphi_{11}(\sigma_k)$ using the first curve in Fig. 11.9, for which $\varphi_{11}(0) = 1$:

$$\varphi_{11}(\sigma_k) = \frac{f(\varepsilon_u^k)\, э_{12}^k}{f(\varepsilon_u^{(1)})\, э_{12}^{(1)}}\,.$$

The corresponding points are shown in Fig. 11.10b. The approximation formula for $\varphi_1(\sigma,T)$ is

$$\varphi_1(\sigma,T) = (1 - A_2\,|\,\sigma\,|^{\alpha_2})(1 + B(\Delta T\,/\,T_{ïë})^{\gamma}\,\mathrm{sgn}\,\Delta T), \tag{11.24}$$

where $\Delta T = T - T_0$.

The conformity of design curves based on formula (11.24) to experimental values is illustrated in Figs. 11.8–11.11. The experimental curves for PTFE creep at shear ($s_{12}{}^0$ = 4 MPa) with superposition of hydrostatic pressure (Fig. 11.11a—$p = 20$ MPa; Fig. 11.11b—$p = 100$ MPa) at various temperatures (curve *1*—$T = 303^0$ K, *2*—$T = 313$ K,

3—$T$ = 323 K) were not used in calculations of the constants included in the approximation formula (11.24), that is they were the check ones. According to the works by Goldman and Shukov, the volume creep of PTFE makes up less than 5% of the shear creep, so it is neglected.

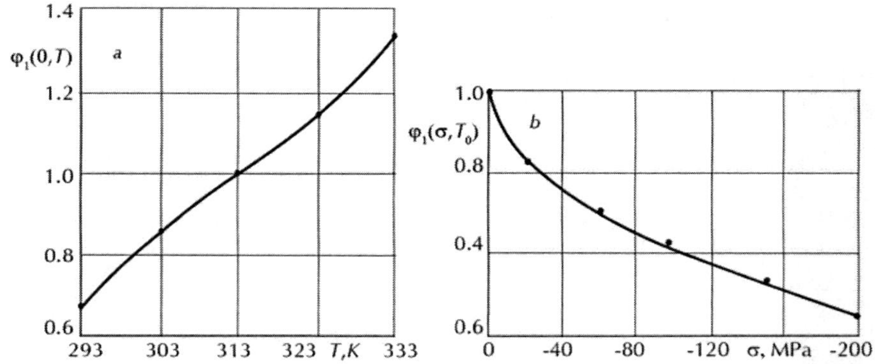


**Figure 11.10.**

The mechanism of a 3D behavior of polymeric materials at positive mean stresses differs qualitatively and quantitatively from that at a uniform compression, for which we have no reliable experimental data as yet. Therefore, the function of nonlinearity $\varphi_2(\sigma, T)$ will be determined only in the area $\sigma < 0$:

$$\varphi_2(\sigma,T)=\begin{cases} 1, & p \geq p_0, \\ A_3\sigma^{\alpha_3}, & p < p_0. \end{cases} \tag{11.25}$$

where $p_0$ is the minimal pressure able healing all to internal defects.

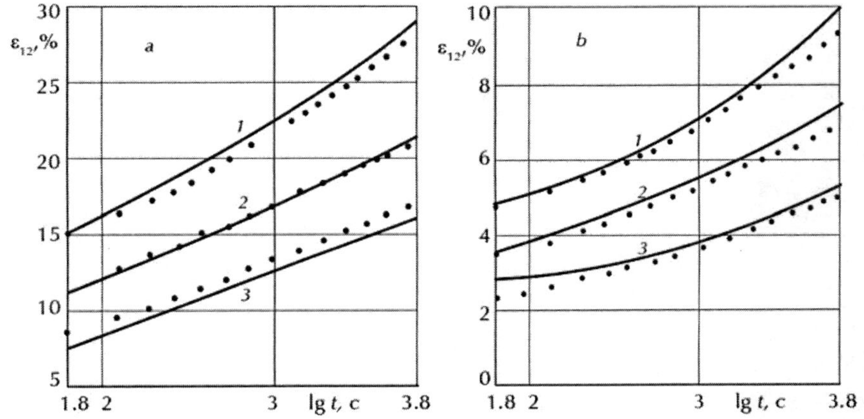


**Figure 11.11.**

To determine the volume strain modulus $K$ and parameters of nonlinearity $A_3$, $\alpha_3$, we use curve $p \sim \theta$ (Fig. 11.12) received at $T_1 = 303^0$ K. For some $k$-th and $n$-th points of this curve we should meet according to (11.25) the relation

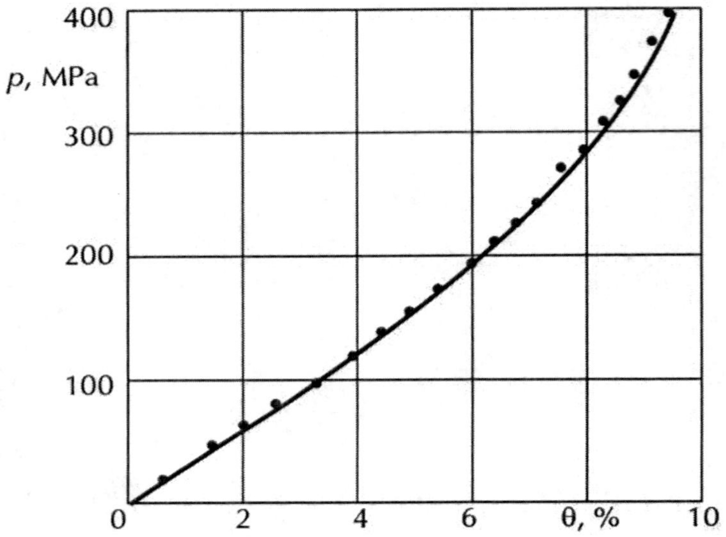


**Figure 11.12.**

$$K\left|\theta_k\right| = A_3\left|\sigma_k\right|^{\alpha_3+1},$$

$$K\left|\theta_n\right| = A\left|\sigma_n\right|^{\alpha_3+1}.$$

Where from,

$$\alpha_3 = \frac{\ln(\theta_k/\theta_n)}{\ln(\sigma_k/\sigma_n)} - 1.$$

Factor $A_3$ is determined from the condition of continuity on the boundaries of the zones of physical linearity and nonlinearity:

$$f(\sigma_0) = A_3\sigma_0^{\alpha_3} = 1, \quad A_3 = 1/\sigma_0^{\alpha_3}.$$

then,

$$K = A|\sigma|^{\alpha+1}/\theta_k.$$

The factor of volume strain $\alpha_V$ is calculated on the base of experimental curves of the relative volume strain $\theta$ dependence on temperature (Fig. 11.13). It bears a clear-cut linear character, so by accepting

$$\varphi_2(\sigma,T) \equiv \varphi_2(\sigma),\ \alpha_v = 5{,}33 \cdot 10^{-5}\,\mathrm{K}^{-1},$$

we shall reach good enough approximation for the function $\varphi_2(\sigma)$ (Fig. 11.14).

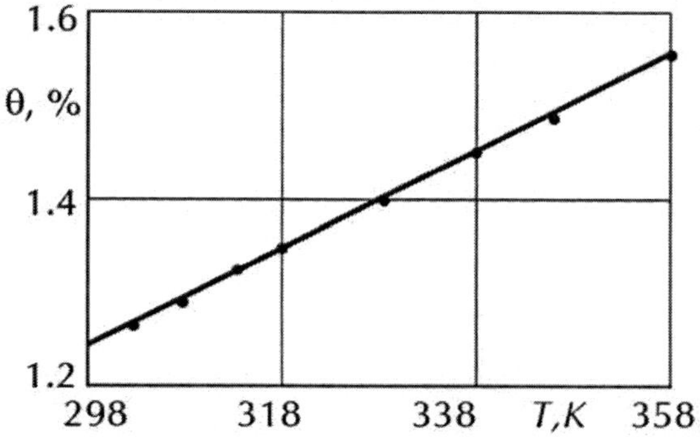


**Figure 11.13.**

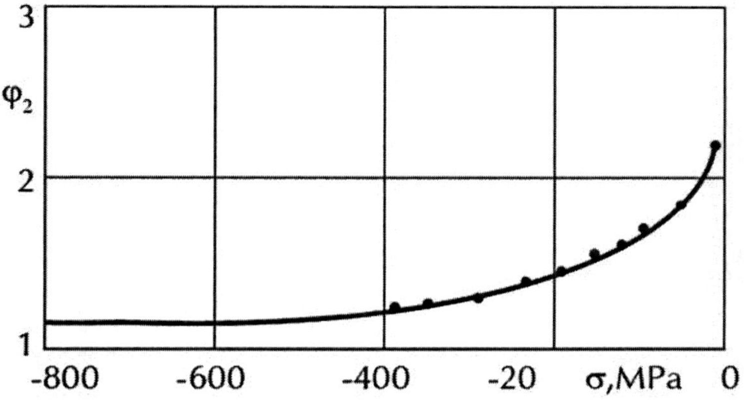


**Figure 11.14.**

There often arises a necessity to use a mechanical characteristic like the yield point or conditional yield strength at various engineering strength calculations of design elements from polymeric materials. These values are varying with temperature too. The dependencies of the conditional yield point of fluoroplastics at tension on time and temperature (Fig. 11.15, curve *1*—$T = 293^0$ K, *2*—$T = 313^0$ K, *3*—$T = 333^0$ K, *4*—$T = 353^0$ K) are described quite accurately by the relation

$$\sigma_y = \sigma_{y1}(1 - A_4((T - T_1)/T_m)^{\alpha_4})(1 - k\lg t) \tag{11.26}$$

The averaged constants for fluoroplastics included in equations (11.16)–(11.26) and reduced to temperature $T_0$ are shown in Table 11.3.

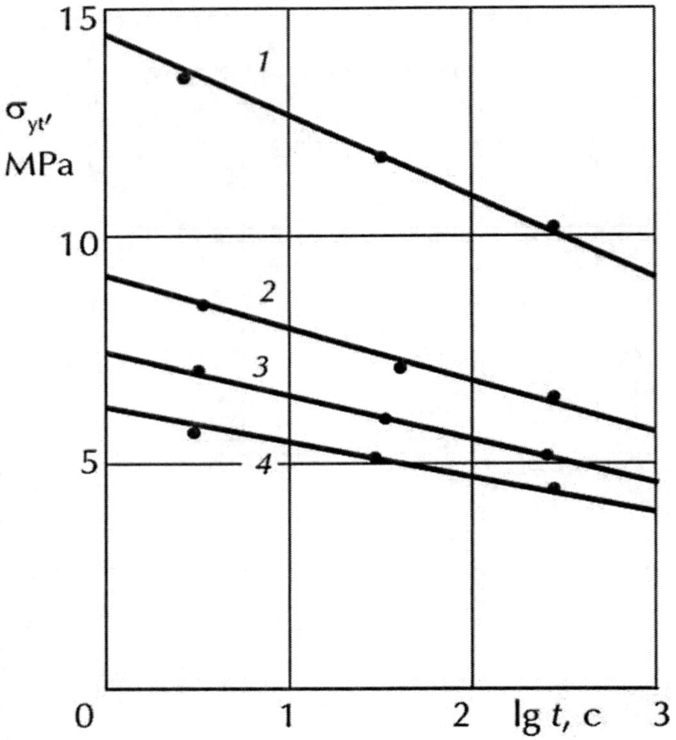


**Figure 11.15.**

**Table 11.3.**

| Parameter | Value | Parameter | Value |
|---|---|---|---|
| $A_1$ | 0.905 | $T_0$, K | 313 |
| $\alpha_1$ | 1.48 | $T_1$, K | 293 |
| $\varepsilon_s$, % | 3.3 | $A_3$, $Pa^{-\alpha 3}$ | 63.1 |
| $A$, $s^{-\alpha}$ | 0.02366 | $\alpha_3$ | –0.2 |
| $\beta$, $s^{-1}$ | $3.33 \cdot 10^{-4}$ | $A_4$, $Pa^{-\alpha 4}$ | $2.525 \cdot 10^{-19}$ |
| $\alpha$ | 0.05 | $\alpha_4$ | 3.226 |
| $A_2$, $Pa^{-\alpha 2}$ | $1.98 \cdot 10^{-5}$ | $p_0$, MPa | 800 |
| $\alpha_2$ | 0.536 | $\alpha_0$,1/K | $1.78 \cdot 10^{-5}$ |
| $B$ | 24.44 | $\sigma_{y1}$, MPa | 14.54 |
| $\gamma$ | 1.27 | $A_5$ | 1.394 |
| $G_0$, MPa | 90 | $\alpha_5$ | 0.388 |
| $K_0$, MPa | 4700 | $k$, 1/lg$C$ | 0.13 |
| $T_m$, K | 600 | $\lambda$, W/m·K | 0.25 |
| $C$, J/kg·K | 2300 | $\rho$, kg/m³ | 2150 |

## RADIATION EFFECT ON MECHANICAL PROPERTIES OF MATERIALS

When exposed to neutrons, ions or electrons mechanical properties of materials of design elements, such as hardness, yield point, plasticity, and creep undergo variations. *Neutron irradiation* is of special interest for us. According to experimental data, intensification neutron flux $I = \varphi t$ ($\varphi$—intensity of radiation, $t$—time) within the limits of low deformations results usually in increased radiation-induced hardening of the material and the yield point growth. This is visualized on the example of aluminum alloy 356. Experimental values of its yield point are shown by dots in Fig. 11.16 as dependent on the neutron flux value at a constant temperature. To analyze the radiation-induced increase of the material yield point, the following formula is suggested:

$$\sigma_y = \sigma_{y0}\left[1 + A(1 - \exp(-\xi I))^{1/2}\right], \tag{11.27}$$

where $\sigma_y$, $\sigma_{y0}$ are yield points of the exposed and unexposed materials; $A$, $\xi$—constants of the material determined experimentally.

It is also taken into account that with the radiation doze growth there occurs a saturation that ceases hardening of the material. In the absence of radiation ($I = 0$), equation (11.27) does not shows any hardening ($\sigma_y = \sigma_{y0}$), which distinguishes it from Makin and Minter[1] basic formulas for steel alloys. The constants accepted for aluminum alloy $A = 1.09$; $\xi = 9.73 \cdot 10^{-26}$ m²/neutron make the design curve to agree well with the experimental data (Fig. 11.16). The influence of radiation on elasticity constants, Young modulus, Poisson's ratio is insignificant and is not taken into account.

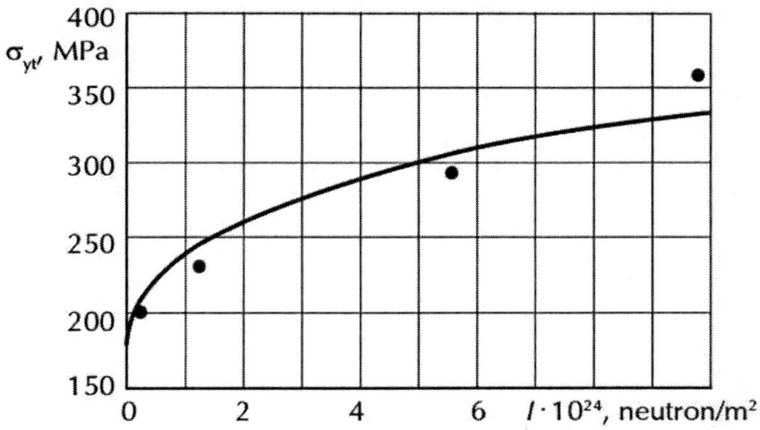


**Figure 11.16.**

The first data on the influence of radiation on creep were derived during testing of uranium in a reactor. The amount of microdefects increasing at irradiation accelerates creep of uranium from 50 till hundreds of times in spite that radiation hardening of the material results in a reduced velocity of dispositions.

There is very scarce information on creep of nonfissionable materials due to, first of all, difficulties in experimentation with the measurements of low deformations at exposure in the reactor. The creep curves of zirconium alloy "Zircaloy-2" are shown in Fig. 11.17 (Pleskatshevsky et al., 2004) (curve *1*—unexposed; *2*—exposed in reactor at intensity $\varphi_0 = 5{\cdot}10^{12}$ neutron/(cm$^2{\cdot}$s); the dotted line denotes calculations). The load during the tests was 14 kg/m$^2$, temperature—300°C. The extremes on the curves marked by dashes correspond to short-term changes in the temperature.

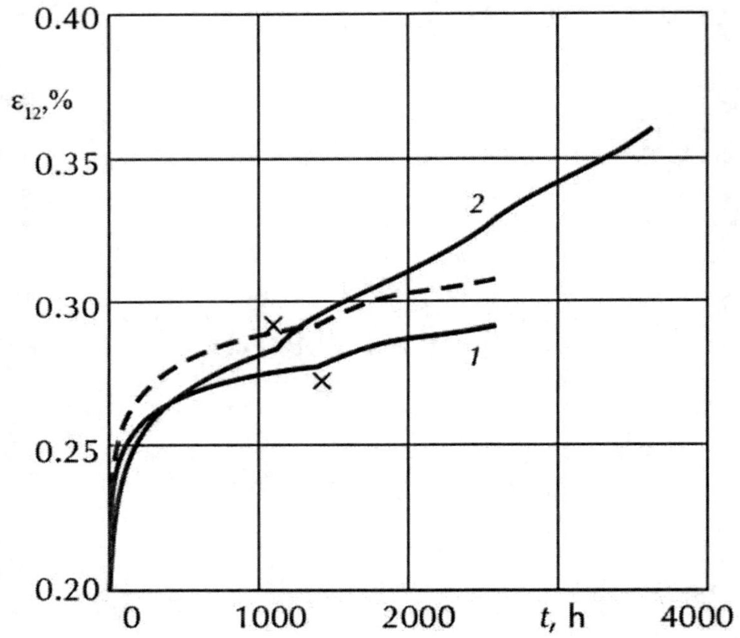


**Figure 11.17.**

To take into account the neutron irradiation effect upon viscous properties of the relaxation kernel of the material, we shall accept the relaxation in the form

$$R_\varphi(t, \varphi) = g(\varphi)R(t),$$

where *R (t)*—relaxation kernel of unexposed material. In this case, if the neutron flux intensity does not eventually change, the value of function g(φ) will be constant and it can be taken outside the integral sign when defining the relations like (11.1)

$$s_{ij} = 2G(T)\left( f_1(\varepsilon_u, T, I)\ni_{ij} - g(\varphi)\int_0^t R(t-\tau) f_2(\varepsilon_u, T)\ni_{ij}(\tau)\,d\tau \right).$$

In the absence of physical nonlinearity ($f_1 = f_2 = 1$), it follows that

$$g(\varphi)=\frac{э_{12}^{(2)}-s_{12}/2G}{э_{12}^{(1)}-s_{12}/2G}=\frac{э_{12}^{(2)}-\varepsilon_{12}(0)}{э_{12}^{(1)}-\varepsilon_{12}(0)}.$$

The superscript is the curve number, $\varepsilon_{12}(0)$ is the instant deformation value. In the case considered we may accept for the zirconium alloy that $g(\varphi_0) = 1.18$.

## CALCULATION OF TEMPERATURE FIELDS IN A THREE-LAYER PLATE

At numerical research of analytical solutions of the boundary problems for three-layered elements of constructions found in thermal flows it is often necessary to know the temperature distribution through their thickness. In this connection, we must have an approximation solution of the corresponding problem of heat conductivity for a three-layer plate.

Let us consider an unconstrained three-layer plate of the total thickness $H$ exposed to a thermal flow with intensity $q$ falling perpendicularly to the outer plane $z = c + h_1$. The surface of the plate $z = - c - h_2$ is supposed to be heatproof.

Under above conditions of heat exchange the nonstationary unidimensional temperature field $\theta(z)$ is described by a differential equation of heat conductivity

$$\theta_{,zz}=\dot{\theta}/a_k \tag{11.28}$$

at the initial condition $\theta=0$ ($t = 0$) and conditions on the outer planes of the plate

$$\lambda_1\theta_{,z}=-q \text{ at } z=c+h_1,\ \theta_{,z}=0 \text{ at } z=-c-h_2. \tag{11.29}$$

Here, $a_k = \lambda_k/(C_k\rho_k)$—thermal diffusivity of the $k$-th layer; $\lambda_k$, $C_k$—heat conductivity and heat capacity coefficients; $\rho_k$—density of the material.

Since characteristics $\lambda_k$, $C_k$, $\rho_k$ change across thickness of the three-layer pack discontinuously, equation (11.28) should be solved inside each homogeneous area (layer) independently by setting additional conditions of heat exchange and temperature parity on the splices of the layers at exact statement of the problem on the temperature field. It is hard to solve this problem. To simplify the problem, we may average thermophysical parameters over thickness of the plate and bring it finally to the problem of the temperature field in a homogeneous plate with modified characteristics. Let us introduce parameter $a=\lambda/C$, where

$$\lambda=\sum_{k=1}^{3}\lambda_k h_k/H,\quad C=\sum_{k=1}^{3}C_k h_k/H.$$

The problem on finding the temperature field in the plate follows in this case from (11.28), (11.29) after substitution of $a_k$ for $a$ and $\lambda_1$ by $\lambda$.

Let us introduce a dimensionless coordinate $s = z/H$ and reduce time $\tau = at/H$ (Fourier[2] number). Then, the equation for heat conductivity becomes:

$$\theta_{,ss}=\theta_{,\tau}. \tag{11.30}$$

The initial conditions will be zero ($\theta=0$ at $\tau=0$). We should meet the condition on the outer planes of the plate

$$\frac{\lambda}{h}\theta,_s=-q \text{ at } s=(c+h_1)/H\ ;\ \theta,_s=0 \text{ at } s=-(c+h_2)/H\ . \tag{11.31}$$

To solve Eq. (11.30), we shall use the operational method based on Laplace transform. Determination of the transform is reduced in this case to solution of the equation

$$\theta^*,_{ss}-p\theta^*=0 \tag{11.32}$$

under the conditions following from (11.31):

$$\frac{\lambda}{H}\theta^*,_s=-\frac{q}{p} \text{ at } s=(c+h_1)/H\ ;\ \theta^*,_s=0 \text{ at } s=-(c+h_2)/H\ . \tag{11.33}$$

The general solution of equation (11.32) can be written in hyperbolic functions:

$$\theta^*=C_1\text{ch}\sqrt{p}s+C_2\text{sh}\sqrt{p}s\ . \tag{11.34}$$

Upon finding integration constants $C_1$, $C_2$ from conditions (11.33) and substituting their values into solution (11.34), we obtain a transform for the plate temperature:

$$\theta^*=A(p)/B(p), \tag{11.35}$$

where

$$A(p)=\frac{qH}{\lambda}\text{ch}\sqrt{p}\left(\frac{c+h_2}{H}+s\right),\ B(p)=p^2\frac{\text{sh}\sqrt{p}}{\sqrt{p}}\ .$$

The roots of equation $B(p)=0$ are the double root $p_1=0$ and roots $p_{n+1}=-\pi^2n^2$ ($n$=1, 2, ...) of the equation

$$\frac{\text{sh}\sqrt{p}}{\sqrt{p}}=\frac{\sin i\sqrt{p}}{\sqrt{p}}=0\ .$$

Based on the theorems on the inverse transform using the transform presented by the polynomial relations with account of multiplicity of the roots, we have from (11.35)

$$\theta=\frac{qH}{\lambda}\left\{\lim_{p\to 0}\left(\frac{d}{dp}\frac{\sqrt{p}\text{ch}\sqrt{p}\,(s+(c+h_2)/H)}{\text{sh}\sqrt{p}}e^{p\hat{o}}\right)\right.$$

$$\left.+\sum_{n=1}^{\infty}\frac{\text{ch}\sqrt{p_{n+1}}\,(s+(c+h_2)/H)}{\left[\frac{d}{dp}(p\sqrt{p}\text{sh}\sqrt{p})\right]_{p=p_{n+1}}}e^{p_{n+1}\hat{o}}\right\}. \tag{11.36}$$

By applying L'Hospital[3] rule for disclosing uncertainties, we transform the first item in braces:

$$\lim_{p\to 0}\frac{d}{dp}\left[\frac{\sqrt{p}\,\mathrm{ch}\sqrt{p}\left(s+\frac{c+h_2}{H}\right)}{\mathrm{sh}\sqrt{p}}e^{p\delta}\right]=\tau+\lim_{p\to 0}\frac{\left(s+\frac{c+h_2}{H}\right)\mathrm{sh}\sqrt{p}\left(s+\frac{c+h_2}{H}\right)}{2\mathrm{sh}\sqrt{p}}$$

$$+\lim_{p\to 0}\frac{\mathrm{ch}\sqrt{p}\left(s+\frac{c+h_2}{H}\right)\mathrm{sh}\sqrt{p}-\sqrt{p}\,\mathrm{ch}\sqrt{p}}{2\sqrt{p}\,\mathrm{sh}^2\sqrt{p}}=\tau+\frac{1}{2}\left(s+\frac{c+h_2}{H}\right)^2-\frac{1}{6}.$$

Upon substitution of this expression in (11.36) and taking into account that

$$\left[\frac{d}{dp}(p\sqrt{p}\,\mathrm{sh}\sqrt{p})\right]_{p=p_{n+1}}=\left(\frac{3}{2}\sqrt{p}\,\mathrm{sh}\sqrt{p}+\frac{1}{2}p\,\mathrm{ch}\sqrt{p}\right)_{p=p_{n+1}}=(-1)^{n+1}\frac{\pi^2n^2}{2},$$

$$\mathrm{ch}\sqrt{p_{n+1}}\left(s+\frac{c+h_2}{H}\right)=\cos\pi n\left(s+\frac{c+h_2}{H}\right),$$

we receive an expression for the sought temperature field

$$\theta=\frac{qH}{\lambda}\left\{\tau+\frac{1}{2}\left(s+\frac{c+h_2}{H}\right)^2-\frac{1}{6}-\frac{2}{\pi^2}\sum_{n=1}^{\infty}\frac{(-1)^n}{n^2}\cos\left[\pi n\left(s+\frac{c+h_2}{H}\right)\right]e^{-n^2\delta^2\delta}\right\}, \tag{11.37}$$

where

$$\tau=at/H^2;s=z/H;a=\lambda/C;\lambda=\lambda_kh_k/H;C=\rho_kC_kh_k/H.$$

Notice that the solution of (11.37) at $h_1=h_2=0$ coincides with the known one for a homogeneous plate.

We shall not consider here the nature of the heat current formation. Nevertheless, at a severe heating of the external (internal) surface of the plate in the presence of a liquid or gaseous flow round it, one of the probable reasons of the surface damage may be a hydrodynamic removal of the metal (ablation) untransformed yet into the liquid or gaseous state. The intensive ablation of the solid substance from the surface begins from the moment the dynamic head of the gas or liquid $\rho v^2/2$ reaches the yield point $\sigma_y$ of the heated metal surface. A key parameter describing ablation has been introduced in the monograph by Ilyushin and Ogibalov (1960)

$$\Gamma=\frac{\rho v^2}{2\sigma_y}.$$

A dangerous state may arise at $\Gamma\geq 1$.

In practice, ablation appears at the entrance of space vehicles into the atmosphere, in the rocket engine chambers, during laser effect on solids, in the barrels of artillery guns, and so forth.

The removal velocity of a substance from a considered shell surface is conditioned by the value of the heat flow intensity, dynamic head of the inflowing gas, and thermomechanical characteristics of the surface material. Ablation brings about changes in the temperature field of the plate and reduces thickness of the external layer.

## NOTES

1. *Makin M.J., Minter F.J.*, Acta Met., 8, 691 (1960).
2. *Fourier Jean Batiste Joseph* (1768–1830), French mathematician and physicist, worked in algebra, differential equations and mathematical physics.
3. *L'Hopital Guillaume Francois Antoine* (1661–1704), French mathematician, marquis, author of the first printed textbook in differential calculus based on *I. Bernoulli's* lectures (1667–1748). The latter being in fact the author of "L'Hopital's rule."

## KEYWORDS

- **Cordierite**
- **Neutron irradiation**
- **Polytetrafluoroethylene**
- **Thermoviscoelastoplastic**
- **Zirconium alloy**

# Chapter 12

## Dynamic Problems of the Elasticity Theory

The general static equations of the elasticity theory and corresponding boundary conditions as well as statement of the problem on the elasticity theory were dwelt upon in Ch. 3. Notice, however, that some effects on constructions bear an expressed dynamic character. Nevertheless, the resultant displacements are usually negligible, the velocity and especially acceleration may reach the values hazardous for the constructions. Seismic impacts, wind gusts, as well as various dynamic influences of technological origin: movement of trains, cranes, unbalanced parts of machines and mechanisms belong to such loads.

It is known from theoretical mechanics that accelerated or slowed down motion of masses generates the inertial forces influencing in their turn the elements of constructions in the way the static loads do. The dynamic loads result in most cases in vibrations while and periodic recurrence of low dynamic actions leads under certain conditions to accumulation of the system energy. The peak-to-peak value of vibrations and the intensity of inertial forces gradually increase up to very high values. This phenomenon, called *resonance*, is hazardous for the constructions resistant to usual static loads since their failure may occur even under slight affects.

### STATEMENT OF DYNAMIC PROBLEMS OF THE ELASTICITY THEORY

Based on *D'Alembert*[1] *principle* we may derive the equations for motion of an elastic body from equilibrium equation (3.1) by adding the forces of inertia to the current bulk forces $\rho F_i$:

$$\rho F_i \to \rho F_i - \rho \ddot{u}_i \,.$$

Hence, the equations of motion will be:

$$\sigma_{ij},_j + \rho(F_i - \ddot{u}_i) = 0 \quad (i, j = 1, 2, 3). \tag{12.1}$$

The rest equations (equilibrium, compatibility of deformations, Cauchy) (3.2)–(3.4) and boundary conditions (3.5), (3.7) are retained and we add the initial conditions to them

$$u_i = u_{0i}, \quad \dot{u}_i = \dot{u}_{0i}, \quad t = 0 \,. \tag{12.2}$$

Leaving aside the question on proving the existence and uniqueness of the solution, we should note that the motivation of uniqueness remains the same as for the static problem (see 3.5).

Based on Lame's equation (3.6), the equations of motion in displacements of a homogeneous isotropic elastic continuum are obtained similarly to (12.1)

$$\rho \ddot{u}_i = (\lambda + \mu)\theta_{,i} + \mu \Delta u_i + \rho F_i .$$

After formulation of the general statement of the dynamic problem of the theory of elasticity, we may consider the problems on free (natural) and forced (harmonic) vibrations, as well as the problems on progressive waves.

**Free (natural) vibrations.** Let an elastic body be free from the effect of external forces

$$F_i = 0, \sigma_{ij} l_j = 0 \text{ on } S_\sigma . \tag{12.3}$$

A part of surface $S_u$ can be fixed and its $u_i = 0$. We set the initial conditions (12.2), which initiate motion of the body by imparting the initial distribution of displacements and velocities.

**Forced harmonic vibrations.** In this case, volume forces $F_i$, surface forces $R_i$, and specified displacements of surface points $u_{0i}$ are periodic functions of time, such that

$$F_i = F_i^0 \varphi(t), \quad R_i = R_i^0 \varphi(t), \quad u_{0i} = u_i^0 \varphi(t) . \tag{12.4}$$

The values with "zero" superscript are independent of time, so for a typical representative of function $\varphi(t)$ in (12.4) we can accept

$$\phi(t) = \exp(ipt) = \cos pt + i \sin pt .$$

Actually, any periodic function can be presented by Fourier's series. Based on solution for one member of the series, we can use the principle of superposition for construction of a full solution.

## VARIATIONAL PRINCIPLE IN DYNAMICS

The variational equation for an elastic body at dynamic loads can be found using Lagrangian principle (4.5) by adding the force of inertia to mass forces:

$$\int_{S_\sigma} R_i \delta u_i \, \mathrm{d}S + \int_V \rho (F_i - \ddot{u}_i) \delta u_i \, \mathrm{d}V = \int_V \sigma_{ij} \delta \varepsilon_{ij} \, \mathrm{d}V . \tag{12.5}$$

Equation (12.5) expresses the *principle of virtual works in dynamics*. It is true for both elastic and inelastic bodies, as well as for the linear and nonlinear relations between stresses and strains. Introduction of Hooke's law constrains the principle of virtual works till the linearly elastic bodies. Introduction of strain energy $W$ (12.4) converts equation (12.5) into the form

$$\int_{S_\sigma} R_i \delta u_i \, \mathrm{d}S + \int_V \rho (F_i - \ddot{u}_i) \delta u_i \, \mathrm{d}V = \delta W . \tag{12.6}$$

It follows from equations (12.5), (12.6) that the virtual work of external forces and the forces of inertia is equal to the strain energy variation. Notice that in the case of mixed boundary conditions, the surface integral included into the variational equations is taken only on the part of surface $S_\sigma$ where stresses have been specified.

From the principle of virtual works (12.6), we can deduce the minimal principle for the field of displacements called *Hamilton's*[2] (Novackii, 1975) *principle*, according to which the transition of the system from one possible state into another at any time interval $[t_1, t_2]$ occurs so that the functional of the action takes a stationary value according to Hamilton, that is

$$\delta S = \int_{t_1}^{t_2} (\delta K - \delta W_v - \delta A)\mathrm{d}t = 0 ,$$

where $K$ is kinetic energy of the whole system, $W_v$—potential energy of deformation of the whole system, $A$—work of external forces.

## FREE VIBRATIONS OF ELASTIC BODIES

In the case of free (natural) vibrations (12.3), the solution of the equations of motion takes the form

$$u_k = U_k \exp(i\omega t) .$$

$U_k$ is here a function of coordinates only, but not of time. The components of strain and stress are represented similarly.

For convenience we shall denote the amplitudes of displacements and stresses as $u_i$, $\sigma_{ij}$, meaning that instead of $U_i$ we shall further write $u_i$. Then, after substitution of the sought solution into the equations of motion (12.1), we shall obtain a system of equations for amplitudes

$$\sigma_{ij},_j - \rho\omega^2 u_i = 0 \qquad (12.7)$$

The external forces and boundary conditions for free vibrations should be zeroth, so the multiplier $\exp(i\omega t)$ is reduced. Leaving aside the question on the initial conditions for the present, the relationship between the amplitudes of stresses and strains will keep the form of a usual Hooke's law

$$\sigma_{ij} = E_{ijkl}\, u_{k},_{l}.$$

The system of equations (12.7) for the homogeneous boundary conditions may have an obviously trivial solution $u_i \equiv 0,\ \sigma_{ij} \equiv 0$. However, at certain values of parameter $\omega = \omega_k$ a nonzero solution is also possible

$$u_i \equiv u_i^k, \quad \sigma_{ij} \equiv \sigma_{ij}^k .$$

The corresponding values of parameter $\omega_k$ are called natural *frequencies* of the elastic body, and functions $u_i^k$ determine *eigenmodes of vibrations*. Remember that the squares of natural frequencies will enter in (12.7), which will be retained in all further calculations, so root $\omega_k$ will always correspond to the second root $\omega_k$ equal in magnitude and opposite in sign. We shall not introduce a special numbering for these negative roots but we should remember that along with solution $u_i \exp(i\omega t)$, there is always another solution $u_i \exp(-i\omega t)$. This remark allows us to form real combinations from them, being the only ones having a mechanical sense.

The equations relating values $u_i^k$ and $\sigma_{ij}^k$ follow from (12.7):

$$\sigma_{ij,j}^k - \rho\omega_k^2 u_i^k = 0. \tag{12.8}$$

It is evident that owing to homogeneity of the system of equations and boundary conditions, the sought functions included in (12.8) are determined till a random multiplier.

Equations (12.1) can be treated as the equations of a static problem of the elasticity theory with mass forces $-\rho\omega^2 u_i$. Let $\omega = \omega_k$ be any of the natural frequencies. Then, $u_i = u_i^k$ is a displacement caused by the action of forces $P_i^k = -\rho\omega_k^2 u_i^k$ distributed over the volume. Similarly, forces $P_i^s = -\rho\omega_s^2 u_i^s$ cause displacements $u_i^s$ at frequency $\omega_s$. However, from Betti's theorem (4.4) follows that

$$\int_V P_i^s u_i^k \, dV = \int_V P_i^k u_i^s \, dV,$$

Or

$$\omega_s^2 \int_V \rho u_i^s u_i^k \, dV = \omega_k^2 \int_V \rho u_i^k u_i^s \, dV.$$

Since $\omega_s \neq \omega_k$ the latter equality is true if the integral is equal to zero. Hence,

$$\int_V \rho u_i^k u_i^s \, dV = 0 \quad (k \neq s).$$

This equality expresses a *property of orthogonality of eigenmodes* of vibrations. It follows from the condition of orthogonality, in particular, that frequencies $\omega_k$ are always valid.

To prove this, we shall on the contrary assume that $\omega_1 = \alpha + i\beta$. The equation for natural frequencies will have one more complex-conjugate root $\omega_2 = \alpha - i\beta$. The corresponding eigenmodes will be also complex-conjugate:

$$u_i^1 = p_i + iq_i, u_i^2 = p_i - iq_i.$$

From the condition of orthogonality follows

$$\int_V \rho u_i^1 u_i^2 \, dV = \int_V \rho (p_i^2 + q_i^2) dV = 0.$$

This equality is, however, possible only when $p_i \equiv 0$, $q_i \equiv 0$. Thus, there cannot be motions with complex frequencies in the linear elasticity theory. It is obvious that it also excludes the case of purely imaginary frequencies.

Since $u_i^k$ are determined only till a random constant multiplier, they can be normalized arbitrarily. It is usually accepted that

$$\int_V \rho u_i^k u_i^s \, \mathrm{d}V = \delta_{ks}, \tag{12.9}$$

where $\delta_{ks}$ are Kronecker's symbols. Relations (12.9) express the conditions of normalizing and simultaneously repeat the conditions of orthogonality of eigenmodes.

The principle of superposition (Fourier method) allows us to present a general expression of displacements at free vibrations of the elastic body in the form:

$$u_i = \sum_{k=1}^{\infty} (A_k \sin \omega_k t + B_k \cos \omega_k t) u_i^k (x_s) \cdot \tag{12.10}$$

Here, $A_k$, $B_k$ are indefinite constants. By differentiating (12.10) in time, we find

$$\dot{u}_i = \sum_{k=1}^{\infty} (A_k \omega_k \cos\omega_k t - B_k \omega_k \sin\omega_k t) u_i^k (x_s) \, .$$

After equating the values of displacements and velocities at $t = 0$ to their preset initial values $u_{0i}$ and $v_{0i}$, we shall have

$$\sum_{k=1}^{\infty} B_k u_i^k (x_s) = u_{0i}(x_s) \, , \quad \sum_{k=1}^{\infty} A_k \omega_k u_i^k (x_s) = v_{0i}(x_s) \, .$$

Then, each of these equalities is multiplied by $\rho u_i^k$ and integrated by volume. Proceeding from (12.9), only one member from each series remains in the left-hand side

$$B_k = \int_V \rho u_{0i} u_i^k \, \mathrm{d}V \, , \quad A_k = \frac{1}{\omega_k} \int_V \rho v_{0i} u_i^k \, \mathrm{d}V \, . \tag{12.11}$$

Notice that relations (12.11) do not presume any expansion of functions $u_{0i}$ and $v_{0i}$ in a series by the eigenmodes or fundamental functions $u_i^k$. The initial distribution of velocities in cannot general be continuous while and speaking about convergence it can be only the convergence on the average.

## FORCED VIBRATIONS OF ELASTIC BODIES

Let a body be affected by periodic forces with angular frequency $p$. For simplicity we shall accept mass forces to be absent ($F_i = 0$) and the external forces on the whole surface of the body will be presented as

$$R_i = R_i^0 \exp(ipt) \, .$$

There are no additional difficulties in considering a more general case. By taking displacements and stresses also proportional $\exp(ipt)$ and retaining notations $u_i$ and $\sigma_{ij}$ for the amplitudes of displacements and stresses, we derive from (12.1) the following equations:

$$\sigma_{ij},_j - \rho p^2 u_i = 0 \, . \tag{12.12}$$

The boundary conditions on surface $S_\sigma$ should be met

$$\sigma_{ij} l_j = R_i^0 . \tag{12.13}$$

Let us present the sought decision as

$$u_i = u_i' + u_i^0 , \sigma_{ij} = \sigma_i' + \sigma_{ij}^0 . \tag{12.14}$$

Here, $u_i^0$, $\sigma_{ij}^0$ is the solution of the static problem of the elasticity theory corresponding to the equilibrium equations

$$\sigma_{ij}^0,_j = 0,$$

the equations of coupling and boundary conditions (12.13). Then, upon substitution of (12.14) into (12.12), we find that the first part of solution $u'_i$, $\sigma'_{ij}$ meets the following equations of motion

$$\sigma_{ij}',_j - \rho p^2 u_i' = \rho p^2 u_i^0 \tag{12.15}$$

under the homogeneous boundary conditions.

We accept that:

$$u_i' = \sum_{k=1}^{\infty} a_k u_i^k , \sigma_{ij}' = \sum_{k=1}^{\infty} a_k \sigma_{ij}^k .$$

After substitution of these relations into (12.15) and accepting the derivatives from the amplitudes of stresses using (12.8), we have

$$\sum_{k=1}^{\infty} a_k \rho (\omega_k^2 - p^2) u_i^k = \rho p^2 u_i^0 .$$

If we multiply this equality by $u_i^m$ and integrate by volume under condition of orthonormal eigenfunctions (12.9), we come to

$$a_m (\omega_m^2 - p^2) = p^2 \int_V \rho u_i^0 u_i^m \, \mathrm{d}V .$$

where from,

$$a_m = \frac{p^2}{\omega_m^2 - p^2} \int_V \rho u_i^0 u_i^m \, \mathrm{d}V .$$

If $p = \omega_m$, then the frequency of the perturbing force will coincide with that of the natural frequencies of the elastic body and the corresponding factor in (12.5) will convert into infinity, that is, turn into a *resonance*.

In the case of a nonperiodic influence of external forces, to describe forced vibrations of an elastic body, we present the surface load and the sought solution as expansions in a series following the fundamental eigenfunctions. Substitution of these series into the equations of motion allows us to receive the equations for the unknown time function. This method will be described further on the example of a three-layer plate.

## RAYLEIGH'S INEQUALITY AND RITZ METHOD

Let us multiply both parts of equation (12.7) by $u_i$, integrate by volume and solve the received equality relative to $\omega^2$. Then,

$$\omega^2 = \frac{\int_V \sigma_{ij},_j u_i \, \mathrm{d}V}{\int_V \rho u_i^2 \, \mathrm{d}V}. \tag{12.16}$$

After transforming the numerator by integration in parts at homogeneous boundary conditions (the integral on the surface is zero), we obtain

$$\omega^2 = \frac{\int_V E_{ijkl} \varepsilon_{ij} \varepsilon_{kl} \, \mathrm{d}V}{\int_V \rho u_i^2 \, \mathrm{d}V}. \tag{12.17}$$

If $u_i = u_i^k$, equation (12.7) is fulfilled at $\omega = \omega_k$ and formula (12.16) or (12.17) will give an exact value of a squared natural frequency with number $k$. If $u_i$ are random functions, equation (12.7) cannot be satisfied, and formula (12.17) determines some number $\omega^2$, which does not present, generally speaking, the squared frequency of any free vibrations of the system.

We may prove that the functional appearing in the right-hand side of equations (12.16), (12.17) gives us an estimate of at least the least natural frequency. We shall number the natural frequencies in the ascending order

$$\omega_1 < \omega_2 < \omega_3 < \ldots$$

Let us choose an arbitrary system of three differentiated and continuous functions satisfying the kinematic boundary conditions for $u_i$. and decompose them into a series by the system of eigenfunctions

$$u_i = \sum a_k u_i^k.$$

Proceeding from (12.8),

$$\sigma_{ij},_j = \sum a_k \sigma_{ij}^k,_j = \rho \sum a_k \omega_k^2 u_i^k,$$

then, if we multiply by $u_i = \sum a_m u_i^m$ and integrate we shall have

$$\int \sigma_{ij},_j u_i dV = \sum \sum a_k a_m \omega_k^2 \int \rho u_i^k u_i^m dV = \sum a_k^2 \omega_k^2 .$$

Similarly, in view of orthogonality of the principal forms,

$$\int \rho u_i^2 \, \mathrm{d}V = \sum a_k^2.$$

Now, the relation (12.16) can be rewritten as:

$$\omega^2 = \frac{a_1^2\omega_1^2 + a_2^2\omega_2^2 + \ldots}{a_1^2 + a_2^2 + \ldots},$$

or

$$\omega^2 = \omega_1^2 \frac{a_1^2 + a_2^2(\omega_2/\omega_1)^2 + \ldots}{a_1^2 + a_2^2 + \ldots}.$$

Each member of the series in the numerator is not less then the corresponding member of the series in the denominator since $(\omega_k/\omega_1)^2 > 1$, so $\omega_1^2 \leq \omega^2$ and (12.17) can be substituted by inequality

$$\omega_1^2 \leq \widetilde{U}/\widetilde{T}. \tag{12.18}$$

where $\widetilde{U}$—elastic potential calculated for a specified system of displacements $u_i$; $\widetilde{T}$—expression for kinetic energy in which the velocities are substituted for displacements $u_i$.

Rayleigh's inequality (12.18) gives us the upper estimation for the lowest frequency of vibrations of the elastic body. If functions $u_i$ contain a certain number of indefinite parameters $c_s$, then $\widetilde{U} = \widetilde{U}(c_s)$ and $\widetilde{T} = \widetilde{T}(c_s)$. The best approximation for $\omega_1^2$ will be $c_s$ values minimizing the fraction in the right-hand side of (12.18), therefore it should be that

$$\frac{\partial}{\partial c_s}\left(\frac{\widetilde{U}}{\widetilde{T}}\right) = 0.$$

Where from,

$$\frac{1}{\widetilde{T}^2}\left(\frac{\partial \widetilde{U}}{\partial c_s}\widetilde{T} - \widetilde{U}\frac{\partial \widetilde{T}}{\partial c_s}\right) = 0$$

or assuming that $\widetilde{U}/\widetilde{T} = \omega^2$,

$$\frac{\partial \widetilde{U}}{\partial c_s} - \omega^2 \frac{\partial \widetilde{T}}{\partial c_s} = 0. \tag{12.19}$$

The simplest result is obtained when parameters $c_s$ are included into relation $u_i$ linearly, namely:

$$u_i = \sum_{s=1}^{\infty} c_s f_i^s. \tag{12.20}$$

Equation (12.19) is linear and homogeneous, therefore to satisfy a nontrivial solution, we should make the determinant of the system equal to zero. This condition results in an algebraic equation of the $k$th power about $\omega^2$. In terms of Rayleigh's inequality, the least root of the equation will give the upper estimation for $\omega^2$, which only improves with $k$ increase. With increasing $k$, the root of the equation with number $m$ will tend to $\omega_m^2$, and we cannot say for sure whether from top or from bottom. We do not present the proof for this theorem here, since that to satisfy it, the system of functions $f_i^s$ is to be complete, that is, there is a possibility of presenting any admissible system of displacements $u_i$ in the form of (12.20). Above-described approximated procedure of determining frequencies is known as *Ritz method.*

## VIBRATIONS OF AN ELASTIC CIRCULAR THREE-LAYER PLATE

Let us consider axisymmetric transverse vibrations of an elastic three-layer plate of a circular shape asymmetric through thickness (Fig. 6.11). The statement of the problem and its solution is presented similarly to statics (6.14) in cylindrical coordinates $r$, $\varphi$, $z$. However, the filler is light in this case, so we neglect its work in tangential direction [item $2cG_3\psi$ in the second equation of system (6.62)]. The external vertical load is $q = q(r, t)$. There is a rigid diaphragm on the plate contour that hinders relative shear of the layers. By a virtue of the problem symmetry, the tangential displacements in the layers are absent [$u_\varphi^{\ k} = 0$ ($k$—number of the layer)], and deflection of the plate $w$, relative shear in the filler $\psi$ and radial displacement of the coordinate surface $u$ are independent of coordinate $\varphi$, that is, $u(r, t)$, $\psi(r, t)$, $u(r, t)$. These functions will be considered further as the sought ones. All displacements and linear values of the plate are related to its radius $r_0$, force characteristics—to 1 Pa, thickness of the $k$th layer is denoted as $h_k$.

The equations of transverse vibrations of the plate without account of rotation inertia of the normal in the layers can be deduced from the equilibrium equations by adding the forces of inertia $-M_0\ddot{w}$ to the external load, where

$$M_0 = (\rho_1 h_1 + \rho_2 h_2 + \rho_3 h_3) r_0^{\ 2},$$

$\rho_k$—density of the material of the $k$th layer.

As a result, we have a system of differential equations in the partial derivatives describing forced transversal vibrations of the circular three-layer plate:

$$\begin{aligned}
&\mathrm{L}_2(a_1 u + a_2\psi - a_3 w,_r) = 0\,,\\
&\mathrm{L}_2(a_2 u + a_4\psi - a_5 w,_r) = 0\,,\\
&\mathrm{L}_3(a_3 u + a_5\psi - a_6 w,_r) - M_0\ddot{w} = -q\,, \qquad (12.21)
\end{aligned}$$

where factors $a_i$ and differential operators $\mathrm{L}_2$, $\mathrm{L}_3$ are found from equation (6.63).

The problem of functions $u(r, t)$, $\psi(r, t)$, $u(r, t)$ is closed by adding boundary and initial conditions to (12.21)

$$w(r,0) \equiv f(r), \quad \dot{w}(r,0) \equiv g(r)\,. \qquad (12.22)$$

Let us take first a homogeneous system of differential equations describing free vibrations of a plate. It follows from (12.21) at $q = 0$ that

$$\begin{aligned}
&\mathrm{L}_2(a_1 u + a_2\psi - a_3 w,_r) = 0\,,\\
&\mathrm{L}_2(a_2 u + a_4\psi - a_5 w,_r) = 0\,,
\end{aligned}$$

$$L_3(a_3u+a_5\psi-a_6w,_r)-M_0\ddot{w}=0 .$$

Using the first two equations and double integration, this system is transformed into the form

$$u=b_1w,_r+C_1r+C_2/r ,$$

$$\psi=b_2w,_r+C_3r+C_4/r ,$$

$$L_3(w,_r)+M^4\ddot{w}=0 . \quad (12.23)$$

Here,

$$b_1=\frac{a_3a_4-a_2a_5}{a_1a_4-a_2^2} , \quad b_2=\frac{a_1a_5-a_2a_3}{a_1a_4-a_2^2} ,$$

$$M^4=\frac{M_0a_1(a_1a_4-a_2^2)}{(a_1a_6-a_3^2)(a_1a_4-a_2^2)-(a_1a_5-a_2a_3)^2} . \quad (12.24)$$

In view of continuity of the anticipated solution, we should take $C_2 = C_4 = 0$ at the beginning of coordinates for the continuous plates.

The sought deflection is accepted as

$$w(r,t)=v(r)(A\cos\omega t+B\sin\omega t) , \quad (12.25)$$

where $v(r)$ - unknown coordinate function, ω—frequency of the natural vibrations of the considered plate, $A$ and $B$ are the integration constants determined from the initial conditions.

After substitution of equation (12.25) into the last equation of system (12.23), we receive Bessel's equation for coordinate function $v(r)$:

$$L_3(v,_r)-\beta^4v=0 . \quad (12.26)$$

In the decomposed form

$$v,_{rrrr}+\frac{2}{r}v,_{rrr}-\frac{1}{r^2}v,_{rr}+\frac{1}{r^3}v,_r-\beta^4v=0 .$$

The notation accepted here

$$\beta^4=M^4\omega^2. \quad (12.27)$$

The solution of equation (12.26) can be presented as [29]

$$v(\beta r)=C_5J_0(\beta r)+C_6I_0(\beta r)+C_7Y_0(\beta r)+C_8K_0(\beta r) , \quad (12.28)$$

where $J_0$, $Y_0$ are Bessel's functions of zeroth order of the first and second kinds, accordingly; $I_0$, $K_0$—Bessel's modified function and McDonald's function of zeroth orders; $C_5$, ..., $C_8$—integration constants.

Leaving aside the description of specified functions, notice that $Y_0(\beta r)$ and $K_0(\beta r)$ have a feature like a logarithm in the beginning of coordinates, namely, in the center of the plate (Korn and Korn, 1996, Zubchaninov, 1990). Therefore, similarly to the previous cases, we take $C_7 = C_8 = 0$ for the continuous plates. Remember that for the plates with a hole, the constants $C_2$, $C_4$, $C_7$, $C_8$ are determined from additional conditions over the internal contour.

If the edge of the plate is pinched, the requirements at $r = 1$ should be met

$$u = \psi = w = w,_r = 0.$$

After substitution of solution (12.25) into the last two conditions and taking account of coordinate function (12.28), we have

$$C_5 J_0(\beta)+C_6 I_0(\beta)=0 \; ,$$
$$-C_5 J_1(\beta)+C_6 I_1(\beta)=0, \tag{12.29}$$

where $J_1$, $Y_1$ are specified earlier Bessel's functions of the first order.

The homogeneous system of equation (12.29) has a nontrivial solution for the integration constants $C_5$, $C_6$, providing its determinant is equal to zero.

Hence,

$$I_1(\beta)J_0(\beta)+I_0(\beta)J_1(\beta)=0. \tag{12.30}$$

Above transcendental equation (12.30) gives eigenvalues $\beta_n$ ($n = 0, 1, 2, \ldots$). After calculations, the frequencies of natural vibrations follow from equation (12.27) with account of (12.24):

$$\omega_n^2 = \frac{\beta_n^4}{M^4}. \tag{12.31}$$

We should underline that equation (12.30) coincides with a similar equation for a circular single-layer plate pinched over its contour.

At a simply supported plate contour and the presence of a rigid diaphragm on it, the condition $u = \psi = w = M_r = 0$ is satisfied. Then, the equation for eigenvalues is:

$$J_0(\beta)[a_7(\beta I_0(\beta)-I_1(\beta)\ +a_8 I_1(\beta)]+I_0(\beta)[a_7(\beta J_0(\beta)-J_1(\beta)\ +a_8 J_1(\beta)]=0. \tag{12.32}$$

Where

$$a_7 = a_6 - a_3 b_1 - a_5 b_2\ ,\quad a_8 = a_{60} + a_3 b_1 + a_5 b_2,\ a_{60} = a_6\left\{K_k^- \rightarrow K_k^+\right\}.$$

After solving equation (12.32) we may find the frequencies of natural vibrations by (12.31). A transcendental equation for eigenvalues and a free-supported contour is built similarly.

In a general case, to describe a deflection of a circular three-layer plate pinched over the boundary by one of above-named means, we should use at transverse vibrations *a system of orthonormal eigenfunctions* $v_n \equiv v(\beta_n, r)$:

$$v_n \equiv \frac{1}{d_n}\left[J_0(\beta_n r) - \frac{J_0(\beta_n)}{I_0(\beta_n)} I_0(\beta_n r)\right]. \tag{12.33}$$

We take into account here the condition $w = 0$ at $r = 1$ following from the relation between integration constants

$$C_6 = -C_5 J_0(\beta)/I_0(\beta).$$

The constants $d_n$ are determined from the requirement of normalizing:

$$d_n^2 = \int_0^1 \left[J_0(\beta_n r) - \frac{J_0(\beta_n)}{I_0(\beta_n)} I_0(\beta_n r)\right]^2 r\,\mathrm{d}r = \frac{1}{2}\left[J_0^2(\beta_n) + J_1^2(\beta_n)\right]$$

$$-\frac{J_0(\beta_n)}{\beta_n I_0(\beta_n)}\left[J_1(\beta_n) I_0(\beta_n) + J_0(\beta_n) I_1(\beta_n)\right] + \frac{J_0^2(\beta_n)}{2 I_0^2(\beta_n)}\left[I_0^2(\beta_n) - I_1^2(\beta_n)\right].$$

Finally, the sought deflection is presented by decomposition into a series based on the fundamental system of orthonormal eigenfunctions (12.33):

$$w(r,t) = \sum_{n=0}^{\infty} v_n (A_n \cos\omega_n t + B_n \sin\omega_n t)\cdot \tag{12.34}$$

The radial displacement and relative shear we obtain by using the first two equations from (12.23) and the boundary condition on the contour $\psi(1, t) = u(1, t) = 0$:

$$u(r,t) = b_1 \sum_{n=0}^{\infty} \varphi_n (A_n \cos\omega_n t + B_n \sin\omega_n t),$$

$$\psi(r,t) = b_2 \sum_{n=0}^{\infty} \varphi_n (A_n \cos\omega_n t + B_n \sin\omega_n t). \tag{12.35}$$

Here, $\varphi_n \equiv \varphi_n(\beta_n, r)$:

$$\varphi_n = \frac{\beta_n}{d_n}\left[J_1(\beta_n) r - J_1(\beta_n r) + \frac{J_0(\beta_n)}{I_0(\beta_n)}(I_1(\beta_n) r - I_1(\beta_n r))\right]. \tag{12.36}$$

Factors $A_n$, $B_n$ in formulas (12.34), (12.35) follow from the initial conditions of motion (12.22)

$$A_n = \int_0^1 f(r) v_n r\,\mathrm{d}r, \quad B_n = \frac{1}{\omega_n}\int_0^1 g(r) v_n r\,\mathrm{d}r. \tag{12.37}$$

All the functions are assumed to have a continuity of derivatives till the sixth order inclusive.

To describe forced vibrations of the considered plate, the external load $q(r, t)$ and the sought solution $u(r, t)$, $\psi(r, t)$ and $w(r, t)$ are represented as the following decompositions into series based on the systems of functions (12.33), (12.36):

$$q(r,t)=M_0\sum_{n=0}^{\infty}v_n q_n(t),\quad w(r,t)=\sum_{n=0}^{\infty}v_n T_n(t),$$

$$\psi(r,t)=b_2\sum_{n=0}^{\infty}\varphi_n T_n(t),\quad u(r,t)=b_1\sum_{n=0}^{\infty}\varphi_n T_n(t). \tag{12.38}$$

A uniform convergence of the series (12.38) is provided by the completeness of the used system of fundamental functions. The expressions for functions $q_n(t)$ are derived by multiplying the first relation from (12.38) by r $v_n$ and integrating over the plate radius:

$$\int_0^1 q(r,t)v_n r\,dr = M_0\int_0^1\sum_{m=0}^{\infty}v_m q_m(t)v_n r\,dr = M_0\sum_{m=0}^{\infty}q_m(t)\int_0^1 v_m v_n r\,dr.$$

In view of the orthonormal property of the systems of eigenfunctions $v_n$, we have

$$q_n(t)=\frac{1}{M_0}\int_0^1 q(r,t)v_n r\,dr,$$

$$\int_0^1 v_m v_n r\,dr=\begin{cases}1, & m=n\\ 0, & m\neq n\end{cases}.$$

The equation for the unknown function $T_n(t)$ can be found in this case from the third equation of system (12.21) after substitution of equation (12.38) and using a linear relation of functions $v_n$, $\varphi_n$:

$$\ddot{T}_n+\omega_n^2 T_n=q_n. \tag{12.39}$$

A general solution of equation (12.39) can be accepted as

$$T_n(t)=A_n\cos\omega_n t+B_n\sin\omega_n t+\frac{1}{\omega_n}\int_0^t \sin\omega_n(t-\tau)q_n(\tau)\,d\tau. \tag{12.40}$$

Thus, the parameters of the forced transverse vibrations of the three-layer plate are determined by equations (12.38), (12.40). Factors $A_n$, $B_n$ are found from formulas (12.37) since the integrated summand in (12.40) and its derivative at the initial moment turns into zero.

The transcendental equation for eigenvalues (12.30) is independent of the geometrical and elastic characteristics of the materials of the layers, which was numerically confirmed over the interval of numerical axis 0–50. The first 15 roots were calculated till 0.001 accuracy and are given in Table 12.1. The first four of them coincide with the known in literature roots for the pinched single-layer plates. It is seen from the other values that distribution of eigenvalues within a specified interval is quite uniform.

**Table 12.1.**

| Number n | Eigenvalue $\beta_n$ | Number n | Eigenvalue $\beta_n$ |
|---|---|---|---|
| 0 | 3.196 | 8 | 28.279 |
| 1 | 6.306 | 9 | 31.420 |
| 2 | 9.439 | 10 | 34.561 |
| 3 | 12.577 | 11 | 37.702 |
| 4 | 15.716 | 12 | 40.844 |
| 5 | 18.857 | 13 | 43.985 |
| 6 | 21.997 | 14 | 47.126 |
| 7 | 25.138 | | |

The dependence for the first two frequencies of natural vibrations of a pinched over its contour plate with the materials of the layers D-16T–fluoroplastic–D-16T versus thickness of the external layer ($h_2 = 0.02$, $h_3 = 0.05$) and thickness of the filler ($h_1 = h_2 = 0.02$) is shown in Fig. 12.1a, and 12.1b, accordingly, curve *1* – $\omega_0$, *2* – $\omega_1$. In the first case, the frequencies first decrease and then start to grow. In the second case, the frequencies with incremented thickness relative to the soft filler are decreasing continuously, which proves drop of rigidity of the plate.

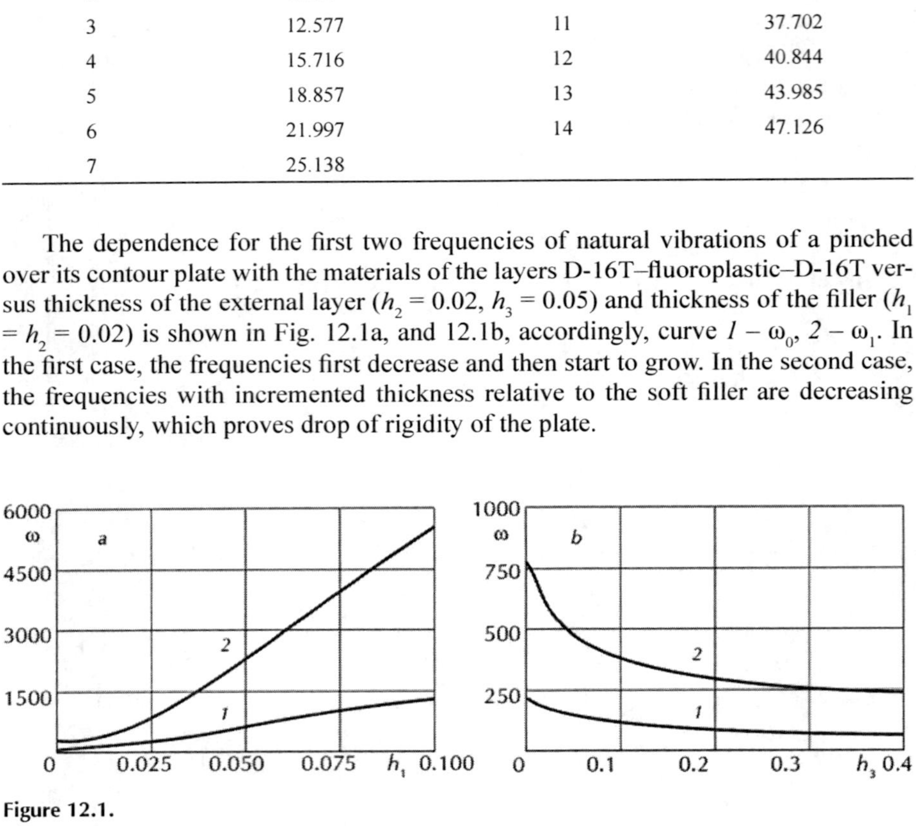


**Figure 12.1.**

The equation for eigenvalues (12.32) corresponding to a hingidly supported plate contour contains rigidity parameters of the layers numerically proved on a pack D-16T–fluoroplastics–D-16T at $h_1 = h_2 = 0.02$, $h_3 = 0.05$. The first 15 roots calculated with up to 0.001 accuracy are given in Table 12.2. Their values are somewhat lower than the pinched three-layer plate has, which is a confirmation to a decreasing rigidity at a different type of fixing over the contour.

**Table 12.2.**

| Number n | Eigenvalue $\beta_n$ | Number n | Eigenvalue $\beta_n$ |
|---|---|---|---|
| 0 | 3,141 | 8 | 27,950 |
| 1 | 6.203 | 9 | 31.069 |
| 2 | 9.293 | 10 | 34.191 |
| 3 | 12.392 | 11 | 37.314 |
| 4 | 15.497 | 12 | 40.439 |
| 5 | 18.605 | 13 | 43.565 |
| 6 | 21.717 | 14 | 47.692 |
| 7 | 24.832 | | |

Figure 12.2 shows the dependence of eigenvalues $\beta_0$ – *1* and corresponding natural frequencies $\omega_0$ – *2* versus thickness of the bearing layer. We should underline differences in the reactions of these parameters: growth of $h_1$ increases rigidity of the plate and the frequencies but decrease eigenvalues.

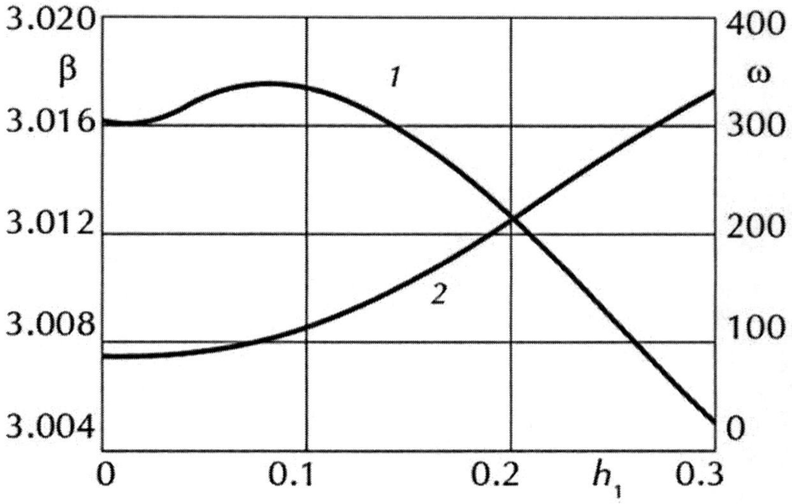


**Figure 12.2.**

Figure 12.3 illustrates the reduction of the first two natural frequencies of the considered circular three-layer plate (*1* – $\omega_0$, *2* – $\omega_1$) with incrementing thickness of the filler. The character of this variation of the curves is similar to the case with pinching over the contour (Fig. 12.1b).

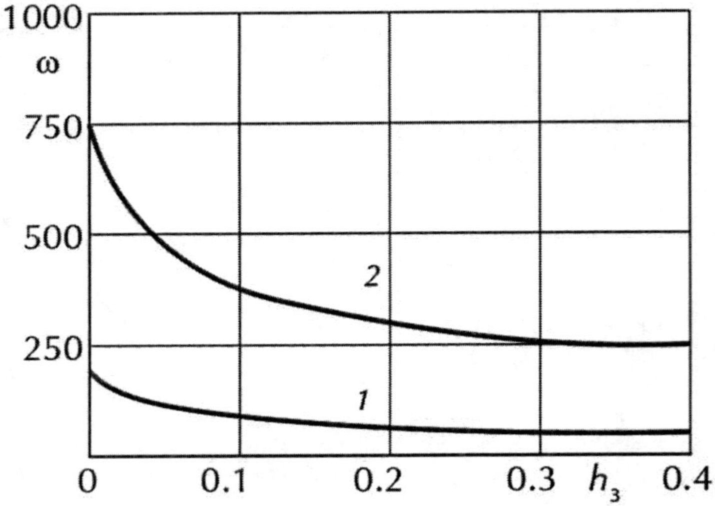


Figure 12.3.

## RESONANCE VIBRATIONS OF A CIRCULAR THREE-LAYERED PLATE

We shall consider forced resonant vibrations of the elastic circular three-layer plates under the action of harmonic resonant loads, that is, the loads whose frequency coincides with one of eigenfrequencies of the system.

**The load uniformly distributed over the entire surface.** Let a harmonic resonant distributed load be acting on the entire external surface of the bearing layer of the considered three-layer plate

$$q(r,t) = q_0(D\cos(\omega_k t) + E\sin(\omega_k t) \ . \tag{12.41}$$

The frequency of the external perturbing force $\omega_k$ coincides with one of the eigenfrequencies $\omega_n$ of the plate, $t$—time, $q_0$, $D$, $E$, $k$—preset loading parameters.

The solution of the corresponding initially boundary problem $u(r, t)$, $\psi(r, t)$, $w(r, t)$ and the load (12.41) can be presented like previously (12.38) by expansion into a series over the orthonormal eigenfunctions $v_n$ (12.33):

$$u(r,t) = b_1 \sum_{n=0}^{\infty} \varphi_n T_n(t) \quad , \quad \psi(r,t) = b_2 \sum_{n=0}^{\infty} \varphi_n T_n(t) \quad ,$$

$$w(r,t) = \sum_{n=0}^{\infty} v_n T_n(t) \quad , \quad q(r,t) = M_0 \sum_{n=0}^{\infty} v_n q_n(t) \quad ,$$

$$q_n(t) = \frac{1}{M_0}\int_0^1 q(r,t) v_n r \mathrm{d}r \ , \quad v_n \equiv \frac{1}{d_n}\left[ J_0(\beta_n r) - \frac{J_0(\beta_n)}{I_0(\beta_n)} I_0(\beta_n r) \right] . \tag{12.42}$$

The parameters $q_n(t)$ corresponding to the $n$th harmonic (12.41) of expanding into a series over eigenfunctions $v_n$:

$$q_n(t)=\frac{1}{M_0}\int_0^1 q(r,t)v_n r\mathrm{d}r = D_n\cos(\omega_k t)+E_n\sin(\omega_k t), \tag{12.43}$$

$$D_n=\frac{Dq_0}{M_0 d_n \beta_n}\left[J_1(\beta_n)-\frac{J_0(\beta_n)}{I_0(\beta_n)}I_1(\beta_n)\right],$$

$$E_n=\frac{Eq_0}{M_0 d_n \beta_n}\left[J_1(\beta_n)-\frac{J_0(\beta_n)}{I_0(\beta_n)}I_1(\beta_n)\right].$$

The differential equation (12.39) for determining unknown time function $T_n(t)$ with account of (12.43) will be in our case:

$$\ddot{T}_n(t)+\omega_n^2 T_n(t) = D_n\cos(\omega_k t)+E_n\sin(\omega_k t). \tag{12.44}$$

Its solution can be presented in the form

$$T_n(t)=A_n\cos(\omega_n t)+B_n\sin(\omega_n t)+y_n(t), \tag{12.45}$$

as a partial solution (12.44) we can take

$$y_n(t)=\begin{cases}\dfrac{D_n}{\omega_n^2-\omega_k^2}\cos(\omega_k t)+\dfrac{E_n}{\omega_n^2-\omega_k^2}\sin(\omega_k t) & n\neq k,\\ -\dfrac{E_k}{2\omega_k}t\cos(\omega_k t)+\dfrac{D_k}{2\omega_k}t\sin(\omega_k t) & n=k.\end{cases}$$

Integration constants are calculated from the initial conditions of motion (12.22)

$$w(r,0)\equiv f(r),\quad \dot{w}(r,0)\equiv g(r).$$

If we substitute here the relation for bending (12.42) with account of a time function (12.45) and orthonormal property of the fundamental system, $v_n$, we shall finally have

$$A_n=\int_0^1 f(r)v_n r dr-\begin{cases}\dfrac{D_n}{\omega_n^2-\omega_k^2}, & n\neq k,\\ 0, & n=k,\end{cases}$$

$$B_n=\frac{1}{\omega_n}\left[\int_0^1 g(r)v_n r dr-\begin{cases}\dfrac{\omega_k E_n}{\omega_n^2-\omega_k^2}, & n\neq k\\ -\dfrac{E_k}{2\omega_k}, & n=k\end{cases}\right]. \tag{12.46}$$

Hence, the forced transversal vibrations of the three-layer circular plate under a resonant load (12.41) are described by the displacements (12.42) in which time function $T_n(t)$ and integration constants are found from equations (12.45), (12.46). All displacements and linear dimensions of the plate are related to its radius $r_0$.

The numerical calculations here and further in this chapter are made for a plate of a unit radius pinched over its contour and having the layers made of D16T-fluoroplastic-D16T materials. The natural frequencies $\omega_n$ were calculated by the formula (12.31) using eigenvalues from Table 12.1 and parameters of the layers: $h_1 = h_2 = 0.01$, $c = 0.05$. The initial conditions of motion were assumed homogeneous $w(r,0) \equiv \dot{w}(r,0) \equiv 0$ , which simplifies integration constants $A_n$, $B_n$ according to (12.46) and makes possible to accept that

$$A_n = -\begin{cases} \dfrac{D_n}{\omega_n^2 - \omega_k^2}, & n \neq k, \\ 0, & n = k, \end{cases} \quad B_n = -\frac{1}{\omega_n}\left[\begin{cases} \dfrac{\omega_k E_n}{\omega_n^2 - \omega_k^2}, & n \neq k \\ -\dfrac{E_k}{2\omega_k}, & n = k \end{cases}\right]. \tag{12.47}$$

Figure 12.4 shows that the amplitude of the resonant vibrations (flexure in the center of the plate) grows with time under the external load frequency $\omega_k$ that coincides with one of the frequencies of natural vibrations: (*a*) – $\omega_k = \omega_0 = 94$ c$^{-1}$, (*b*) – $\omega_k = \omega_1 = 368$ c$^{-1}$, (*c*) – $\omega_k = \omega_2 = 824$ c$^{-1}$, (*d*) – $\omega_k = \omega_3 = 1463$ c$^{-1}$.

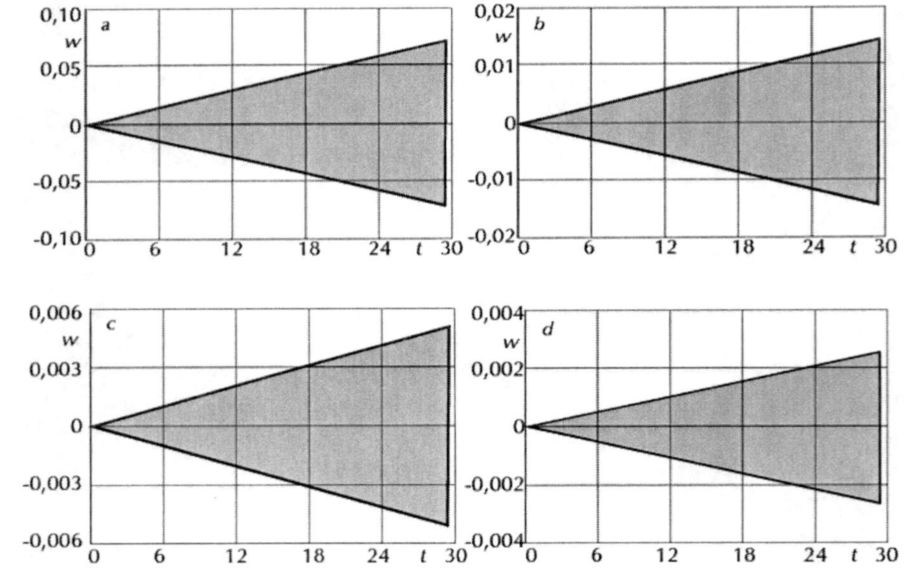


**Figure 12.4.**

The amplitude of oscillations grows with time and the higher the natural frequency at which the frequency of the perturbing force coincides, the lower the gain velocity of the amplitude per accepted time interval. For instance, the flexure (*a*) ratio to the maximal flexures presented in Fig. 12.4b–d makes up roughly 5, 13 and 27 times. The amplitude of a uniformly distributed load is $q_0 = 50$ Pa. The number of resonant vibrations of the plate in a given time interval is rather large, therefore the process illustrated in Fig. 12.4 is graphically indiscernible.

Figure 12.5 is devoid of this drawback due to a 200-fold times reduced interval. Curve *1* complies with the case $\omega_k = \omega_0$, *2* – $\omega_k = \omega_1$. We may also the observe intensification of the resonant amplitude with a noticeable advance in the first case.

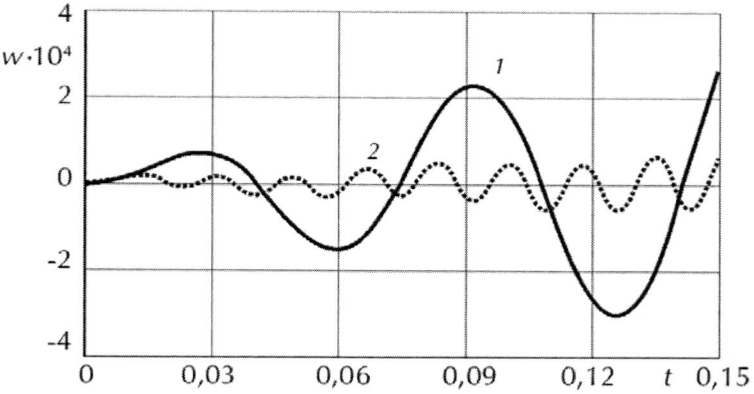


**Figure 12.5.**

Figure 12.6 illustrates convergence of the series (12.42) in calculations of the plate bending (*a*) and relative shear in the filler (*b*) under $q_0 = 10^4$ Pa loading. Curve *1* corresponds to a single term of the series ($n = 0$), *2* – $n = 5$, *3* – $n = 14$. Curves 2 and 3 are overlapping.

Further increment in the number of terms does not in fact change the sum of the series. This is why, we have restricted to the first six summands. We shall now consider some examples of the local axisymmetric resonant action on the three-layer circular plate. The analytical recording of the load will be made like previously using the Heaviside function $H_0(x)$ (6.74).

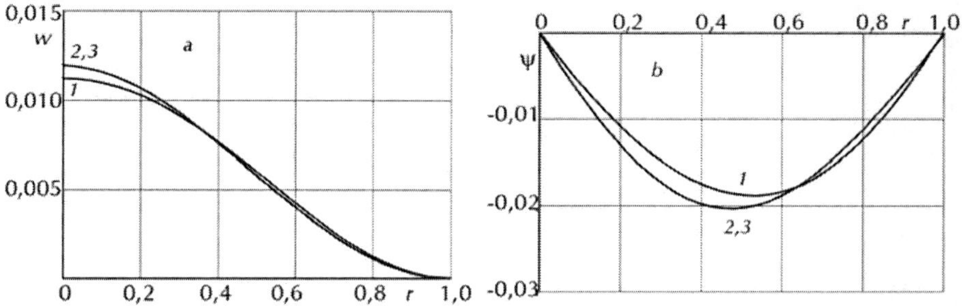


**Figure 12.6.**

**The load uniformly distributed over a circle [0, *b*].** Let us assume that a local resonant harmonic load is exerted on the surface of the plate under study, which is evenly distributed over a circle of a relative radius $b \leq 1$. Then, imposing this restriction on equation (12.41), we obtain:

$$q(r,t) = q_0 H_0 (b-r)\big(D\cos(\omega_k t) + E\sin(\omega_k t)\big). \qquad (12.48)$$

Some integrals contained in Bessel's and Heaviside's functions have been calculated and are presented in the Appendix (see Ch. 14.4). Correspondingly, the parameters of the load expanding (12.48) into a series over eigenfunctions are

$$D_n = \frac{q_0 Db}{M_0 d_n \beta_n}\left(J_1(\beta_n b) - \frac{J_0(\beta_n)}{I_0(\beta_n)} I_1(\beta_n b)\right),$$

$$E_n = \frac{q_0 Eb}{M_0 d_n \beta_n}\left(J_1(\beta_n b) - \frac{J_0(\beta_n)}{I_0(\beta_n)} I_1(\beta_n b)\right). \qquad (12.49)$$

Thus, the time function is found from relation (12.45), where integration constants $A_n$, $B_n$ are calculated by equation (12.47) with account of (12.49). The resonant load at $b = 1$ is distributed over the total surface of the plate, so the solution coincides with the previous one.

Figure 12.7 presents the dependence of the central flexure of the plate on the radius of the resonant load spot at the intensity amplitude $q_0 = 10^4$ Pa. These curves were calculated for moments $t_k = \pi/\omega_k$ corresponding to different resonant frequency values: *1* – $\omega_k = \omega_0$, *2* – $\omega_k = \omega_1$, *3* – $\omega_k = \omega_2$, *4* – $\omega_k = \omega_3$. Notice that the maximal flexure is observed at the resonance of the eigentone $\omega_0$ frequency, in which case the loading spot radius increase brings about a nonlinear growth of the central flexure of the plate. The rest flexures are by far less.

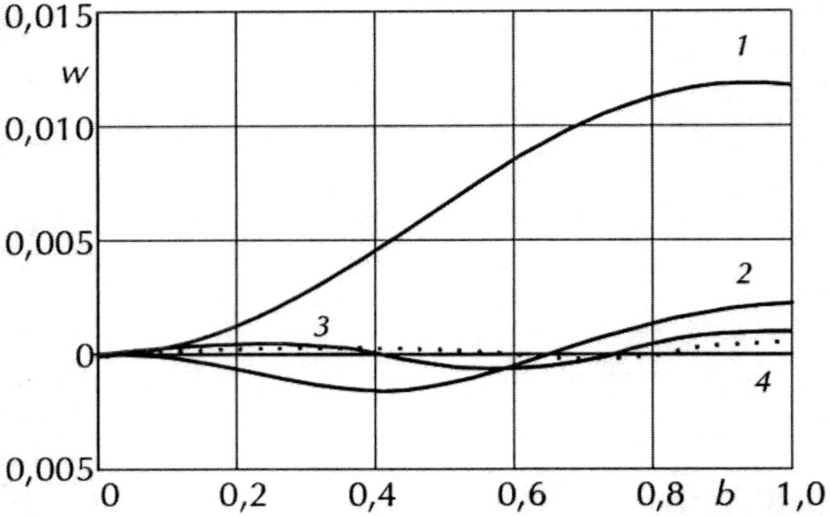


**Figure 12.7.**

Figure 12.8 shows variations in shears of the filler (*a*), (*c*) and flexures (*b*), (*d*) over the radius of the circular plate. Figures 12.8a and 12.8b correspond to moment $t_1 = \pi/\omega_0$ at the perturbing load frequency coinciding with that of the eigentone $\omega_k = \omega_0$, (*c*) and (*d*) – $t_2 = \pi/\omega_1$ at $\omega_k = \omega_1$. The number of the curves indicates the spot radius of the distributed intensity load $q_0 = 10^4$ Pa: *1* – $b = 0.25$, *2* – $b = 0.50$, *3* – $b = 0.75$, *4* – $b = 1.0$.

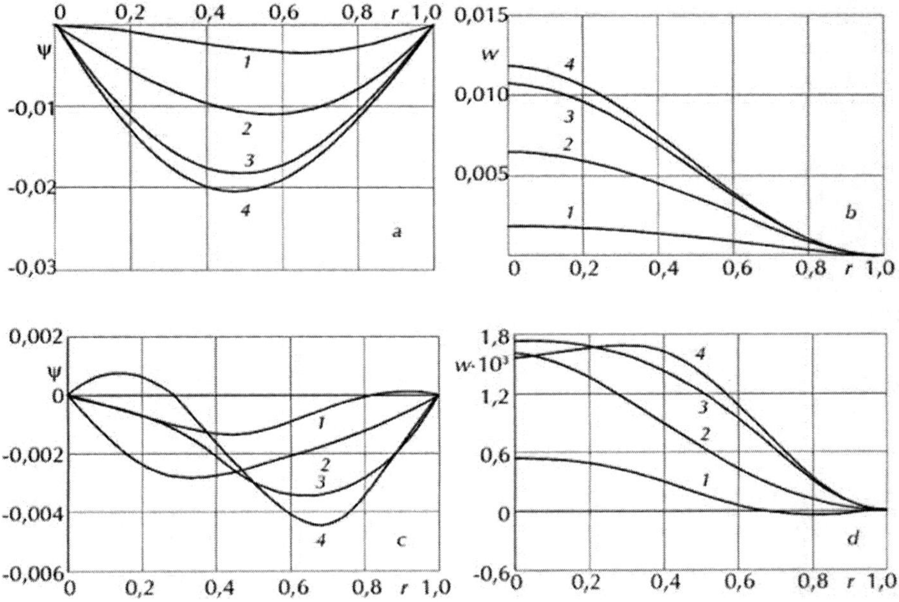


**Figure 12.8.**

The flexures and relative shear at $\omega_k = \omega_0$ surpass the corresponding resonant curves of the second case. As the load spot radius increments, they increase over modulus reaching the maximum at $\omega_k = \omega_0$ in cross-sections $r = 0.5$ and $r = 0$, respectively.

The load uniformly distributed over a circle $[a, b]$. We assume that the harmonic resonant load uniformly locally distributed in the interval $a \leq r \leq b$ over the surface of a circular three-layer plate, which can be expressed in the form

$$q(r,t) = q_0(H_0(b-r) - H_0(a-r))(D\cos(\omega_k t) + E\sin(\omega_k t)) \ . \qquad (12.50)$$

The parameters of load decomposition (12.50) into a series following orthonormal eigenfunctions (12.33) are

$$D_n = \frac{q_0 D}{M_0 d_n \beta_n}\left( bJ_1(\beta_n b) - aJ_1(\beta_n a) - \frac{J_0(\beta_n)}{I_0(\beta_n)}(bI_1(\beta_n b) - aI_1(\beta_n a))\right),$$

$$E_n = \frac{q_0 E}{M_0 d_n \beta_n}\left( bJ_1(\beta_n b) - aJ_1(\beta_n a) - \frac{J_0(\beta_n)}{I_0(\beta_n)}(bI_1(\beta_n b) - aI_1(\beta_n a)) \right). \tag{12.51}$$

The corresponding time function $T_n(t)$ follows from equation (12.45), where integration constants $A_n$, $B_n$ are found from formulas (12.47) with account of (12.51).

Figure 12.9 shows a central flexure of the plate in moment $t = \pi/\omega_0$ and frequency of the resonant load ($q_0 = 10^5$ Pa) coinciding with the eigentone $\omega_k = \omega_0$ depending on the location of the force ring of width $d = b - a = 0.25$. The displacement of this ring from the center towards the contour at first increases the flexure till a maximum and then decreases to its minimal value.

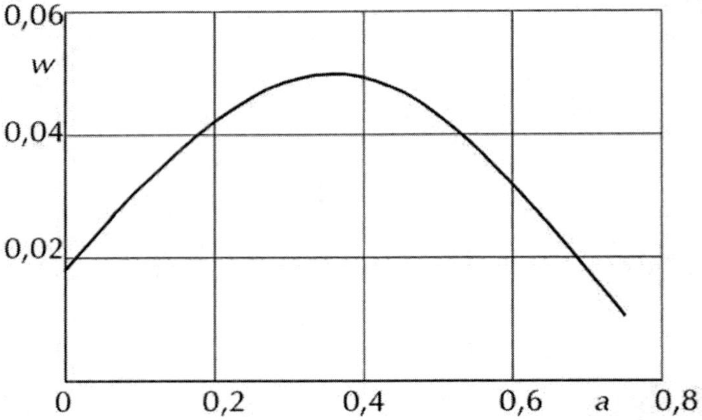


**Figure 12.9.**

Variations in the flexure and relative shear in the filler over the plate radius are shown in Fig. 12.10 at the same moment, load and frequency. These curves correspond to different locations of the load ring: *1* – $a = 0$, $b = 0.25$, *2* – $a = 0.25$, $b = 0.5$, *3* – $a = 0.5$, $b = 0.75$, *4* – $a = 0.75$, $b = 1$. The flexure and relative shear in this case are maximal when the load ring contacts by its outer contour the median radius of the plate (*2*).

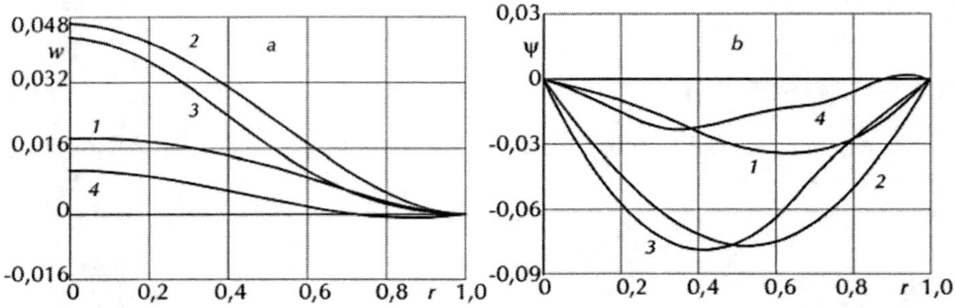


**Figure 12.10.**

**Transverse force.** Let us consider forced vibrations of a circular three-layer plate under the action of a resonant transverse linear force $Q(r, t)$ applied over the circle of radius $a$. The load can be presented analytically in the form

$$Q(r,t)=Q_0H_0(r-a)H_0(a-r)(D\cos\omega_k t+E\sin\omega_k t). \quad (12.52)$$

To solve this problem, we may use the parameters for expanding the surface load into the series (12.51) for the case of its distribution over the ring $[a-\xi, a+\xi]$. Let us substitute $q_0 = Q_0/(2\xi)$ and turn $\xi$ to zero. As a result,

$$D_n=\frac{Q_0D}{M_0d_n\beta_n}\lim_{\xi\to 0}\left[\frac{1}{2\xi}\left((a+\xi)J_1(\beta_n(a+\xi)\right.\right. -(a+\xi)\frac{J_0(\beta_n)}{I_0(\beta_n)}I_1(\beta_n(a+\xi)$$

$$\left.\left.-(a-\xi)J_1(\beta_n(a-\xi)+(a-\xi)\frac{J_0(\beta_n)}{I_0(\beta_n)}I_1(\beta_n(a-\xi)\right)\right]$$

$$=\frac{Q_0aD}{M_0d_n}\left(J_0(\beta_n a)-\frac{J_0(\beta_n)}{I_0(\beta_n)}I_0(\beta_n a)\right),$$

$$E_n=\frac{Q_0aE}{M_0d_n}\left(J_0(\beta_n a)-\frac{J_0(\beta_n)}{I_0(\beta_n)}I_0(\beta_n a)\right). \quad (12.53)$$

Time function $T_n(t)$ for the case of load (12.52) is derived from equation (12.45), where integration constants $A_n$, $B_n$ are calculated by (12.47) with account of (12.53).

Let the *resultant* of the linear transverse force remain *constant* as the radius of the force circumference changes. Its expansion coefficients into series following the system of eigenfunctions can be calculated by (12.53) upon substitution of $Q_0 = Q_1/(2\pi a)$. Hence,

$$D_n=\frac{Q_1D}{2\pi M_0d_n}\left(J_0(\beta_n a)-\frac{J_0(\beta_n)}{I_0(\beta_n)}I_0(\beta_n a)\right),$$

$$E_n=\frac{Q_1E}{2\pi M_0d_n}\left(J_0(\beta_n a)-\frac{J_0(\beta_n)}{I_0(\beta_n)}I_0(\beta_n a)\right). \quad (12.54)$$

Equation (12.45) remains true for the function $T_n(t)$ with account of integration constants (12.47) and coefficients (12.54).

Figure 12.11 illustrates variations in the central flexure of the plate at $t = \pi/\omega_0$ depending on the radius of the force circumference $a$: where (*a*)—flexure calculated using (12.53) for the transverse linear intensity forces $Q_0 = 10^5$ N/m; (*b*)—flexure corresponding to formulas (12.54) for the case of the linear forces with a constant resultant $Q_1 = 10^5$ N. Curves *1* correspond to the resonance at the perturbing force frequency complying with the eigentone frequency of the plate $\omega_k = \omega_0$, *2* – $\omega_k = \omega_1$. When $a = 0$, the linear forces in the first case degenerate and the flexure equals to zero.

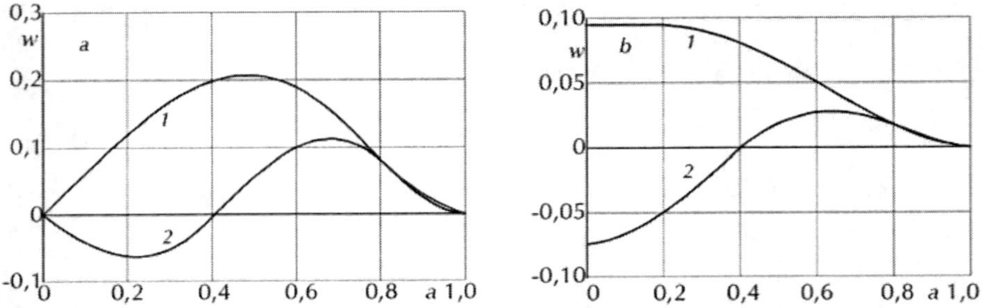


**Figure 12.11.**

Figure 12.12 illustrates changes in the flexure (*a*) and shear in the filler (*b*) along the plate radius under the action of linear forces with a constant resultant $Q_1 = 10^5$ N. The calculations were based on (12.54) under the external load frequency coinciding with that of the eigentone $\omega_k = \omega_0$ in moment $t = \pi/\omega_0$. The curves are related to various radius values of the force circumference: *1* – $a = 0$, *2* – $a = 0.25$, *3* – $a = 0.5$, *4* – $a = 0.75$. These values reach their maxima when a concentrated force is applied in the center of the plate. As the force circumference moves towards the contour of the plate the absolute value of both flexure and shear reduce.

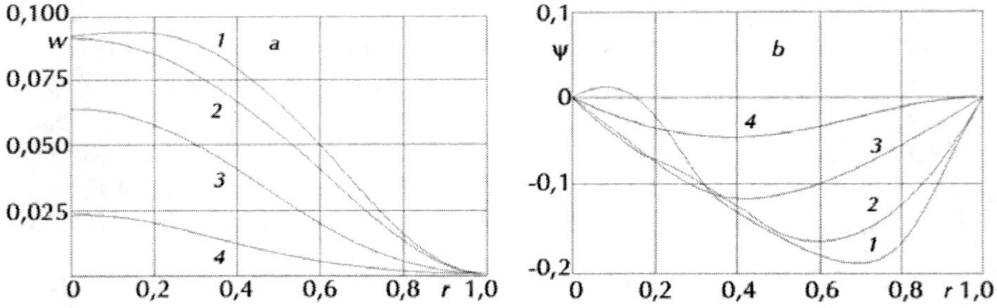


**Figure 12.12.**

**Moment:** Let us assume that the linear resonant moments distributed over a circumference of radius *a* are acting on a circular three-layer plate

$$m(r,t) = m_0 H_0(r-a) H_0(a-r)\left(D\cos\omega_k t + E\sin\omega_k t\right). \tag{12.55}$$

The corresponding forced vibrations will be studied in terms of the sum of two equal in magnitude transverse linear forces (12.52) applied over the circumferences of radii $a - \xi$, $a + \xi$ and directed in the opposite sides.

We shall make substitution of $Q_0 = m_0/(2\xi)$ for parameters $D_n$, $E_n$ (12.53) and turn $\xi$ to zero. As a result, the parameters of the load expansion (12.55) into series over the eigenfunctions will be:

$$D_n = \frac{m_0 D}{M_0 d_n \beta_n} \lim_{\xi \to 0} \left[ \frac{1}{2\xi} \left( (a+\xi) J_0(\beta_n(a+\xi) \right. - (a+\xi) \frac{J_0(\beta_n)}{I_0(\beta_n)} I_0(\beta_n(a+\xi) \right.$$

$$-(a-\xi)J_0(\beta_n(a-\xi) + (a-\xi)\frac{J_0(\beta_n)}{I_0(\beta_n)}I_0(\beta_n(a-\xi)\Bigg)\Bigg]$$

$$=\frac{m_0 D}{M_0 d_n}\left(J_0(\beta_n a)-a\beta_n J_1(\beta_n a)-\frac{J_0(\beta_n)}{I_0(\beta_n)}(I_0(\beta_n a)+a\beta_n I_1(\beta_n a)\right),$$

$$E_n=\frac{m_0 E}{M_0 d_n}\left(J_0(\beta_n a)-a\beta_n J_1(\beta_n a)-\frac{J_0(\beta_n)}{I_0(\beta_n)}(I_0(\beta_n a)+a\beta_n I_1(\beta_n a)\right). \qquad (12.56)$$

If the resultant of the moments is constant at different radii of the instant circumference, then we may obtain the equations for parameters $D_n$, $E_n$ from (12.56) by substitution of $m_0 = m_1/(2\pi a)$:

$$D_n=\frac{m_1 D}{2\pi M_0 d_n}\left(J_0(\beta_n a)-\frac{\beta_n J_1(\beta_n a)}{a}-\frac{J_0(\beta_n)}{I_0(\beta_n)}(I_0(\beta_n a)+\frac{\beta_n I_1(\beta_n a)}{a})\right),$$

$$E_n=\frac{m_1 E}{2\pi M_0 d_n}\left(J_0(\beta_n a)-\frac{\beta_n J_1(\beta_n a)}{a}-\frac{J_0(\beta_n)}{I_0(\beta_n)}(I_0(\beta_n a)+\frac{\beta_n I_1(\beta_n a)}{a})\right). \qquad (12.57)$$

The solution of (12.57) is valid in all cases except for the center of the plate, where $a = 0$.

Figure 12.13 illustrates variations in the central flexure of the circular three-layer plate at $t = \pi/\omega_0$ versus the instant circle radius $a$: ($a$)—flexure derived from (12.56) for the linear intensity moments $m_0 = 10^4$ N, ($b$)—flexure corresponding to (12.57) for the linear moments of intensity $m_1 = 10^4$ N×m.

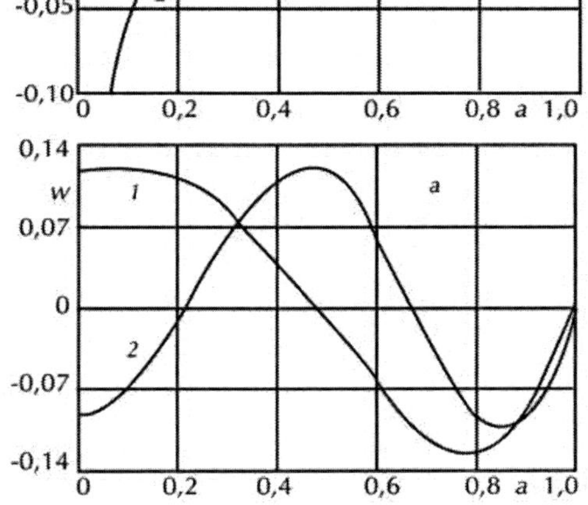

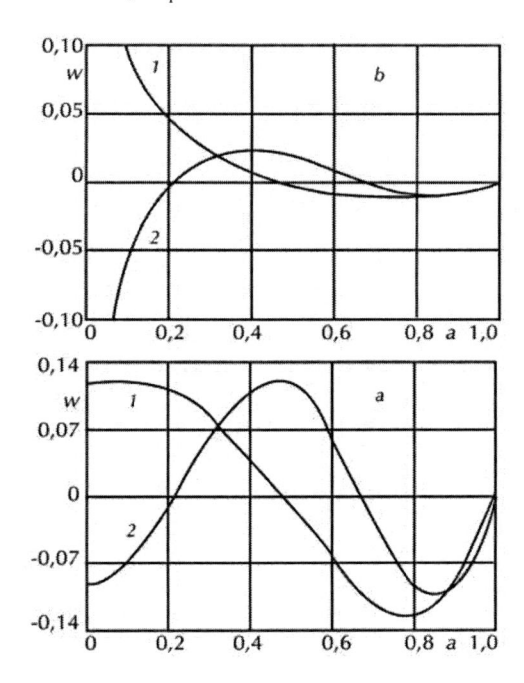

**Figure 12.13.**

Curve *1* answers the resonance when the perturbing load frequency conforms to the eigentone frequency of the plate $\omega_k = \omega_0$, *2* – $\omega_k = \omega_1$. When $a = 0$, then in the first case ($a$) the linear moment is zeroth and the flexure is finite. In the second case ($b$) the flexure tends to infinity.

Figure 12.14 illustrates variations in the flexure ($a$) and shear in the filler ($b$) along the plate radius at dynamic interaction of the moments with the constant resultant $m_1 = 10^4$ N×m. Equations (12.57) were used for calculations at the external load frequency complying with that of the eigentone $\omega_k = \omega_0$ in the instant $t = \pi/\omega_0$. The curves correspond to different radii values of the instant circumference: *1* – $a = 0.25$, *2* – $a = 0.50$, *3* – $a = 0.75$. These values reach their maxima at $a = 0.25$. Further on the flexure reduces in the modulus and changes its sign.

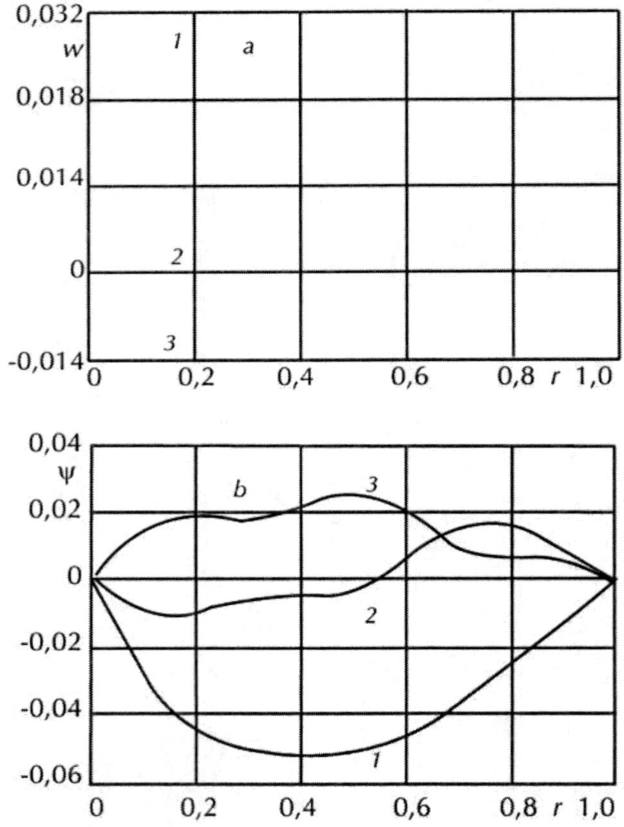


**Figure 12.14.**

Consequently, if the frequency of the perturbing load surpasses that of the eigentone but conforms to one of the higher eigenfrequencies of the circular three-layer plate, then the amplitude of vibrations will augment essentially. This may lead with time to undesirable outcomes during operation of engineering structures in conditions of harmonic affects.

## NOTES

1. *D'Alembert, J.L.* (1717–1783), French mathematician, mechanics; he formulated the rules for compilation of differential equations of motion; he substantiated the perturbation theory of plants; he is a follower of sensualism and scepticism.
2. *Hamilton, W.R.* (1805–1865), Irish mathematician; works in the theory of complex numbers, in mechanics—principle of least action.

## KEYWORDS

- **Axisymmetric**
- **Eigenfunctions**
- **Eigenmodes**
- **Orthogonality**
- **Orthonormal**
- **Principle of superposition**
- **Sensualism and scepticism**

# Appendix

## TENSORS IN CARTESIAN COORDINATES

To solve specific problems in mechanics of continua, we introduce various physical values: scalar, vector, tensor, and other. *Scalar values* are independent of the coordinate system. *Vectors* and *tensors* are characterized by their components, which vary at transition from one coordinate system to another.

The quantitative evaluation of phenomena is usually connected with introduction of coordinates. In many cases, the Cartesian (rectilinear orthogonal) coordinates are quite enough. Introduction of arbitrary curvilinear coordinates demands tensor calculations in a general form, which are not included in this book. For more details on the tensor analysis, we recommend such basic editions as (Gorshkov et al., 2000, Koltunov et.al., 1983).

We shall use the index ("tensor") form of notation for Cartesian coordinates, which presets the equations in a compact form. For this aim, we shall also use a symbolic (vector) form of recording.

*A coordinate system* is a reference frame in a physical three-dimensional Euclidean (uncurved) space relative to which the position and displacement of material particles is determined.

*A vector* is a directed value characterized by an absolute value (modulus) and direction or by giving its components. This definition one can find in the course of mathematics for secondary school, while in this edition we are going to give a stricter variant.

According to Fig. 13.1, the Cartesian system of coordinates is defined by the basic vectors $e_x$, $e_y$, $e_z$, which form the right-hand system (if to look from the side of axis *z*, the rotation from axis *x* to axis *y* should be counter-clockwise). The basic vectors are the unit vectors possessing a property $|e_x| = |e_y| = |e_z| = 1$.

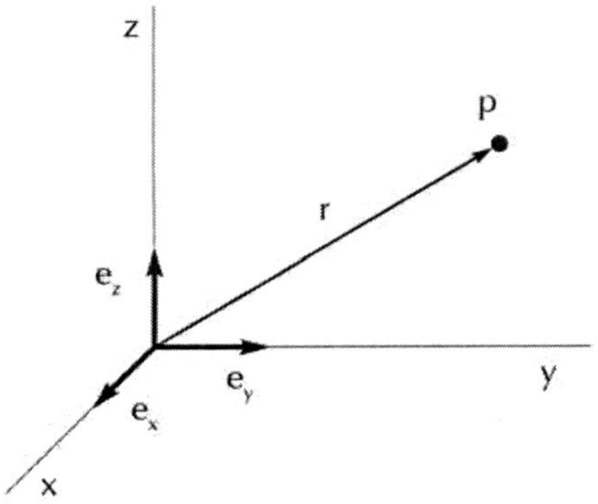


**Figure 13.1.**

The position of a point in space is characterized by a radius-vector **r** whose components are coordinates $x, y, z$, which means that

$$r = e_x x + e_y y + e_z z\,.$$

The representation of a random vector **A** in the coordinate form is generally valid if:

$$\mathbf{A} = e_x A_x + e_y A_y + e_z A_z\,, \tag{13.1}$$

where components $A_x$, $A_y$, $A_z$ are vector projections on the coordinate axes. The modulus of the vector is

$$|\,\mathbf{A}\,| = A = +\sqrt{A_x^2 + A_y^2 + A_z^2}\,. \tag{13.2}$$

We may introduce a product of vectors characterized by some (not all) properties of a usual product of numbers. A *scalar product* of two vectors **A** and **B** gives a scalar value equal to the product of the modules of these vectors by the cosine of the angle between them:

$$A \times B = C = AB\cos(A,\, B)\,, \tag{13.3}$$

The equality $A \times B = B \times A$ is valid. Further, the scalar product of two orthogonal vectors is equated to zero. Therefore, the relations for the basic vectors are satisfied

$$e_x \times e_x = e_y \times e_y = e_z \times e_z = 1\,,$$

$$e_x \times e_y = e_y \times e_z = e_z \times e_x = 0\,.$$

If to present vectors **A** and **B** of the scalar product (13.3) in the coordinate form (13.1), we shall have

$$\mathbf{A} \cdot \mathbf{B} = A_x B_x + A_y B_y + A_z B_z\,. \tag{13.4}$$

*The vector product* of two vectors **A** and **B** is a vector calculated by the formula

$$A \times B = C = \begin{vmatrix} e_x & e_y & e_z \\ A_x & A_y & A_z \\ B_x & B_y & B_z \end{vmatrix}$$

$$= e_x(A_y B_z - A_z B_y) + e_y(A_z B_x - A_x B_z) + e_z(A_x B_y - A_y B_x)\,. \tag{13.5}$$

Vector **C** is perpendicular to a parallelogram formed by vectors **A** and **B** and directed to the side from which the nearest rotation from **A** to **B** is seen as counterclockwise. This means that vectors **A**, **B**, and **C** form in a sequence the right-hand coordinates. The modulus of the vector product is

$$|C| = C = AB\sin(A, B).$$

The equality A × B = B × A. is satisfied. It follows from the definition (13.5) that the product of two parallel vectors tends to zero. Using these relations for the basic vectors, we come to

$$e_x \times e_x = e_y \times e_y = e_z \times e_z = 0$$
,

$$e_x \times e_y = e_z, \quad e_y \times e_z = e_x, \quad e_z \times e_x = e_y.$$

When speaking about *a tensor form of* recording coordinates, we mean the following. For Cartesian coordinates we use

$$x = x_1, \quad y = x_2, \quad z = x_3.$$

In the relation

$$a_1 x_1 + a_2 x_2 + a_3 x_3 = \sum_{i=1} a_i x_i = a_i x_i$$

we may omit the summation symbol if to accept that the summation should be done for a twice met index. The summation index (so-called *dummy index*) may be taken arbitrarily,

$$a_i x_i = a_n x_n = a_m x_m \text{ and so on.}$$

An expression like $a_i b_i x_i$ is indeterminate according to this rule. If the summation is meant in this case, the summation symbol should be applied. It will always be specified further if the rule of summation on the repeating index is not valid. Sometimes, the summation is stipulated only for the repeating Latin indices leaving aside the Greek indices.

For a double sum, the equality is similarly true

$$\sum_{i=1}^{n}\sum_{j=1}^{n} a_{ij} x_i x_j = a_{ij} x_i x_j$$

at $n = 3$ we have only 9 items.

With the help of a scalar product of the basic vectors, we introduce *a metric tensor* $\delta_{ij}$ determined in the Euclidean space as

$$\delta_{ij} = e_i \times e_j, \; \delta_{ij} = \begin{cases} 1, & i = j, \\ 0, & i ^1 j. \end{cases} \tag{13.6}$$

Its components are also termed as *Kronecker's symbols*. Using the metric tensor, we can substitute indices of *"juggling"*, for example.

$$\delta_{ik} a_i = a_k, \quad \delta_{ik} a_{kj} = a_{ij}.$$

When the tensor forms of recording are used, one should observe some rules, most important of which are touched upon below.

If the relation $x_i = B_{im}y_m$ is substituted in $a_i = A_{im}x_m$, then indices $i \to m$ and $m \to k$ should be renamed in the former one. Then, $x_m = B_{mk}y_k$, hence,

$$a_i = A_{im}B_{mk}y_k .$$

This is also true for the product of vectors $a = a_i x_i$ and $b = b_i y_i$, to be written as **a** × b = $a_i x_i b_j y_j$.

Similarly, for the scalar product of two vectors A = $e_i A_i$ and B = $e_i B_i$ with account of (13.4) and (13.6), we obtain

$$A \times B = (e_i A_i)(e_j B_j) = (e_i \times e_j) A_i B_j = A_i B_i .$$

If we put vector $n_i$ outside the parentheses in

$$T_{ij}n_j - \lambda n_i = 0 , \tag{13.7}$$

where λ is a scalar multiplier, we may write $n_i = \delta_{ij}n_i$, then it follows from (13.7)

$$T_{ij}n_j - \lambda\delta_{ij}n_j = (T_{ij} - \lambda\delta_{ij})n_j = 0 .$$

*Tensor contraction or convolution* is a frequently employed operation. The convolution means that the indices are equated to each other in tensor value of the second rank and summed up. This implies, for example, that the value $T_{ij}$ at $i = j$ $(i, j = 1, 2, 3)$ will be equal to $T_{ii} = T_{11} + T_{22} + T_{33}$. For the metric tensor the convolution gives $\delta_{ii} = \delta_{11} + \delta_{22} + \delta_{33} = 3$.

Proceeding from above stated, and having a correct tensor form in the equation, one and the same index in each member should be met only twice and the summation should be made over this index. If some index appears in a certain item only once (free index), it can also be met only once in all other members.

A scalar triple product (*mixed product*) of three vectors is the value

$$(A \times B) \times C = (A_i e_i \times B_j e_j) C_k e_k = A_i B_j C_k (e_i \times e_j) e_k = A_i B_j C_k \varepsilon_{ijk} . \tag{13.8}$$

There is a *Levi-Civita's* triple-index, and *symbol* $\varepsilon_{ijk}$ (pseudotensor) on the right-side of equation (13.8), which is determined as follows:

$$\varepsilon_{ijk} = \begin{cases} 1, & \text{if substitution } (i, j, k) \text{ is even} \\ -1, & \text{if substitution } (i, j, k) \text{ is not even;} \\ 0, & \text{if not all indices } (i, j, k) \text{ are different,} \end{cases} \tag{13.9}$$

that is,

$$\varepsilon_{123} = \varepsilon_{231} = \varepsilon_{312} = - \varepsilon_{132} = - \varepsilon_{213} = - \varepsilon_{321} = 1.$$

The following an important relation is valid

$$\varepsilon_{ijk}\varepsilon_{mnl} = \begin{vmatrix} \delta_{im} & \delta_{in} & \delta_{il} \\ \delta_{jm} & \delta_{jn} & \delta_{jl} \\ \delta_{km} & \delta_{kn} & \delta_{kl} \end{vmatrix}.$$

In particular, $\varepsilon_{ijk}\varepsilon_{mjk} = 2\delta_{im}$, and also, $\varepsilon_{ijk.}\varepsilon_{ijk} = 6$.

The differential of function $f = f(x_1, x_2, \ldots, x_n)$ is $n$ of independent variables

$$\mathrm{d}f = \frac{\partial f}{\partial x_1}\mathrm{d}x_1 + \frac{\partial f}{\partial x_2}\mathrm{d}x_2 + \ldots + \frac{\partial f}{\partial x_n}\mathrm{d}x_n = \frac{\partial f}{\partial x_i}\mathrm{d}x_i\,,\ i = 1, \ldots, n.$$

A contracted recording is commonly used for partial derivatives

$$\frac{\partial f}{\partial x_i} = f,_i\,; \quad \frac{\partial^2 f}{\partial x_i \partial x_j} = f,_{ij}$$

then, a 3D Laplacian operator can be written as

$$\Delta f = \left(\frac{\partial^2}{\partial x_1^2} + \frac{\partial^2}{\partial x_2^2} + \frac{\partial^2}{\partial x_3^2}\right) f = \frac{\partial}{\partial x_i}\frac{\partial}{\partial x_i} f = f,_{ii} \quad i = 1, 2, 3.$$

Among the important differential vector operations are the *gradient*, *divergence*, and *rotor (curl)*. We may write in the tensor form the relations

$$grad\, f = \left(e_1\frac{\partial}{\partial x_1} + e_2\frac{\partial}{\partial x_2} + e_3\frac{\partial}{\partial x_3}\right) f = e_i f,_i\,,$$

$$div\, A = \frac{\partial A_1}{\partial x_1} + \frac{\partial A_2}{\partial x_2} + \frac{\partial A_3}{\partial x_3} = A_i,_i\,,$$

$$rot\ A = \begin{vmatrix} e_1 & e_2 & e_3 \\ \frac{\partial}{\partial x_1} & \frac{\partial}{\partial x_2} & \frac{\partial}{\partial x_3} \\ A_x & A_y & A_z \end{vmatrix} = \varepsilon_{ijk}\frac{\partial A_k}{\partial x_j} e_i = \varepsilon_{ijk} A_k,_j\, e_i\,.$$

where $\varepsilon_{ijk}$ are Levi-Civita[1] symbols (13.9).

Let us consider transformation of coordinates at the turn from $x_i$ towards $x_i^*$ (Fig. 13.2), $x_i^* = x_i^*(x_1, x_2, x_3)$. We can set it as follows:

$$x_i^* = \sum_k x_k \cos(x_i^*, x_k) = c_k\, x_k\,, \tag{13.10}$$

the transformation factors are so-called direction cosines $c_{ik} = c_{ki}$.

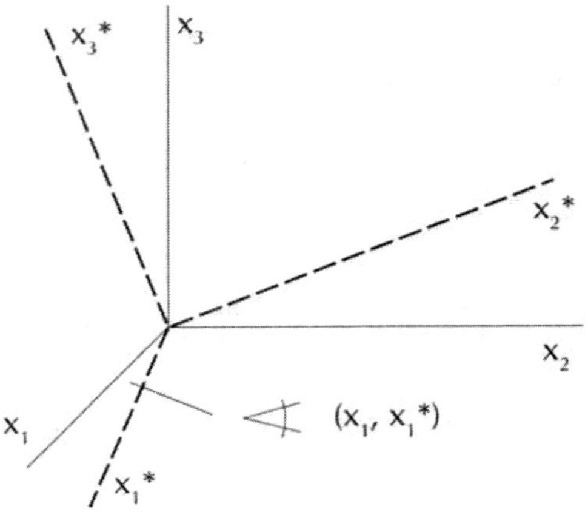


**Figure 13.2.**

The law of transformation corresponding to (13.10) for a differential of the coordinates looks like

$$dx_i^* = \frac{\partial x_i^*}{\partial x_k} dx_k = x_i^*,_k dx_k = c_k \, dx_k .$$

We may determine a more exact vector if to proceed from the variation regularity of the vector components at rotation of coordinates. The advantage of this method consists in its expansion on the higher-order non-scalar values (which is impossible using the definition of the vector as a directed segment).

We shall further accept *vector A* as some object whose components are transformed at rotation of coordinate system similarly to the differentials of coordinates. This implies that the next equality is true

$$A_i^* = \frac{\partial x_i^*}{\partial x_k} A_k = c_{ik} A_k \cdot \tag{13.11}$$

This vector is also termed as the *tensor of the first rank*. Accordingly, the *tensor of the second rank* in Cartesian coordinates is defined as some object having components $T_{ij}$ transformed at the turn of the coordinates system as a product of two vectors:

$$T_{ij}^* = \frac{\partial x_i^*}{\partial x_k} \frac{\partial x_j^*}{\partial x_l} T_{kl} = c_{ik} c_{jl} T_{kl} . \tag{13.12}$$

Similarly to (13.11), (13.12), the transformation formulas are obtained for the components of the third and higher rank tensors.

To describe a particular tensor, we preset its values relative to the basis. A combination of components of the second rank tensor are usually written in the form of a matrix, for example, in a 3D case they are

$$T_{ij} = \begin{vmatrix} T_{11} & T_{12} & T_{13} \\ T_{21} & T_{22} & T_{23} \\ T_{31} & T_{32} & T_{33} \end{vmatrix}. \quad (13.13)$$

In contrast to the vector, the tensor cannot have an absolute value or a modulus like (13.2). Nevertheless, the tensor has its invariants, that is, the values invariable with rotation of coordinates, for example, the matrix determinant (13.13).

We differentiate between *symmetric tensors* $T_{ij} = T_{ji}$ and *antisymmetric (skew-symmetric) tensors* $T_{ij} = -T_{ji}$. Each random tensor of the second rank can be expanded into symmetric and antisymmetric parts:

$$T_{ij} = \tfrac{1}{2}(T_{ij} + T_{ji}) + \tfrac{1}{2}(T_{ij} - T_{ji}).$$

The tensors of a higher rank are formed by differentiating a tensor in Cartesian coordinates. Generally, the scalar differentiation (zeroth-rank tensor) in vector coordinates (tensor of the first rank) gives

$$\frac{\partial U}{\partial x_i} = U,_i = \mathrm{grad}U.$$

This corresponds to a gradient formation operation.

Accordingly, vector differentiation (the first-rank tensor) results in a tensor of the second rank

$$\frac{\partial A_i}{\partial x_j} = A_i,_j,$$

and so on. This is, however, true only for Cartesian coordinates.

To transform a surface integral into the volume one and vice versa, we may use an important *Gauss-Ostrogradskii's theorem*. This theorem was first formulated in different variants by Lagrange (1762), then Gauss (1813), Green (1828), and Ostrogradskii (1831), and is called accordingly.

Its formulation we accept in the form: for the continuous vector functions $\boldsymbol{a}$ with continuous partial derivatives in a constrained and spatially simply-connected volume $V$ of a three-dimensional Euclidean space and on a closed regular surface $S$ restricting this volume, the next integrated formula is true

$$\int_V diva dV \circ \int_V a_i,_i \, dV = \int_S a \times n dS = \int_S a_i l_i dS.$$

This formula for the tensor components of the second rank, for example, for the stress tensor is

$$\int_V \sigma_{ij},_j \, \mathrm{d}V = \int_S \sigma_{ij} l_j \, \mathrm{d}S.$$

This theorem is sometimes termed a *theorem of divergence.*

## BOUNDARY CONDITIONS AND DISPLACEMENTS IN AIRY'S FUNCTIONS

To close the mathematical problem on determining Airy's function, let us express boundary conditions and displacements.

The boundary conditions in stresses for the first boundary problem we shall express through Airy's stress function. The stresses on the boundaries $p_x(s)$ and $p_y(s)$ are set as a coordinate function of contour $s$. The components of the normal unit vector (direction cosines) obey the relations

$$\cos(\nu,x)=\cos(t,y)=\frac{dy}{ds},$$

$$\cos(\nu,y)=-\cos(t,x)=-\frac{dx}{ds}. \tag{13.14}$$

Upon substitution of (13.14) in the boundary conditions (5.7), we have

$$\sigma_{xx}dy - \sigma_{xy}dx = p_x ds,\ \sigma_{yy}dx - \sigma_{xy}dy = p_y ds.$$

Using Airy's function of stresses (5.11), we come to

$$\frac{\partial^2 \hat{O}}{\partial y^2}dy+\frac{\partial^2 \hat{O}}{\partial x \ddot{a} y}dx=p_x ds,\quad \frac{\partial^2 \hat{O}}{\partial x^2}dx+\frac{\partial^2 \hat{O}}{\partial x \ddot{a} y}dy=-p_y ds.$$

Since they are total differentials, these expressions can be rewritten as

$$\frac{\partial}{\partial x}\left(\frac{\partial \hat{O}}{\partial y}\right)dx+\frac{\partial}{\partial y}\left(\frac{\partial \hat{O}}{\partial y}\right)dy=d\left(\frac{\partial \hat{O}}{\partial y}\right)=p_x ds,$$

$$\frac{\partial}{\partial x}\left(\frac{\partial \hat{O}}{\partial x}\right)dx+\frac{\partial}{\partial y}\left(\frac{\partial \hat{O}}{\partial x}\right)dy=d\left(\frac{\partial \hat{O}}{\partial x}\right)=-p_y ds.$$

By integrating over the contour, we obtain

$$\frac{\partial \hat{O}}{\partial x}=-\int_0^s p_y ds=g_x(s)+C_1,\quad \frac{\partial \hat{O}}{\partial y}=-\int_0^s p_x ds=g_y(s)+C_2.$$

Here, $g_x(s)$, $g_y(s)$ are the given functions.

Now, the plane elastic problem can be formulated as follows: a biharmonic function is to be found in a domain if the derivative functions for both coordinates are given on the boundary.

Another definition can be given if we calculate the stress function $\Phi$, and its normal derivative $d\Phi/d\nu$ via functions $g_x(s)$, $g_y(s)$. Based on

$$d\hat{O} = \frac{\partial \hat{O}}{\partial x} dx + \frac{\partial \hat{O}}{\partial y} dy = g_x(s)dx + g_y(s)dy$$

we come to

$$\hat{O} = \int_0^s \left[ g_x(s)\frac{dx}{ds} + g_y(s)\frac{dy}{ds} \right] ds = g(s) + \text{const}$$

The boundary conditions will be

$$\frac{d\hat{O}}{d\nu} = \frac{\partial \hat{O}}{\partial x}\cos(\nu, x) + \frac{\partial \hat{O}}{\partial y}\cos(\nu, y) = \frac{\partial \hat{O}}{\partial x}\frac{dy}{ds} - \frac{\partial \hat{O}}{\partial y}\frac{dx}{ds} = h(s) + \text{const}$$

The constant is insignificant for determination of stresses.

Hence, the first boundary problem can be formulated also in the form: we seek for a biharmonic function in a domain using the values of the function itself given on the boundary, and its normal derivative.

Named conditions on the boundary can be given in a static interpretation. The function of stresses corresponds to a resultant moment of the external forces operating on a section of the domain contour, where the integration takes place. Its normal derivative corresponds to a projection of these external forces on the tangent direction to the boundary (so-called flux of forces in tangential direction).

To determine displacements through Airy's function of stresses let us use a generalized Hooke's law for a plane strain state

$$\varepsilon_{xx} = \frac{1}{E}\left[\sigma_{xx} - \nu(\sigma_{yy} + \sigma_{zz})\right],$$

$$\varepsilon_{yy} = \frac{1}{E}\left[\sigma_{yy} - \nu(\sigma_{zz} + \sigma_{xx})\right],\ \varepsilon_{xy} = \frac{\sigma_{xy}}{2G}.$$

where from, since $\varepsilon_{zz} = 0$ и $\sigma_{zz} = \nu(\sigma_{xx} + \sigma_{yy})$,

$$\varepsilon_{xx} = \frac{1}{2G}\left[\sigma_{xx} - \nu(\sigma_{xx} + \sigma_{yy})\right],\ \varepsilon_{yy} = \frac{1}{2G}\left[\sigma_{yy} - \nu(\sigma_{xx} + \sigma_{yy})\right].$$

By expressing strains through displacements using kinematic relations (5.2), we get

$$\frac{\partial u}{\partial x} = \frac{1}{2G}\left[\sigma_{xx} - \nu(\sigma_{xx} + \sigma_{yy})\right],$$

$$\frac{\partial v}{\partial y} = \frac{1}{2G}\left[\sigma_{yy} - \nu(\sigma_{xx} + \sigma_{yy})\right],\ \frac{\partial v}{\partial x} + \frac{\partial u}{\partial y} = \frac{\sigma_{xy}}{G}.$$

After introduction of Airy's function, we obtain

$$\frac{\partial u}{\partial x} = \frac{1}{2G}\left[\frac{\partial^2 \Phi}{\partial y^2} - \nu\Delta\Phi\right] = \frac{1}{2G}\left[(1-\nu)\Delta\Phi - \frac{\partial^2 \Phi}{\partial x^2}\right],$$

$$\frac{\partial v}{\partial y}=\frac{1}{2G}\left[\frac{\partial^2\Phi}{\partial x^2}-\nu\Delta\Phi\right]=\frac{1}{2G}\left[(1-\nu)\Delta\Phi-\frac{\partial^2\Phi}{\partial y^2}\right],$$

$$\frac{\partial v}{\partial x}+\frac{\partial u}{\partial y}=-\frac{1}{G}\frac{\partial^2\Phi}{\partial x\partial y}. \qquad (13.15)$$

Integration of (13.15) results in general relations for displacements in the form

$$u=\frac{1}{2G}\int\left[(1-\nu)\Delta-\frac{\partial^2}{\partial x^2}\right]\Phi(x,y)\,\mathrm{d}x+f_1(y),$$

$$v=\frac{1}{2G}\int\left[(1-\nu)\Delta-\frac{\partial^2}{\partial y^2}\right]\Phi(x,y)\,\mathrm{d}y+f_2(x),$$

where $f_1(y)$ and $f_2(x)$ are integration functions determined by the third equation (13.15).

In the case of a plane SSS, we proceed from Hooke's law similarly to (5.8). As a result, we obtain kinematic relations

$$\frac{\partial u}{\partial x}=\frac{1}{E}\left[\sigma_{xx}-\nu\sigma_{yy}\right],\ \frac{\partial v}{\partial y}=\frac{1}{E}\left[\sigma_{yy}-\nu\sigma_{xx}\right],\ \frac{\partial v}{\partial x}+\frac{\partial u}{\partial y}=\frac{\sigma_{xy}}{G},$$

which integration results in equations

$$Eu=\int(\sigma_{xx}-\nu\sigma_{yy})\,\mathrm{d}x+g_1(y),$$

$$Ev=\int(\sigma_{yy}-\nu\sigma_{xx})\,\mathrm{d}y+g_2(x),$$

where $g_1(y)$ and $g_2(x)$—integration functions, which can be received from the third kinematic relation.

## GENERALIZED FUNCTIONS

Let us consider in more detail the examples of the generalized functions used earlier and borrowed from (Korn and Korn, 1961).

*A step function* of argument $x$ is a function that alters its values only in a discrete sequence of break points (of the first kind). The values of the function in break points can be either determined, or undetermined. Most often, the following step functions are used:

- symmetrical unit function

$$U(x)=\begin{cases}0 & \text{at}\quad x<0,\\ 1/2 & \text{at}\quad x=0,\\ 1 & \text{at}\quad x>0,\end{cases}$$

- asymmetric unit functions

$$U_-(x)=\begin{cases}0 & \text{at}\quad x<0,\\ 1 & \text{at}\quad x\geq 0,\end{cases}\quad U_+(x)=\begin{cases}0 & \text{at}\quad x\leq 0;\\ 1 & \text{at}\quad x>0.\end{cases} \qquad (13.16)$$

Each step function can be presented (except for its values in break points $x = x_k$) as a sum

$$\sum_k a_k U(x-x_k), \quad \sum_k a_k U_-(x-x_k), \quad \sum_k a_k U_+(x-x_k).$$

*The step functions* can be approximated by the following continuous functions:

$$U(x)=\lim_{\alpha\to\infty}\left[\frac{1}{2}+\frac{1}{\pi}\operatorname{arctg}(\alpha x)\right], \quad U(x)=\lim_{\alpha\to\infty}\frac{1}{2}[\operatorname{erf}(\alpha x)+1],$$

$$U(x)=\lim_{\alpha\to\infty}2^{-e^{-\alpha x}}, \quad U(x)=\lim_{\alpha\to\infty}\frac{1}{\pi}\int_{-\infty}^{\infty}\frac{\sin t}{t}\mathrm{d}t.$$

For convenience of analytical notation of the local loading, the zeroth-order Heavyside function is used as one of step function (13.16):

$$H_0(x)=\begin{cases}1, & x\geq 0;\\ 0, & x<0.\end{cases}$$

The whole Heavyside's family of functions $H_n(z)$ is found from:

$$H_n(z)=\begin{cases}\dfrac{z^n}{n!}, & z\geq 0;\\ 0, & z<0,\end{cases} \tag{13.17}$$

where $n$ is a non-negative integer.

The next property of functions $H_n(z)$ is true:

$$\int_0^z H_n(z-a)\mathrm{d}z = H_{n+1}(z-a) \quad (a>0).$$

The latter relation presumes continuation of the sequence of functions $H_n(z)$ in the domain of the negative indices. So, for the case $n = -1$, this function coincides with Dirac's delta-function $\delta(z)$. This function can be used in mechanics to describe the effect of a concentrated force on a strained solid. Function $H_{-2}(z)$ expresses the influence of a lumped moment.

*The Symmetric unit pulse function* or *Dirac's function* $\delta(x)$ of a real variable $x$ is determined from condition

$$\int_a^b f(\tau)\delta(\tau-x)\mathrm{d}\tau=\begin{cases}f(x) & \text{at } a<x<b;\\ \frac{1}{2}f(x) & \text{at } x=a \text{ or } x=b;\\ 0 & \text{at } x<a \text{ or } x>b,\end{cases} \tag{13.18}$$

where $a < b$; $f(\tau)$ is a random function continuous in point $\tau = x$.

Dirac's function $\delta(x)$ is not a function in its usual sense because of incompatible conditions following from (13.18)

$$\delta(x) = 0\ (x \neq 0),\quad \int_{-\infty}^{\infty} \delta(\tau)\mathrm{d}\tau = 1.$$

A symbolic (generalized) function $\delta(x)$ helps to present formally the functional transformation $f(\tau) \to f(x)$ as an integral transformation. The formal use of $\delta(x)$ results in convenient notations that suggest generalizations of a number of mathematical relations. Although, there are no functions repeating the properties of (13.18), we may consider $\delta(x)$ as the limits of usual functions.

Another variant of the symmetrical pulse function can be the following:

$$\int_a^b f(\tau)\delta(\tau - x)\,\mathrm{d}\tau = \begin{cases} f(x) & \text{at} \quad a \le x \le b; \\ 0 & \text{at} \quad x < a,\ x > b. \end{cases}$$

The formal relations containing $\delta(x)$ are:

$$\delta(ax) = \frac{1}{a}\delta(x) \quad (a > 0),\ \delta(-x) = \delta(x),$$

$$f(x)\delta(x - a) = \frac{1}{2}\left[f(a - 0) + f(a + 0)\right]\delta(x - a),\ x\delta(x) = 0,$$

$$\delta(x^2 - a^2) = \frac{1}{2a}\left[\delta(x - a) + \delta(x + a)\right] \quad (a > 0),$$

$$\int_{-\infty}^{\infty} \delta(a - x)\delta(x - b)dx = \delta(a - b).$$

The derivatives of the step and pulse functions

$$\delta(x) = \frac{\mathrm{d}}{\mathrm{d}x}U(x),$$

$$\delta^{(r)}(x) = (-1)^r r!\frac{\delta(x)}{x^r} \quad (r = 0, 1, 2, \ldots).$$

The approximation of the pulse functions by continuously differentiated functions is

$$\delta(x, \alpha) = \frac{\alpha}{\pi(\alpha^2 x^2 + 1)} \quad \text{at} \quad \alpha \to \infty,$$

$$\delta(x, \alpha) = \frac{\alpha}{\sqrt{\pi}}e^{-\alpha^2 x^2} \quad \text{at} \ \alpha \to \infty,$$

$$\delta(x, \alpha) = \frac{\alpha}{\pi} \frac{\sin(\alpha x)}{\alpha x} \quad \text{at } \alpha \to \infty,$$

meaning that $\lim_{\alpha \to \infty} \delta(x, \alpha) = 0 \ (x \neq 0)$, and

$$\lim_{\alpha \to \infty} \int_{-\infty}^{\infty} f(\xi)\delta(x - \xi, \alpha) d\xi = \frac{1}{2}[f(x-0) + f(x+0)]$$

in provision that $f(x - 0)$ and $f(x + 0)$ exist.

Let us refer also to the relation

$$\lim_{\alpha \to \infty} \int_{-\infty}^{\infty} \delta(\xi, \alpha) d\xi = 1.$$

*Asymmetric pulse functions* $\delta_+(x)$, $\delta'_+(x)$, …, $\delta^{(r)}{}_+(x)$ are determined by the relations

$$\int_{a+0}^{b} f(\tau)\delta_+(\tau - x)\,d\tau = \begin{cases} f(x+0) & \text{at } a \le x \le b, \\ 0 & \text{at } x < a \text{ or } x \ge b; \end{cases}$$

$$\int_{a+0}^{b} f(\tau)\delta_+^{(r)}(\tau - x)\,d\tau = \begin{cases} (-1)^r f^{(r)}(x+0) & \text{at } a \le x \le b, \\ 0 & \text{at } x < a \text{ or } x \ge b. \end{cases}$$

Multivariate delta functions are introduced similarly.

## COEFFICIENTS OF LINEARLY VISCOELASTIC THREE-LAYER PLATE

The values of the coefficients introduced in Ch. 9.8 to describe the analytical solution for a round linearly viscoelastic three-layer plate bending are the next.

$$\alpha_0 = \frac{2(a_{51}a_{11} - a_{21}a_{31})}{3k\omega_3\omega_4(a_{62}a_{12} - a_{32}^2)}, \quad \alpha_1 = \frac{a_{11}\beta_1 - a_{12}}{\omega_1\omega_2(a_{62}a_{12} - a_{32}^2)(\beta_1 - \beta_2)},$$

$$\alpha_2 = \frac{a_{12} - a_{11}\beta_2}{\omega_1\omega_2(a_{62}a_{12} - a_{32}^2)(\beta_1 - \beta_2)}, \quad \alpha_3 = \frac{2(s_1\beta_3^2 - s_2\beta_3 + s_3)}{3K\omega_3\omega_4(a_{62}a_{12} - a_{32}^2)(\beta_4 - \beta_3)},$$

$$\alpha_4 = \frac{2(s_1\beta_4^2 - s_2\beta_4 + s_3)}{3K\omega_3\omega_4(a_{62}a_{12} - a_{32}^2)(\beta_3 - \beta_4)}, \quad \omega_1 = -d_1 + \sqrt{d_1^2 - d_2},$$

$$\omega_2 = -d_1 - \sqrt{d_1^2 - d_2}, \quad \omega_3 = -d_4 + \sqrt{d_4^2 - d_3}, \quad \omega_4 = -d_4 - \sqrt{d_4^2 - d_3},$$

$$d_1 = \frac{a_{61}a_{12} + a_{62}a_{11} - 2a_{31}a_{32}}{2(a_{62}a_{12} - a_{32}^2)}, \quad d_2 = \frac{a_{61}a_{11} - a_{31}^2}{a_{62}a_{12} - a_{32}^2},$$

$$d_3 = \frac{a_{61}a_{11} - a_{12}^2}{a_{62}a_{12} - a_{32}^2}, \quad d_4 = d_1,$$

$$s_1 = a_{51}a_{11} - a_{21}a_{31},\ s_2 = a_{51}a_{12} + a_{52}a_{11} - a_{21}a_{32} - a_{22}a_{31},$$

$$s_3 = a_{52}a_{12} - a_{22}a_{32},\ \beta_i = \omega_i^{-1} \quad (i = 1, ..., 8) \cdot$$

Coefficients $\alpha_5$, ..., $\alpha_{11}$ are in the process of decomposing into simple fractions of the relation

$$\frac{b_2^*}{b_3^* G_3^* (8+\beta^{*2})} == \frac{s_4(s_1 + s_2\omega^* + s_3\omega^{*2})(s_5 + s_6\omega^* + s_7\omega^{*2} + s_8\omega^{*3} + s_9\omega^{*4})}{\omega^*(1+\beta_3\omega^*)(1+\beta_4\omega^*)(1+\beta_5\omega^*)(1+\beta_6\omega^*)(1+\beta_7\omega^*)(1+\beta_8\omega^*)},$$

where

$$s_4 = 2/(3K\omega_3\omega_4\omega_5\omega_6\omega_7\omega_8(a_{62}a_{12} - a_{32}^2)(8s_9 + s_{13})),$$

$$s_5 = (a_{11}a_{41} - a_{21}^2)(a_{11}a_{61} - a_{31}^2) - (a_{11}a_{51} - a_{21}a_{31})^2,$$

$$s_6 = (a_{11}a_{41} - a_{21}^2)(a_{11}a_{62} + a_{12}a_{61} - 2a_{31}a_{32})$$

$$+(a_{11}a_{61} - a_{31}^2)(a_{11}a_{42} + a_{12}a_{41} - 2a_{21}a_{22})$$

$$-2(a_{11}a_{51} - a_{21}a_{31})(a_{11}a_{52} + a_{12}a_{51} - a_{21}a_{32} - a_{22}a_{31}),$$

$$s_7 = (a_{11}a_{41} - a_{21}^2)(a_{12}a_{62} - a_{32}^2) + (a_{12}a_{42} - a_{22}^2)(a_{11}a_{61} - a_{31}^2)$$

$$+(a_{11}a_{42} + a_{12}a_{41} - 2a_{21}a_{22})(a_{11}a_{62} + a_{12}a_{61} - 2a_{31}a_{32})$$

$$-(a_{11}a_{52} + a_{12}a_{51} - a_{21}a_{32} - a_{22}a_{31})^2 - 2(a_{11}a_{51} - a_{21}a_{31})(a_{12}a_{52} - a_{22}a_{32}),$$

$$s_8 = (a_{11}a_{42} + a_{12}a_{41} - 2a_{21}a_{22})(a_{12}a_{62} - a_{32}^2) +$$

$$(a_{12}a_{42} - a_{22}^2)(a_{11}a_{62} + a_{12}a_{61} - 2a_{31}a_{32})$$

$$-2(a_{12}a_{52} - a_{22}a_{32})(a_{11}a_{52} + a_{12}a_{51} - a_{21}a_{32} - a_{22}a_{31}),$$

$$s_9 = (a_{12}a_{42} - a_{22}^2)(a_{12}a_{62} - a_{32}^2) - (a_{12}a_{52} - a_{22}a_{32})^2,$$

$$s_{10} = 3Kca_{11}(a_{11}a_{61} - a_{31}^2),$$

$$s_{11} = 3Kc(a_{11}^2a_{62} + 2a_{11}a_{12}a_{61} - a_{31}^2a_{12} - 2a_{11}a_{32}a_{31}),$$

$$s_{12} = 3Kc(a_{12}^2 a_{61} + 2a_{11}a_{12}a_{62} - a_{32}^2 a_{11} - 2a_{12}a_{31}a_{32}),$$

$$s_{13} = 3Kca_{12}(a_{12}a_{62} - a_{32}^2).$$

Values $-_{\omega 5}, \ldots, -_{\omega 8}$ are here the roots of the polynomial

$$8s_5 + (8s_6 + s_{10})\omega + (8s_7 + s_{11})\omega^2 + (8s_8 + s_{12})\omega^3 + (8s_9 + s_{13})\omega^4.$$

As far as the coefficients of the polynomial are positive, then according to the of Descartes[2] rule, all actual roots of the polynomial are negative, that is, numbers $\omega_5, \ldots, \omega_8$ (as well as $\omega_1, \ldots, \omega_4$) are positive.

Values $\alpha_{12}, \ldots, \alpha_{16}$ are the expansion coefficients into the simple and multiple fractions of the relation

$$\frac{s_{14}(s_1 + s_2\omega^* + s_3\omega^{*2})^2}{\omega^*(1+\beta_3\omega^*)^2(1+\beta_4\omega^*)^2},\ s_{14} = 2/(3K\omega_3^2\omega_4^2(a_{62}2a_{12} - a_{32}^2)^2),$$

and $\alpha_{17}, \ldots, \alpha_{25}$ are the coefficients of a similar expansion of the relation

$$\frac{s_{15}(s_1 + s_2\omega^* + s_3\omega^{*2})^2(s_5 + s_6\omega^* + s_7\omega^{*2} + s_8\omega^{*3} + s_9\omega^{*4})}{\omega^*(1+\beta_3\omega^*)^2(1+\beta_4\omega^*)^2(1+\beta_5\omega^*)(1+\beta_6\omega^*)(1+\beta_7\omega^*)(1+\beta_8\omega^*)},$$

$$s_{15} = s_{14}/(\omega_5\omega_6\omega_7\omega_8(8s_9 + s_{13})),$$

$$\alpha_{26} = \frac{a_{31}\beta_1 - a_{32}}{\omega_1\omega_2(a_{62}a_{12} - a_{32}^2)(\beta_1 - \beta_2)},\ \alpha_{27} = \frac{a_{32} - a_{31}\beta_2}{\omega_1\omega_2(a_{62}a_{12} - a_{32}^2)(\beta_1 - \beta_2)},$$

where $\alpha_{28}, \ldots, \alpha_{32}$ are expansion coefficients into the simple fractions of the equation

$$s_{16}(s_1 + s_2\omega^* + s_3\omega^{*2})\frac{(a_{31} + a_{32}\omega^*)(a_{51} + a_{52}\omega^*) - (a_{21} + a_{22}\omega^*)(a_{61} + a_{62}\omega^*)}{\omega^*(1+\beta_1\omega^*)(1+\beta_2\omega^*)(1+\beta_3\omega^*)(1+\beta_4\omega^*)},$$

$$s_{16} = 2/(3K\omega_1\omega_2\omega_3\omega_4(a_{62}a_{12} - a_{32}^2)^2).$$

Values $\alpha_{33}, \ldots, \alpha_{41}$ are determined as the expansion coefficients into simple fractions of the latter equation multiplied by a fraction

$$\frac{s_5 + s_6\omega^* + s_7\omega^{*2} + s_8\omega^{*3} + s_9\omega^{*4}}{\omega_5\omega_6\omega_7\omega_8(8s_9 + s_{13})(1+\beta_5\omega^*)(1+\beta_6\omega^*)(1+\beta_7\omega^*)(1+\beta_8\omega^*)}.$$

Thus, the described constants close the analytical solution of the bending problem for a linearly viscoelastic three-layer plate based on Ilyushin's experimental-theoretical method of approximation.

## SPECIAL FUNCTIONS

This chapter deals with transcendental functions used in calculations of the boundary problems for three-layered elements of various designs, namely, Besselian, Lommel's, Kelvin's, and some other functions. To this effect, corresponding definitions, major properties, as well as differentiation and integration operations are given. Some integration formulas for the products of Bessel functions, trigonometric functions, and polynomials are original and were not described in literature earlier.

### Bessel functions

Bessel functions (cylindrical functions or cylindrical harmonics) are the solutions $Z_\nu(z)$ of the differential Besselian equation

$$z^2 \frac{d^2 w}{dz^2} + z\frac{dw}{dz} + (z^2 - \nu^2)w = 0, \tag{13.19}$$

where $z$ is a complex variable with a real part $x$; number $\nu$ is called an order and can also be complex, however if it is a real integer, we must write $\nu = n$. We shall confine further only to the real values of $\nu$, for which it is suffice to consider region $z > 0$. In case $\nu$ is not an integer and we do not restrict to the region $z > 0$, we accept a single-valued function in a complex $z$ plane that is cut along a beam from point $z = 0$ for $z^\nu$. The same relates to the logarithmic members.

Notice that solutions of the differential equation (13.19) are *Besselian functions* of the first order $J_\nu(z)$ presented by expansion into the series

$$J_\nu(z) = \left(\frac{z}{2}\right)^\nu \sum_{k=0}^{\infty} \frac{(-z^2/4)^k}{k!\Gamma(k+\nu+1)}, \tag{13.20}$$

where $\nu$ is not equal to any negative number. The series converges at any $z$ value. The solutions are also presented by Bessel's functions of the second order (Neumann function) $Y_\nu(z)$, found by noninteger $\nu$ in the form

$$z^2 \frac{d^2 w}{dz^2} + z\frac{dw}{dz} + (z^2 - \nu^2)w = 0.$$

If the index is non-negative integer, it will be justified to present it in the form of a power series

$$Y_n(z) = \frac{2}{\pi} I_n(z) \ln\frac{\gamma z}{2} - \frac{1}{\pi}\sum_{k=0}^{n-1} \frac{(-1)^k (n-k-1)!}{k!}\left(\frac{z}{2}\right)^{2k-n}$$

$$-\frac{1}{\pi}\sum_{k=0}^{\infty} \frac{(-1)^k (z/2)^{n+2k}}{k!(k+n)}\left[\frac{\Gamma'(n+k+1)}{(n+k)} + \frac{\Gamma'(k+1)}{(k)}\right], \tag{13.21}$$

$\gamma = 0.5772$—Euler's constant; $\Gamma(z)$—gamma function;

$$\Gamma(z)=\int_0^\infty e^{-t}t^{z-1}\,dt\,,\quad \Gamma(n+1)=n!\,,$$

the upper prime means a derivative per $z$.

Bessel function $J_\nu(z)$ is restricted in case $z \to 0$ in any bounded region of arg $z$ variation; $J_\nu(z) \to 0$ at a real $x \to +\infty$. Function $Y_\nu(z)$ is multiple-valued $z = 0$; $Y_\nu(z) \to 0$ and has a branch point at $x \to +\infty$.

Bessel's (13.20) and Neumann's functions (13.21) of the real argument $x > 0$ are the real functions $J_\nu(z)$ and $Y_\nu(z)$. They are linearly independent at any $v$ values. In any case, cylindrical functions

$$Z_\nu(z)=C_1J_\nu+C_2Y_\nu, \tag{13.22}$$

($C_1$,$C_2$—random constants) exhaust all solutions of equation (13.19), being transcendental functions.

*Bessel functions of the third order or Hankel functions* are presented by

$$H_\nu^{(1)}(z)=J_\nu(z)+iY_\nu(z),\ H_\nu^{(2)}(z)=J_\nu(z)-iY_\nu(z). \tag{13.23}$$

The Hankel function is important since it is the only solution of Bessel function obeying the zero boundary conditions at infinity

$$\lim_{r\to\infty} H_\nu^{(1)}(re^{i\theta})=0,\ \lim_{r\to\infty} H_\nu^{(2)}(re^{-i\theta})=0,\ (\varepsilon\le\theta\le\pi-\varepsilon,\ \varepsilon>0).$$

In contrast to functions $J_\nu(z)$ and $Y_\nu(z)$, Hankel functions (13.23) of the real order $v$ take complex values for the real argument $x > 0$.

Each $J_\nu(z)$, $J'_\nu(z)$, $Y_\nu(z)$, $Y'_\nu(z)$ function has an infinite number of real nulls. The diagrams of functions $(x)$, $J_1(x)$, $Y_0(x)$, $Y_1(x)$ are shown in Fig. 13.3.

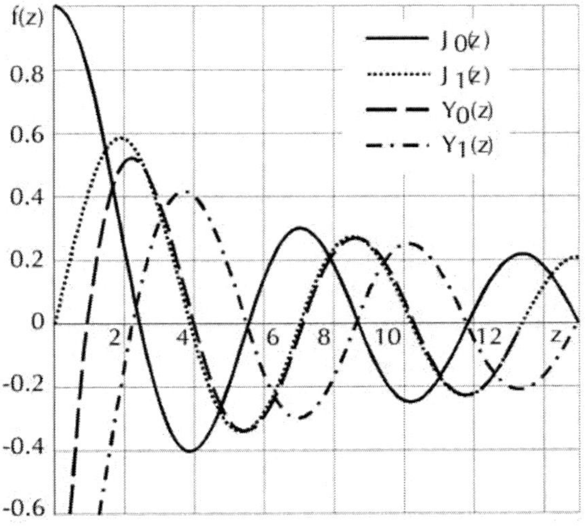


**Figure 13.3.**

In case $\beta_m$ and $\beta_k$ are two real nulls of the function $J_n(z)$, we are to meet the condition of orthogonality of these functions

$$\int_0^1 J_n(\beta_m z)J_n(\beta_k z)z\,\mathrm{d}z = \begin{Bmatrix} 0, & \text{если} \quad m \neq k \\ \frac{1}{2}[J_n'(\beta_k)]^2, & \text{если} \quad m = k \end{Bmatrix}.$$

Notice also that,

$$\alpha\int_0^\infty J_n(\alpha z)J_n(\beta z)z\,\mathrm{d}z = \delta(\alpha-\beta),$$

where $\delta(z)$—impulse delta function introduced further.

For cylindrical functions $Z_\nu(z)$ (13.22) and, consequently, for above-considered Bessel functions $J_\nu(z)$, $Y_\nu(z)$, $H_\nu^{(1)}(z)$, $H_\nu^{(2)}(z)$ the following relations are valid

$$Z_{-n}(z) = (-1)^n Z_n(z),\ Z_{\nu-1}(z) + Z_{\nu+1}(z) = \frac{2\nu}{z}Z_\nu(z)$$

$$Z_\nu'(z) = -\frac{\nu}{z}Z_\nu(z) + Z_{\nu-1}(z),\ Z_\nu'(z) = \frac{\nu}{z}Z_\nu(z) - Z_{\nu+1}(z)$$

$$Z_0'(z) = -Z_1(z),\ Z_1'(z) = Z_0(z) - \frac{1}{z}Z_1(z)$$

$$\frac{\mathrm{d}}{\mathrm{d}z}\left[z^\nu Z_\nu(\alpha z)\right] = \alpha z^\nu Z_{\nu-1}(\alpha z),\ \frac{\mathrm{d}}{\mathrm{d}z}\left[z^{-\nu} Z_\nu(\alpha z)\right] = -\alpha z^{-\nu} Z_{\nu+1}(\alpha z)$$

$$\left(\frac{1}{z}\frac{\mathrm{d}}{\mathrm{d}z}\right)^k \left[z^\nu Z_\nu(z)\right] = z^{\nu-k} Z_{\nu-k}(z)$$

$$\left(\frac{1}{z}\frac{\mathrm{d}}{\mathrm{d}z}\right)^k \left[z^{-\nu} Z_\nu(z)\right] = (-1)^k z^{-\nu-k} Z_{\nu+k}(z),$$

where $k = 0, 1, 2, \ldots$ —serial number of the derivative.

Wronskian determinants of indicated systems are written as

$$W\left\{J_\nu(z), Y_\nu(z)\right\} = \begin{vmatrix} J_\nu & Y_\nu \\ J_\nu' & Y_\nu' \end{vmatrix} = J_{\nu+1}(z)Y_\nu(z) - J_\nu(z)Y_{\nu+1}(z) = \frac{2}{\pi z},$$

$$W\left\{J_\nu(z), J_{-\nu}(z)\right\} = J_{\nu+1}(z)J_{-\nu}(z) + J_\nu(z)J_{-(\nu+1)}(z) = -\frac{2\sin(\nu\pi)}{\pi z},$$

$$W\left\{H_\nu^{(1)}(z)\ H_\nu^{(2)}(z)\right\} = H_{\nu+1}^{(1)}(z)H_\nu^{(2)}(z) - H_\nu^{(1)}(z)H_{\nu+1}^{(1)}(z) = -\frac{4i}{\pi z}.$$

There are recurrent relations between the functions $J_\nu(z)$, $Y_\nu(z)$, $H_\nu^{(1)}(z)$, $H_\nu^{(2)}(z)$

$$J_\nu(z)J_{-\nu+1}(z)+J_{\nu-1}(z)J_{-\nu}(z)=\frac{2\sin(\nu\pi)}{\pi z},$$

$$J_{\nu-1}(z)H_\nu^{(1)}(z)-J_\nu(z)H_{\nu-1}^{(1)}(z)=\frac{2}{\pi iz},$$

$$J_\nu(z)H_{\nu-1}^{(2)}(z)-J_{\nu-1}(z)H_\nu^{(2)}(z)=\frac{2}{\pi iz}.$$

*The modified Bessel functions* are presented by solutions of the differential Bessel equation

$$z^2\frac{d^2w}{dz^2}+z\frac{dw}{dz}-(z^2+\nu^2)w=0.$$

The solutions of this equation are the modified Bessel function $I_\nu(z)$ and Macdonald function $K_\nu(z)$ or Hankel modified function. Their relationship and expansion into the power series looks like:

$$I_\nu(z)=\sum_{k=0}^{\infty}\frac{(z/2)^{\nu+2k}}{k!\Gamma(k+\nu+1)},\ K_\nu(z)=\frac{\pi}{2}\frac{I_{-\nu}(z)-I_\nu(z)}{\sin(\nu\pi)},$$

$$K_n(z)=(-1)^{n+1}I_n(z)\ln\frac{\gamma z}{2}+\frac{1}{2}\sum_{k=0}^{n-1}\frac{(-1)^k(n-k-1)!}{k!}\left(\frac{z}{2}\right)^{2k-n}$$

$$+\frac{(-1)^n}{2}\sum_{k=0}^{\infty}\frac{(z/2)^{n+2k}}{k!(k+n)!}\left[\frac{\Gamma'(n+k+1)}{(n+k)!}+\frac{\Gamma'(k+1)}{(k)!}\right], \quad (13.24)$$

($n$—non-negative integer, $\gamma$—Euler's constant). At $n = 0$ the first sum $K_n(z)$ in (13.24) is to be equated to zero.

Function $I_\nu(z)$ (Re $\nu \geq 1$) is bounded at $z \to 0$ when the region of arg $z$ variation is restricted; $I_\nu(x) \to 0$ at a real $x \to +\infty$. Function $K_\nu(z)$ is multiple-valued and has a branch point $z = 0$; $K_\nu(z) \to 0$ at $x \to +\infty$.

The modified Bessel and Macdonald's functions of a real argument $x > 0$ are the real functions $I_\nu(x)$ and $K_\nu(x)$. They are linearly independent at any $\nu$.

Wronskian determinants of these systems are

$$W\{K_\nu(z)\ I_\nu(z)\}=\begin{vmatrix} I_\nu & K_\nu \\ I'_\nu & K'_\nu \end{vmatrix}=I_{\nu+1}(z)K_\nu(z)+I_\nu(z)K_{\nu+1}(z)=\frac{1}{z},$$

$$W\{I_\nu(z), I_{-\nu}(z)\}=I_\nu(z)I_{-(\nu+1)}(z)+I_{\nu+1}(z)I_{-\nu}(z)=-\frac{2\sin(\nu\pi)}{\pi z}.$$

At $x >> 1$ function $_\nu(z)$ increases in respect to the demonstrative law, while $K_\nu(z)$ diminishes in respect to this law. The diagrams of the functions $I_0(x)$, $I_1(x)$, $K_0(x)$, $K_1(x)$ are presented in Fig. 13.4. The following relations are obeyed for above functions

$$I_n(z) = i^{-n} J_n(iz),\ K_n(z) = \frac{\pi}{2} i^{n+1} H_n^{(1)}(iz),$$

$$I_{-n}(z) = I_n(z),$$

$$K_{-n}(z) = K_n(z),\ I_n(-z) = (-1)^n I_n(z),$$

$$I_{\nu-1}(z) - I_{\nu+1}(z) = \frac{2\nu}{z} I_\nu(z),$$

$$K_{\nu-1}(z) - K_{\nu+1}(z) = -\frac{2\nu}{z} K_\nu(z),$$

$$2I'_\nu(z) = I_{\nu-1}(z) + I_{\nu+1}(z),$$

$$-2K'_\nu(z) = K_{\nu-1}(z) + K_{\nu+1}(z),\ I'_0(z) = I_1(z),\ K'_0(z) = -K_1(z),$$

$$\left(\frac{1}{z}\frac{d}{dz}\right)^k \left[z^\nu I_\nu(z)\right] = z^{\nu-k} I_{\nu-k}(z),\ \left(\frac{1}{z}\frac{d}{dz}\right)^k \left[z^{-\nu} I_\nu(z)\right] = z^{-\nu-k} I_{\nu+k}(z),$$

$k = 0, 1, 2, \ldots$ —serial number of the derivative.

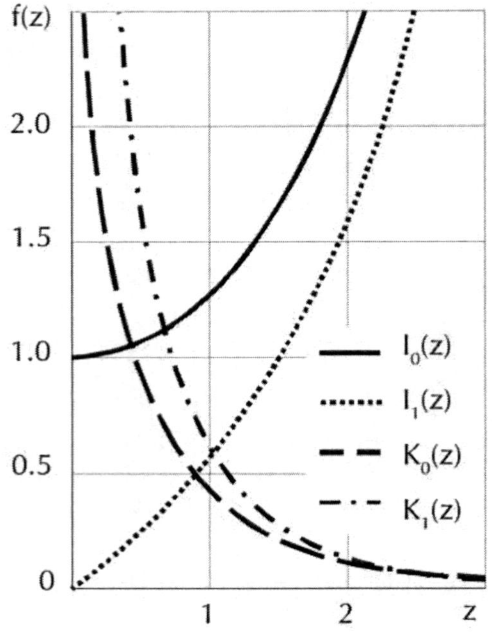


**Figure 13.4.**

For the considered Bessel (13.20) and Neumann's functions (13.21) as well as their modified analogues (13.24) the next *integration formulas* are true [34, 81]:

$$\int J_\nu(z)\mathrm{d}z = 2\sum_{k=0}^{\infty} J_{\nu+2k+1}(z), \quad \int z^{\nu+1}J_\nu(z)\mathrm{d}z = z^{\nu+1}J_{\nu+1}(z),$$

$$\int z^{-\nu+1}Z_\nu(z)\mathrm{d}z = -z^{-\nu+1}Z_{\nu-1}(z),$$

$$\int Z_1(z)\,\mathrm{d}\,z = -Z_0(z), \int zZ_0(z)\,\mathrm{d}\,z = zZ_1(z),$$

$$\int_0^x J_n(\alpha z)J_n(\beta z)z\,\mathrm{d}\,z = \frac{x}{\alpha^2-\beta^2}\left[\alpha J_n(\beta x)J_{n+1}(\alpha x) - \beta J_n(\alpha x)J_{n+1}(\beta x)\right]$$

$$= \frac{x}{\alpha^2-\beta^2}\left[\beta J_{n-1}(\beta x)J_n(\alpha x) - \alpha J_{n-1}(\alpha x)J_n(\beta x)\right], \quad \alpha^2-\beta^2 \neq 0.$$

where from,

$$\int_0^x I_n(\alpha z)I_n(\beta z)z\,\mathrm{d}\,z = \frac{x}{\alpha^2-\beta^2}\left[\alpha I_n(\beta x)I_{n+1}(\alpha x) - \beta I_n(\alpha x)I_{n+1}(\beta x)\right]$$

$$= \frac{x}{\beta^2-\alpha^2}\left[\beta I_{n-1}(\beta x)I_n(\alpha x) - \alpha I_{n-1}(\alpha x)I_n(\beta x)\right], \quad \alpha^2-\beta^2 \neq 0,$$

$$\int_0^x J_0(\alpha z)I_0(\beta z)z\,\mathrm{d}\,z = \frac{x}{\alpha^2+\beta^2}\left[\alpha J_1(\alpha x)I_0(\beta x) + \beta J_0(\alpha x)I_1(\beta x)\right],$$

$$\int_0^1 J_0(\beta_n z)I_0(\beta_n z)z\mathrm{d}z = \frac{1}{2\beta_n}\left[J_1(\beta_n)I_0(\beta_n) + J_0(\beta_n)I_1(\beta_n)\right],$$

$$\int_0^x J_n(\alpha z)J_n(\beta z)z\,\mathrm{d}\,z = \frac{x}{\alpha^2-\beta^2}\left[\alpha J_n(\beta x)J_{n+1}(\alpha x) - \beta J_n(\alpha x)J_{n+1}(\beta x)\right],$$

$$\int_0^x J_0^2(z)z\,\mathrm{d}\,z = \frac{x^2}{2}\left[J_0^2(x) + J_1^2(x)\right], \quad \int_0^x I_0^2(z)z\mathrm{d}z = \frac{x^2}{2}\left[I_0^2(x) - I_1^2(x)\right],$$

$$\int_0^1 J_0^2(\beta_n z)z\mathrm{d}z = \frac{1}{2}\left[J_0^2(\beta_n) + J_1^2(\beta_n)\right], \quad \int_0^1 I_0^2(\beta_n z)z\mathrm{d}z = \frac{1}{2}\left[I_0^2(\beta_n) - I_1^2(\beta_n)\right],$$

$$\int_0^1 \left[J_0(\beta_n r) - \frac{J_0(\beta_n)}{I_0(\beta_n)}I_0(\beta_n r)\right]^2 r\,\mathrm{d}\,r = \frac{1}{2}\left[J_0^2(\beta_n) + J_1^2(\beta_n)\right]$$

$$-\frac{J_0(\beta_n)}{\beta_n I_0(\beta_n)}\left[J_1(\beta_n)I_0(\beta_n) + J_0(\beta_n)I_1(\beta_n)\right] + \frac{J_0^2(\beta_n)}{2I_0^2(\beta_n)}\left[I_0^2(\beta_n) - I_1^2(\beta_n)\right].$$

$$\int_0^a z^{\mu} J_{\nu}(z)\,\mathrm{d}z = (\mu+\nu-1)aJ_{\nu}(a)S_{\mu-1,\nu-1}(a) - aJ_{\nu-1}(a)S_{\mu,\nu}(a)$$

$$+2^{\mu}\Gamma\left(\frac{1+\mu+\nu}{2}\right)\Big/\Gamma\left(\frac{1-\mu+\nu}{2}\right), \quad \mathrm{Re}(\mu+\nu)>-1,$$

$$\int_0^a z^{\nu+1}\sin\left[\frac{\beta}{2}(a^2-z^2)\right]J_{\nu}(z)\,\mathrm{d}z = \beta^{-\nu-1}U_{\nu+2}(a^2\beta, a),$$

$$\int_0^a z\sin\left[\frac{\pi}{2}(a^2-z^2)\right]J_0(\beta z)\,\mathrm{d}z = \frac{1}{\pi}U_2(a^2\pi, a\beta),$$

$$\int_0^a z\sin\left[\frac{\pi}{2}(a^2-z^2)\right]J_0(\beta z)\,\mathrm{d}z = \frac{1}{\pi}U_2(a^2\pi, a\beta),$$

$$\int_0^a z^{\nu+1}\cos\left[\frac{\beta}{2}(a^2-z^2)\right]J_{\nu}(z)\,\mathrm{d}z = \beta^{-\nu-1}U_{\nu+1}(a^2\beta, a),$$

$$\int_0^a z^{\nu+1}\sin\left[b(a^2-z^2)^{1/2}\right]J_{\nu}(z)\,\mathrm{d}z = (\pi/2)^{1/2}a^{\nu+1/2}b(1+b^2)^{-\nu/2-1/4}J_{\nu+3/2}[a(1+b^2)],$$

$$\int_0^a (a^2-z^2)^{-1/2}\cos\left[\beta(a^2-z^2)^{1/2}\right]J_0(z)z\,\mathrm{d}z = (\beta^2+1)^{-1/2}\sin[a(\beta^2+1)^{1/2}].$$

The last four integrals in above formulas should obey the requirement Re $\nu > -1$. This integration uses Lommel's functions of a single $S_{\mu,\nu}(z)$ and two $U_{\nu}(w, z)$ variables. In a general case, these and some other Lommel's functions ($s_{\mu,\nu}$, $V_{\nu}$) as well as related to them hypergeometric functions ${}_mF_n(\alpha_1,\ldots,\alpha_m; \gamma_1, \ldots, \gamma_n; z)$ are found from relations (Korn and Korn, 1961):

$$S_{\mu,\nu}(z) = s_{\mu,\nu}(z) + 2^{\mu-1}\Gamma\left(\frac{\mu-\nu+1}{2}\right)\Gamma\left(\frac{\mu+\nu+1}{2}\right)$$

$$\times\left[\sin\left(\frac{\mu-\nu}{2}\pi\right)J_{\nu}(z) - \cos\left(\frac{\mu-\nu}{2}\pi\right)Y_{\nu}(z)\right],$$

$$s_{\mu,\nu}(z) = \frac{z^{\mu+1}}{(\mu-\nu+1)(\mu+\nu+1)}\,{}_1F_2\left(1; \frac{\mu-\nu+3}{2}; \frac{\mu+\nu+3}{2}; -\frac{z^2}{4}\right),$$

$$U_{\nu}(w,z) = \sum_{m=0}^{\infty}(-1)^m\left(\frac{w}{z}\right)^{\nu+2m}J_{\nu+2m}(z),$$

$$V_{\nu}(w,z) = \cos\left(\frac{w}{2}+\frac{z^2}{2w}+\frac{\nu\pi}{2}\right) + U_{2-\nu}(w,z),$$

$$ {}_mF_n(\alpha_1, ..., \alpha_m; \gamma_1, ..., \gamma_n; z) = \sum_{k=0}^{\infty} \frac{(\alpha_1)_k \cdots (\alpha_m)_k}{(\gamma_1)_k \cdots (\gamma_n)_k} \frac{z^k}{k!}. $$

The *integration formulas* shown below are true for the modified Besselian function and Macdonald's function

$$ \int I_\nu(z)\,\mathrm{d}z = 2\sum_{k=0}^{\infty}(-1)^k I_{\nu+2k+1}(z), \quad \mathrm{Re}\,\nu > -1, $$

$$ \int z^{\nu+1} I_\nu(z)\,\mathrm{d}z = z^{\nu+1} I_{\nu+1}(z), \quad \mathrm{Re}\,\nu > -1, $$

$$ \int_0^a z^{1-\nu} I_\nu(z)\,\mathrm{d}z = a^{1-\nu} I_{\nu-1}(a) - \frac{2^{1-\nu}}{\Gamma(\nu)}, $$

$$ \int_0^x [I_n(\alpha z)]^2 z\,\mathrm{d}z = \frac{x^2}{2}[I_n'(\alpha x)]^2 + \frac{1}{2}\left(x^2 + \frac{n^2}{\alpha^2}\right)[I_n(\alpha x)]^2, \quad n > -1, $$

$$ \int_0^x J_\nu(\alpha z) K_\nu(\beta z) z\,\mathrm{d}z $$

$$ = \frac{1}{\alpha^2 + \beta^2}\left[\left(\frac{\alpha}{\beta}\right)^\nu + \alpha x J_{\nu+1}(\alpha x) K_\nu(\beta x) - \beta x J_\nu(\alpha x) K_{\nu+1}(\beta x)\right], $$

$$ \int_0^a z^{\nu+1}(a^2 - z^2)^{\sigma-1} I_\nu(z)\,\mathrm{d}z = 2^{\sigma-1}\alpha^{\nu+\sigma}\Gamma(\sigma) I_{\nu+\sigma}(a), \quad \mathrm{Re}\,\sigma > 0, \mathrm{Re}\,\nu > -1, $$

$$ \int_0^a z^{\nu+1}(a^2 - z^2)^{-1/2} \cos\left[(a^2 - z^2)^{1/2}\right] I_\nu(z) z\,\mathrm{d}z = \frac{\pi^{1/2} a^{2\nu+1}}{2^{\nu+1}\Gamma(\nu + 1/2)}, \quad \mathrm{Re}\,\nu > -1. $$

$$ \int_0^a z^{\nu+1} K_\nu(z)\,\mathrm{d}z = 2^\nu \Gamma(\nu+1) - a^{\nu+1} K_{\nu+1}(a), \quad \mathrm{Re}\,\nu > -1, $$

$$ \int_0^a z^{1-\nu} K_\nu(z)\,\mathrm{d}z = 2^{-\nu}\Gamma(1-\nu) - a^{1-\nu} K_{\nu-1}(a), \quad \mathrm{Re}\,\nu < 1, $$

$$ \int_{r_0}^1 J_0^2(\beta r) r\,\mathrm{d}r = \frac{1}{2}\left[J_0^2(\beta) + J_1^2(\beta)\right] - \frac{r_0^2}{2}\left[J_0^2(\beta r_0) + J_1^2(\beta r_0)\right], $$

$$ \int_{r_0}^1 J_0(\beta r) I_0(\beta r) r\,\mathrm{d}r = \frac{1}{2\beta}\left[J_1(\beta) I_0(\beta) + J_0(\beta) I_1(\beta)\right] $$

$$ -\frac{r_0}{2\beta}\left[J_1(\beta r_0) I_0(\beta r_0) + J_0(\beta r_0) I_1(\beta r_0)\right], $$

$$\int_{r_0}^{1} J_0(\beta r)Y_0(\beta r)r\,\mathrm{d}r = \frac{1}{2}\left[J_0(\beta)Y_0(\beta)+J_1(\beta)Y_1(\beta)\right]$$

$$-\frac{r_0^2}{2}\left[J_0(\beta r_0)Y_0(\beta r_0)+J_1(\beta r_0)Y_1(\beta r_0)\right],$$

$$\int_{r_0}^{1} J_0(\beta r)K_0(\beta r)r\,\mathrm{d}r = \frac{1}{2\beta}\left[J_1(\beta)K_0(\beta)-J_0(\beta)K_1(\beta)+1/\beta\right]$$

$$-\frac{r_0}{2\beta}\left[J_1(\beta r_0)K_0(\beta r_0)r_0 - J_0(\beta r_0)K_1(\beta r_0)r_0 + 1/\beta\right],$$

$$\int_{r_0}^{1} I_0^2(\beta r)r\,\mathrm{d}r = \frac{1}{2}\left[I_0^2(\beta)-I_1^2(\beta)\right]-\frac{r_0^2}{2}\left[I_0^2(\beta r_0)-I_1^2(\beta r_0)\right],$$

$$\int^{z} I_0(\beta z)Y_0(\beta z)z\,\mathrm{d}z = \frac{z}{2\beta}\left[I_1(\beta z)Y_0(\beta z)+I_0(\beta z)Y_1(\beta z)\right],$$

$$\int_{r_0}^{1} Y_0^2(\beta r)r\,\mathrm{d}r = \frac{1}{2}\left[Y_0^2(\beta)+Y_1^2(\beta)\right]-\frac{r_0^2}{2}\left[Y_0^2(\beta r_0)+Y_1^2(\beta r_0)\right].$$

Let us examine some integrals containing the products of the power and Besselian functions:

$$\int_0^1\left(J_0(\beta_n r)-\frac{J_0(\beta_n)}{I_0(\beta_n)}I_0(\beta_n r)\right)r\,\mathrm{d}r =$$

$$\frac{1}{\beta_n}\left[J_1(\beta_n r)r-\frac{J_0(\beta_n)}{I_0(\beta_n)}I_1(\beta_n r)r\right]_0^1 = \frac{1}{\beta_n}\left[J_1(\beta_n)-\frac{J_0(\beta_n)}{I_0(\beta_n)}I_1(\beta_n)\right],$$

$$\int_0^1\left(J_0(\beta_n r)-\frac{J_0(\beta_n)}{I_0(\beta_n)}I_0(\beta_n r)\right)r^3\,\mathrm{d}r$$

$$=\frac{1}{\beta_n}\left[J_1(\beta_n r)r^3-\frac{2}{\beta_n}J_2(\beta_n r)r^2-\frac{J_0(\beta_n)}{I_0(\beta_n)}\left(I_1(\beta_n r)r^3-\frac{2}{\beta_n}I_2(\beta_n r)r^2\right)\right]_0^1$$

$$=\frac{1}{\beta_n}\left[J_1(\beta_n)-\frac{2}{\beta_n}J_2(\beta_n)-\frac{J_0(\beta_n)}{I_0(\beta_n)}\left(I_1(\beta_n)-\frac{2}{\beta_n}I_2(\beta_n)\right)\right].$$

$$\int_0^1\left(J_0(\beta_n r)-\frac{J_0(\beta_n)}{I_0(\beta_n)}I_0(\beta_n r)\right)r^3 H_0(a-r)\,\mathrm{d}r$$

$$=\frac{1}{\beta_n}\left[J_1(\beta_n r)r^3-\frac{2}{\beta_n}J_2(\beta_n r)r^2-\frac{J_0(\beta_n)}{I_0(\beta_n)}\left(I_1(\beta_n r)r^3-\frac{2}{\beta_n}I_2(\beta_n r)r^2\right)\right]_0^a$$

$$=\frac{1}{\beta_n}\left[a^3J_1(\beta_n a)-\frac{2a^2}{\beta_n}J_2(\beta_n a)-\frac{J_0(\beta_n)}{I_0(\beta_n)}\left(a^3I_1(\beta_n a)-\frac{2a^2}{\beta_n}I_2(\beta_n a)\right)\right]$$

Based on the relations obtained, we may calculate the next definite integral:

$$\int_0^a H_0(a-r)\left(J_0(\beta_n r)-\frac{J_0(\beta_n)}{I_0(\beta_n)}I_0(\beta_n r)\right)\left(\frac{r^2}{a^2}-\frac{2r}{a}+1\right)r\,\mathrm{d}r==\frac{a}{\beta_n}J_1(\beta_n a)$$

$$+\left(\frac{a}{\beta_n}J_1(\beta_n a)-\frac{2}{\beta_n^2}J_2(\beta_n a)\right)-\frac{J_0(\beta_n)}{I_0(\beta_n)}\left(\frac{a}{\beta_n}I_1(\beta_n a)-\frac{2}{\beta_n^2}I_2(\beta_n a)\right.$$

$$\left.\frac{a}{\beta_n}I_1(\beta_n a)\right)$$

$$-\frac{2}{a}\left[\frac{a^2}{\beta_n}J_1(\beta_n a)-\frac{1}{\beta_n^2}\left(-aJ_0(\beta_n a)+\frac{2}{\beta_n}\sum_{m=0}^{\infty}J_{2m+1}(\beta_n a)\right)\right]$$

$$+\frac{2J_0(\beta_n)}{aI_0(\beta_n)}\left[\frac{a^2}{\beta_n}I_1(\beta_n a)-\frac{1}{\beta_n^2}\left(aI_0(\beta_n a)-\frac{2}{\beta_n}\sum_{m=0}^{\infty}(-1)^m I_{2m+1}(\beta_n a)\right)\right]$$

$$=-\frac{2}{\beta_n^2}J_2(\beta_n a)-\frac{2}{\beta_n^2}J_0(\beta_n a)+\frac{4}{\beta_n^3 a}\sum_{m=0}^{\infty}J_{2m+1}(\beta_n a)$$

$$-\frac{J_0(\beta_n)}{I_0(\beta_n)}\left[-\frac{2}{\beta_n^2}I_2(\beta_n a)+\frac{2}{\beta_n^2}I_0(\beta_n a)-\frac{4}{\beta_n^3 a}\sum_{m=0}^{\infty}(-1)^m I_{2m+1}(\beta_n a)\right]$$

$$=-\frac{2}{\beta_n^2}\frac{2}{\beta_n a}J_1(\beta_n a)+\frac{4}{\beta_n^3 a}\sum_{m=0}^{\infty}J_{2m+1}(\beta_n a)$$

$$-\frac{J_0(\beta_n)}{I_0(\beta_n)}\left[\frac{2}{\beta_n^2}\frac{2}{\beta_n a}I_1(\beta_n a)-\frac{4}{\beta_n^3 a}\sum_{m=0}^{\infty}(-1)^m I_{2m+1}(\beta_n a)\right]$$

$$=\frac{4}{\beta_n^3 a}\left\{-J_1(\beta_n a)+\sum_{m=0}^{\infty}J_{2m+1}(\beta_n a)-\frac{J_0(\beta_n)}{I_0(\beta_n)}\left[I_1(\beta_m a)-\sum_{m=0}^{\infty}(-1)^m I_{2m+1}(\beta_n a)\right]\right\}$$

$$=\frac{4}{\beta_n^3 a}\left\{\sum_{m=1}^{\infty}J_{2m+1}(\beta_n a)+\frac{J_0(\beta_n)}{I_0(\beta_n)}\sum_{m=1}^{\infty}(-1)^m I_{2m+1}(\beta_n a)\right\}.$$

*Kelvin's functions*

Kelvin's functions constitute a fundamental system of solutions of the differential equation (Zubchaninov, 1990)

$$x^4 w,_{xxxx}+2x^3 w,_{xxx}-(1+2\nu^2)(x^2 w,_{xx}-xw,_x)$$

$$+(\nu^4-4\nu^2+x^4)w_x=0. \tag{13.25}$$

It can be written in the form

$$\varphi_n(x)=\mathrm{ber}_{\pm\nu}\,x,\,\mathrm{bei}_{\pm\nu}\,x,\,\mathrm{ker}_{\pm\nu}\,x,\,\mathrm{kei}_{\pm\nu}\,x\;. \tag{13.26}$$

In a particular case when ν = 0 it follows from (13.25), (13.26)

$$w_{,xxxx}+\frac{2}{x}w_{,xxx}-\frac{1}{x^2}w_{,x}+\frac{1}{x^3}w_{,x}+w=0, \qquad (13.27)$$

$$w(x)=C_5\,\mathrm{ber}\,x+C_6\,\mathrm{bei}\,x+C_7\,\mathrm{ker}\,x+C_8\,\mathrm{kei}\,x\,.$$

In this case, Kelvin's functions of the zeroth-order (ν = 0 is usually omitted) form a fundamental system of solutions for equation (13.27)

$$\varphi_n(x)=\mathrm{ber}\,x,\ \mathrm{bei}\,x,\ \mathrm{ker}\,x,\ \mathrm{kei}\,x\,.$$

Kelvin's functions can be presented in the power series in the form:

$$\mathrm{ber}_\nu\,x=\left(\frac{x}{2}\right)^{\nu}\sum_{k=0}^{\infty}\frac{\cos\left[\left(\frac{3}{4}\nu+\frac{1}{2}k\right)\pi\right]}{k!\,\Gamma(\nu+k+1)}\left(\frac{x^2}{4}\right)^{k},$$

$$\mathrm{bei}_\nu\,x=\left(\frac{x}{2}\right)^{\nu}\sum_{k=0}^{\infty}\frac{\sin\left[\left(\frac{3}{4}\nu+\frac{1}{2}k\right)\pi\right]}{k!\,\Gamma(\nu+k+1)}\left(\frac{x^2}{4}\right)^{k},$$

$$\mathrm{ker}_n\,x=\frac{1}{2}\left(\frac{x}{2}\right)^{-n}\sum_{k=0}^{n-1}\cos\left[\left(\frac{3}{4}n+\frac{1}{2}k\right)\pi\right]\frac{(n-k-1)!}{k!}\left(\frac{x^2}{4}\right)^{k}-\ln\left(\frac{x}{2}\right)\mathrm{ber}_n\,x$$

$$+\frac{1}{4}\pi\,\mathrm{bei}_n\,x+\frac{1}{2}\left(\frac{x}{2}\right)^{n}\sum_{k=0}^{\infty}\cos\left[\left(\frac{3}{4}n+\frac{1}{2}k\right)\pi\right]\frac{\psi(k+1)+\psi(n+k+1)}{k!(n+k)!}\left(\frac{x^2}{4}\right)^{k},$$

$$\mathrm{kei}_n\,x=-\frac{1}{2}\left(\frac{x}{2}\right)^{-n}\sum_{k=0}^{n-1}\sin\left[\left(\frac{3}{4}n+\frac{1}{2}k\right)\pi\right]\frac{(n-k-1)!}{k!}\left(\frac{x^2}{4}\right)^{k}-\ln\left(\frac{x}{2}\right)\mathrm{bei}_n\,x$$

$$-\frac{1}{4}\pi\,\mathrm{ber}_n\,x+\frac{1}{2}\left(\frac{x}{2}\right)^{n}\sum_{k=0}^{\infty}\sin\left[\left(\frac{3}{4}n+\frac{1}{2}k\right)\pi\right]\frac{\psi(k+1)+\psi(n+k+1)}{k!(n+k)!}\left(\frac{x^2}{4}\right)^{k},$$

$$\Gamma(z)=\int_0^{\infty}e^{-t}t^{z-1}dt;\ \Gamma(n+1)=n!;\ \psi(z)=\frac{\Gamma'(z)}{\Gamma(z)}, \qquad (13.28)$$

where Γ(z)—gamma function, upper prime is a derivative per $z$ variable.

*Differentiation rules* of Kelvin's functions

$$f_\nu'=\frac{1}{2\sqrt{2}}\left(f_{\nu+1}+g_{\nu+1}-f_{\nu-1}-g_{\nu-1}\right), \qquad (13.29)$$

$$\sqrt{2}\,\mathrm{ber}'\,x=\mathrm{ber}_1\,x+\mathrm{bei}_1\,x,\ \sqrt{2}\,\mathrm{bei}'\,x=-\mathrm{ber}_1\,x+\mathrm{bei}_1\,x,$$

$$\sqrt{2}\,\mathrm{ker}'\,x = \mathrm{ker}_1\,x + \mathrm{kei}_1\,x,\ \sqrt{2}\,\mathrm{kei}'\,x = -\mathrm{ker}_1\,x + \mathrm{kei}_1\,x\ ,$$

where

$$\left.\begin{matrix} f_v = \mathrm{ber}_v\,x \\ g_v = \mathrm{bei}_v\,x \end{matrix}\right\};\ \left.\begin{matrix} f_v = \mathrm{bei}_v\,x \\ g_v = -\mathrm{ber}_v\,x \end{matrix}\right\};\ \left.\begin{matrix} f_v = \mathrm{ker}_v\,x \\ g_v = \mathrm{kei}_v\,x \end{matrix}\right\},\ \left.\begin{matrix} f_v = \mathrm{kei}_v\,x \\ g_v = -\mathrm{ker}_v\,x \end{matrix}\right\}.$$

The recurrence relation is

$$f_{v+1} + f_{v-1} = \frac{\sqrt{2}}{x}\left(f_{v+1} + g_{v+1} - f_{v-1} - g_{v-1}\right). \tag{13.30}$$

## NOTES

1. *Levi-Civita Tullio* (1873–1941), Italian mathematician.
2. *Deskartes Rene* (1596–1650), French mathematician, physicist, philosopher; founder of analytical geometry, has put forward the concepts of a function and argument, formulated the law of conservation of momentum, and the concept of the impulse force.

# References


## 1

Amenzade, Yu.A. (1976). *Theory of elasticity* (in Russian). Vysshaya Shkola, Moscow.

Gorshkov, A.G., Rabinskii, L.N., and Tarlakovskii, D.V. (2000). *Fundamental tensor analysis and continuum mechanics* (in Russian). Nauka, Moscow.

Mase, G.E. (1970). *Theory and problems of continuum mechanics*. McGraw-Hill Book Company, New York.

Nadai, A. (1950). *Theory of flow and fracture of solids*, Vol. 1. McGraw-Hill, New York.

Novackii, V. (1975). *Theory of elasticity* (in Russian). Mir, Moscow.

Postnov, V.A. (1977). *Numerical techniques for designing shipboard structures* (in Russian). Sudostroenie, Leningrad.

Sneddon, I.N. and Berry, D.S. (1961). *The classical theory of elasticity* (in Russian). GIFML, Moscow.

Williams, M.L., Landel, R.F., and Ferry, J.D. (1955). The temperature dependence of relaxation mechanisms in amorphous polymers and other glass-forming liquids. *Journal of American Chemical Society,* 77: 370.

## 2

Bell, G. Ph. (1984). *Experimental foundations of deformable body mechanics* (in Russian), Vol. 1 and 2. Nauka, Moscow.

Moskvitin, V.V. (1972). *Strength of viscoelastic materials* (in Russian). Nauka, Moscow.

## 3

Alfutov, N.A. (1991). *Fundamentals of buckling analysis of elastic systems* (in Russian). Mashinostroenie, Moscow.

Pobedrya, B.E. and Georgievskii, D.V. (1999). *Lectures on the theory of elasticity* (in Russian). Editorial URSS, Moscow.

Rzhanicyn, A.R. (1968). *Creep theory* (in Russian). Strojizdat, Moscow.

Trefftz, E. (1934). *Mathematical theory of elasticity* (in Russian). GTTI, Leningrad, Moscow.

## 6

Aleksandrov, A.V. and Potapov, V.D. (1990). *Fundamental theory of elasticity and plasticity* (in Russian). Vysshaya Shkola, Moscow.

Demidov, S.P. (1979). *Theory of elasticity* (in Russian). Vysshaya Shkola, Moscow.

Gorshkov, A.G. and Tarlakovskii, D.V. (1995). *Dynamic contact problems with moving boundaries* (in Russian). Nauka, FIZMATLIT, Moscow.

Green, A.E. and Zerna, W. (1968). *Theoretical elasticity*. Dover Publication, New York.

Koltunov, M.A., Kravchuk, A.S., and Majboroda, V.P. (1983). *Applied mechanics of deformable solid bodies* (in Russian). Vysshaya Shkola, Moscow.

Kupradze, V.D., Gegelia, T.G., Basheleishvili, M.O., and Burchuladze, T.V. (1968). *Three-dimensional problems of mathematical theory of elasticity* (in Russian). Tbilisi State University, Tbilisi.

Novozhilov, V.V. (1958). *Theory of elasticity* (in Russian). Sudpromgiz, Leningrad.

Southwell, R.V. (1948). *An introduction to the elasticity*. Illinois.

Starovoitov, É.I. (1987). Stress state of three-layer metal-polymer plates under variable loads. *International Applied Mechanics*, 4: 351–356.

Starovoitov, É.I. (1988). Description of thermomechanical properties of some structural materials (in Russian). *Journal of Strength of Materials*, 4: 426–431.

Starovoitov, E.I. (2001). *Foundations of elasticity, plasticity and viscoelasticity theory* (in Russian). BelGUT, Gomel.

Starovoitov, E.I. (2002). *Viscous elastoplastic laminated plates and shells* (in Russian). BelGUT, Gomel.

Starovoitov, E.I. (2004). *Strength of materials* (in Russian). BelGUT, Gomel.

Starovoitov, E.I. (2006). *Technical mechanics* (in Russian). BelGUT, Gomel.

Starovoitov, E.I., Yarovaya, A.V., and Gu, Ju. (2002). Vibrations of a circular sandwich-type plate induced by thermal-radiation effect (in Russian). *Problems of Mechanical Engineering and Automation*, 3: 71–77.

Starovoitov, E.I., Yarovaya, A.V., and Leonenko, D.V. (2003). *Local and pulse loading of sandwich structural elements* (in Russian). BelGUT, Gomel.

Starovoitov, E.I., Yarovaya, A.V., and Leonenko, D.V. (2006). *Deformation of three-layer structural elements on the elastic foundation* (in Russian). FIZMATLIT, Moscow.

Starovoitov, E.I. and Zeyad, A.D. (2007). Physical nonlinearity under cyclic loading in neutron flow. *American Journal of Applied Sciences*, 4(9): 653.


## 7


Bezuhov, N.I. (1968). *Fundamentals of elasticity, plasticity and creep theory* (in Russian). Vysshaya Shkola, Moscow.

Filonenko-Borodich, M.M. (1947). *Elasticity theory* (in Russian). Gostechizdat, Moscow.

Galin, L.A. (1953). *Contact problem theory of elasticity* (in Russian). Gostehizdat, Moscow.

Gorshkov, A.G., Starovoitov, E.I., and Tarlakovskii, D.V. (2002). *The theory of elasticity and plasticity* (in Russian). Fizmatlit, Moscow.

Gorshkov, A.G. and Tarlakovskii, D.V. (1995). *Dynamic contact problems with moving boundaries* (in Russian). Nauka, Fizmatlit, Moscow.

Green, A.E. and Zerna, W. (1968). *Theoretical elasticity*. Dover Publ, Inc., New York.

Ilyushin, A.A. (1948). *Plasticity. Part 1. Elastic-plastic deformations* (in Russian). Gostekhizdat, Moscow.

Lechnickiy, S.G. (1977). *Elasticity theory of anisotropic body* (in Russian). Nauka, Moscow.

Lomakin, V.A. (1976). *Elasticity theory of inhomogeneous body* (in Russian). MGU, Moscow.

Prager, W. (1955). *Probleme der plastizitätstheorie*. Birkhäuser, Basel und Stuttgart.

Shtaerman, I.Ya. (1994). *Contact problem theory of elasticity* (in Russian). GITTL, Moscow Leningrad.

Sokolovskii, V.V. (1969). *Theory of plasticity* (in Russian). Vysshaya Shkola, Moscow.

Timoshenko, S.P. and Goodier, J. (1979). *Elasticity theory* (in Russian). Nauka, Moscow.

Tolokonnikov, L.A. (1979). *Mechanics of a deformable solid body* (in Russian). Vysshaya Shkola, Moscow.


## 8


Bauschinger, I. (1881). Uber die Veränderung der Elastizitätsgrenze und des Elastizitätsmoduls verschiedener Metall. *Civilingenieur*, 27: 289–348.

Hencky, H. (1924). Zur Theorie plastischer Deformationen und der deburch im Material hervorgrefenen Nachspanungen. *ZAMM*, 4(4), 323–334.

Ilyushin, A.A. (1963). *Plasticity. Foundations of general mathematical theory* (in Russian). Publ. House of USSR AS, Moscow.

Ilyushin, A.A. and Ogibalov, P.M. (1960). *Elastic-plastic deformations of hollow cylinders* (in Russian). MGU, Moscow.

Ishlinskii, A.Yu. and Ivlev, D. (2001). *Mathematical theory of plasticity* (in Russian). Fizmatlit, Moscow.

Kiiko, I.A. (1978). *Plastic yield theory* (in Russian). MGU, Moscow.

Klyushnikov, V.D. (1979). *Mathematical theory of plasticity* (in Russian). MGU, Moscow.

Lenskiy, V.S. (1969). *Introduction to plasticity theory*, 2nd ed. (in Russian). MGU, Moscow.

Malinin, N.N. (1968). *Applied theory of plasticity and creep* (in Russian). Mashinostroenie, Moscow.

Mises, R. (1913). Mechanik der festen Körper im plastischdeformablen Zustand. Nachrichten d. Geselsch d. Wissensch Zu. *Göttingen Math.-phys. Klasse*, 582–592.

Moskvitin, V.V. (1981). *Cyclic loading of constructional elements* (in Russian). Nauka, Moscow.

Rabotnov, Yu.N. (1966). *Creep of structural elements* (in Russian). Nauka, Moscow.

Starovoitov, É.I. (1987). Stress state of three-layer metal-polymer plates under variable loads. International Journal of Applied Mechanics, 4: 351–356.

Starovoitov, E.I. (2004). *Strength of materials* (in Russian). BelGUT, Gomel.

## 9

Filonenko-Borodich, M.M. (1947). *Elasticity theory* (in Russian). Gostechizdat, Moscow.

Ilyushin, A.A. and Pobedrya, B.E. (1970). *Foundations of mathematical theory of thermoviscoelasticity* (in Russian). Nauka, Moscow.

Koltunov, M.A. (1976). *Creep and relaxation* (in Russian). Vysshaya Shkola, Moscow.

Rabotnov, Yu.N. (1977). *Elements of hereditary mechanics of solid bodies* (in Russian). Nauka, Moscow.

Rabotnov, Yu.N. (1979). *Mechanics of a deformable solid body* (in Russian). Nauka, Moscow.

Rzhanicyn, A.R. (1968). *Creep theory* (in Russian). Strojizdat, Moscow.

## 10

Gorshkov, A.G., Starovoitov, E.I., and Yarovaya, A.V. (2005). *Mechanics of laminated viscous elastic-plastic constructional elements* (in Russian). FIZMATLIT, Moscow.

Moskvitin, V.V. (1972). *Strength of viscoelastic materials* (in Russian). Nauka, Moscow.

Moskvitin, V.V. and Starovojtov, E.I. (1986). To the investigation of stress-strain state of two-layered metal-polymer plates at cyclic loading. *Mechanics of solids*, 1: 116–121.

Pleskatshevsky, Yu.M., Starovoitov, E.I., and Yarovaya, A.V. (2004). *Deformation of metal-polymer systems* (in Russian). Belaruskaya navuka, Minsk.

Pobedrya, B.E. and Georgievskii, D.V. (1999). *Lectures on the theory of elasticity* (in Russian). Editorial URSS, Moscow.

## 11

Gogotsy, G.A. (1982). *Durability of mechanical engineering nitride ceramics*. Preprint Kiev: Institute of problem durability. Academy of Sciences of Ukraine SSR.

Goldman, A.Ya. (1979). *Durability of constructional plastic*. (in Russian). Nauka, Moscow.

Ilyushin, A.A. and Ogibalov, P.M. (1960). *Elastic-plastic deformations of hollow cylinders* (in Russian). Moscow State University, Moscow.

Koltunov, M.A. (1976). *Creep and relaxation* (in Russian). Vysshaya Shkola, Moscow.

Pleskchevsky, Yu.M., Starovoitov, E.I., and Yarovaya, A.V. (2004). *Dynamics of metal-polymer systems* (in Russian). Belaruskaya navuka, Minsk.

Rabotnov, Yu.N. (1977). *Elements of hereditary mechanics of solid bodies* (in Russian). Nauka, Moscow.

Rzhanicyn, A.R. (1968). *Creep theory* (in Russian). Strojizdat, Moscow.

Starovoitov, E.I. (2002). *Viscous elastoplastic laminated plates and shells* (in Russian). BelGUT, Gomel.

Starovoitov, E.I. (2004). *Strength of materials* (in Russian). BelGUT, Gomel.

## 12

Korn, G. and Korn, T.M. (1996). *Mathematics handbook*. McGraw-Hill.

Novackii, V. (1975). *Theory of elasticity* (in Russian). Mir, Moscow.

Zubchaninov, V.G. (1990). *Foundations of elasticity and plasticity theory* (in Russian). Vysshaya Shkola, Moscow.

## APPENDIX

Gorshkov, A.G., Rabinskii, L.N., and Tarlakovskii, D.V. (2000). *Fundamental tensor analysis and continuum mechanics* (in Russian). Nauka, Moscow.

Koltunov, M.A., Kravchuk, A.S., and Majboroda, V.P. (1983). *Applied mechanics of deformable solid bodies* (in Russian). Vysshaya Shkola, Moscow.

Korn, G. and Korn, T.M. (1996). *Mathematics handbook*. McGraw-Hill.

# Index

## D

## H



## I



## J



## K

## R



## S

**T**

## V



## W



## Z